T0315280

Virtual Work Approach to Mechanical Modeling

In memory of Marcel Save (1930–2016)

Virtual Work Approach to Mechanical Modeling

Jean Salençon

WILEY

First published 2018 in Great Britain and the United States by ISTE Ltd and John Wiley & Sons, Inc.

ISTE Ltd
27-37 St George's Road
London SW19 4EU
UK

www.iste.co.uk

John Wiley & Sons, Inc.
111 River Street
Hoboken, NJ 07030
USA

www.wiley.com

Library of Congress Control Number: 2017959464

British Library Cataloguing-in-Publication Data
A CIP record for this book is available from the British Library
ISBN 978-1-78630-295-3

Contents

Preface

This book was triggered by the suggestion from students at the City University of Hong Kong that I should publish a revised version of my textbook[1] as a follow-up to the PhD courses they had been attending. As announced by its title, it comes out in a completely different spirit from the previous work, being centered on the principle of virtual work and the related method for mechanical modeling, with the purpose of showing and enhancing their polyvalence and versatility.

The concepts of virtual velocities and virtual work were progressively formulated through a 2000-year scientific process which started with the quest for a general principle to govern the leverage properties of "simple machines", a process that can be considered as completed by the end of the 18th Century with Lagrange's statement of the principle of virtual velocities, after the concepts of virtual velocities and energy had been specified by Johann Bernoulli. With the introduction of Lagrange's multipliers, the principle led to the definition of internal and external binding forces as the dual variables associated with geometrical constraints, the very concept at the root of the virtual work mechanical modeling method.

The book starts with a brief overview of the genesis of the principle of virtual velocities in relation to the statics of simple machines.

Chapter 2 is then devoted to the complete statement of the principle of virtual work for the dynamics of a system of point masses. It involves external and internal forces and quantities of acceleration, introduces the concept of subsystems and underscores that, for them, external forces are only the result of a thought experiment. Through the dualization process performed on Newton's Laws, linear forms, named virtual (rates of) work, are built up on the vector space of virtual

---

1 [SAL 01].

motions for the system and similar linear forms for any subsystem on its own vector space of virtual motions. Virtual motions are defined by the rate of variation of the parameters (the positions of the point masses in that case) that describe the geometrical state of the system (or subsystem) without complying with any geometrical constraint. These rates of variation are the virtual velocities. Rigid body virtual motions are shown to play a specific role, as the linear forms expressing the virtual rate of work by internal forces, either for the system or any subsystem, are zero in such virtual motions. Conversely, as an introduction to the virtual work method in this particular case, it is shown how, given the geometrical model of point masses, the corresponding force model can be obtained by writing down the linear forms expressing the virtual rates of work in their most general form and then implementing the virtual work statements to specify them and derive the fundamental laws.

In Chapter 3, the principle of virtual work is generalized into a formulation suitable for any general geometrical description encompassing all the parameters which define the state of a system and its subsystems. Their non-constrained variations define the virtual motions generating vector spaces, and the linear forms expressing virtual rates of work are continuous. This leads to the presentation of the virtual work method, making it clear that the "virtuous circle" of mechanical modeling starts from experimentally-induced hypotheses on the geometrical parameters. The method then provides consistent processes to build up associated force models, which must be submitted to validation, the next step in the "circle", in order to assess their domain of relevance. This chapter also takes advantage of the key role played by rigid body virtual motions to establish general results valid for any force modeling. They are expressed in terms of force wrenches.

The three-dimensional continuum model is chosen as a first example. Chapter 4 is devoted to defining the corresponding geometrical modeling from the physical intuition of a continuum gained through experimental observations. The concept is that of a system made of "diluted particles" whose geometrical state is characterized by their spatial position, under the constraint of piecewise continuity and continuous differentiability.

The kinematics of the model is analyzed in Chapter 5. Derived from the velocity field, tensor quantities are defined which measure the deformation rate and mean rigid body rotation rate of each "diluted" point particle. With the definition of convective (time) derivatives, the general theorems established in Chapter 3 are given more explicit expressions in the form of the Euler and Kinetic Energy theorems for the three-dimensional continuum.

In Chapter 6, the construction of the force model for the classical three-dimensional continuum, following the virtual work method, calls first for the expression of the virtual rate of work by external forces for the system, based upon physically-induced assumptions. The same assumptions are heuristically made for subsystems, together with the crucial hypothesis that constituent particles do not exert any action at a distance upon one another. As for the virtual rate of work by internal forces, it is assumed to be described by a volume density independent of the considered subsystem, a linear function of the virtual velocity field first gradient. In order to comply with the principle of virtual work, this expression is specified with a symmetric tensor field for the internal force model: the Cauchy stress tensor. The field and boundary equations of motion are then derived for the system and subsystems, showing that the initial assumptions about the virtual rates of work are consistent with each other and actually describe a physical model where particles only exert contact actions upon one another. The virtual work method thus plays a double role: first in organizing the whole modeling process, then as the final criterion for checking the consistency of the model obtained. The hydrostatic pressure model comes out as a particular case of the Cauchy stress tensor model. It can also be built up directly through the same method with a restricted expression of the virtual rate of work by internal forces on the same vector space of virtual motions; consequently, its relevance for practical applications is reduced.

Chapter 7 develops the same procedure for constructing the one-dimensional curvilinear continuum model. Taking the physical intuition of a slender three-dimensional medium being squeezed onto a director curve as a starting point, a first model is built up exactly in the same way as the three-dimensional continuum: the constituent particles are described as "diluted" material points on the director curve only characterized by their position vector from the geometrical viewpoint. The model obtained that way is physically relevant for wires and cables without stiffness, which turns out not to be sufficient since most practical applications are concerned with rods, beams, arches, etc. For a mechanical model better suited for practical issues, it is necessary to enrich the geometrical description of the constituent particle with parameters that better reflect the physical constitution of the original three-dimensional body. As it comes out from everyday practice that the cross-section of the original slender body plays a crucial role, the constitutive particle in this more sophisticated one-dimensional model is characterized on the director curve by its position vector and the orientation of an underlying microstructure to account for the transverse cross-section. Actual and virtual motions are defined by velocity distributor fields on the director curve. Consistently, external forces are modeled by wrenches and a line density of wrench, while internal forces are represented by a wrench field on the director curve. The field and boundary equations of motion are then derived for the system and subsystems in terms of wrenches, in the form of a conservation law as for the three-dimensional continuum. The internal force wrench field determines contact actions between adjacent particles. The connection between this model and the three-dimensional modeling of the original slender body is established through the principle of virtual work, which identifies the internal force wrench in the one-dimensional model with the wrench of the contact forces exerted on the transverse cross-section in the three-dimensional body.

Based upon a similar physical intuition, the two-dimensional modeling of plates and thin slabs through the virtual work method presented in Chapter 8 follows the same path as the one-dimensional modeling process in Chapter 7, with the director curve being substituted by a director sheet. Again, the constituent particles are diluted material points with an attached oriented microstructure which accounts for the role of the transverse material elements in the original plane and slender body. From the geometrical viewpoint, these particles are characterized by their position vector and the orientation of their attached microstructure with respect to the reference frame. Actual and virtual motions are defined by velocity distributor fields on the director sheet. External forces are modeled by surface and line densities of wrenches, while internal forces are represented by a tensorial wrench field defined on the director sheet. With the use of tensor quantities, the field and boundary equations can be written in a compact form easily identified as a conservation law. They can usually be split into in-plane and out-of-plane equations, which can be compared with their three-dimensional and one-dimensional counterparts, respectively. The internal force tensorial wrench field determines contact actions between adjacent particles in the two-dimensional model in the form of a wrench, in the same way as the Cauchy stress tensor determines the stress vector in the case of the three-dimensional classical continuum. The connection between this two-dimensional continuum modeling and the three-dimensional modeling of the original body through the principle of virtual work identifies the internal force tensorial wrench at the field point: for any given orientation, the wrench of contact actions in the two-dimensional continuum is equal to the wrench of the contact actions exerted on the corresponding element in the original body.

This selection of three typical mechanical continua, with five force models being constructed, first underscores the utmost and fundamental importance of the geometrical modeling that lies at the root of the force modeling process. Non-constrained variations of the chosen geometrical parameters define the virtual motions of the system and subsystems on which the virtual work method is implemented. In a non-axiomatic way, the method enhances the reference to physical intuition when it comes to the crucial step of writing down the linear forms dualizing the external and internal force rates of work and also, after the validation step, when it proves necessary to enrich the geometrical modeling in order to improve the physical and practical relevance of the force model.

From this latter point of view, the analogies and differences between the one-dimensional and three-dimensional continua are enlightening. In both cases, enlarging the practical relevance of the force model calls for a more sophisticated expression of the virtual rate of work by internal forces. For the one-dimensional curvilinear continuum, the inadequacy of the wires and supple cables model to account for beams and arches is obviously due to the fact that the underlying microstructure, namely the transverse cross-section, is not taken into account; thence, the geometrical modeling is enriched and virtual motions are defined by virtual velocity distributor fields instead of virtual velocity fields. Regarding the classical three-dimensional continuum, the hydrostatic pressure model only refers to the trace of the virtual velocity gradient, the virtual volume dilatation rate, in the expression of the virtual rate of work by internal forces. This amounts to taking just a mean vision of the virtual evolution of the constituent particle, thus neglecting its microstructure in some sense.

The analogy may then be pushed further. Considering that the wires and supple cables model is the exact one-dimensional counterpart of the classical three-dimensional continuum, we naturally imagine the three-dimensional counterpart of the beams and arches model, as a matter of fact similar to the two-dimensional model for plates and thin slabs, where the constituent diluted particles are characterized by their position vector and the orientation of an attached microstructure. The virtual work method is the most valuable tool for constructing this "micropolar" three-dimensional continuum in a consistent way.

Throughout these examples, it is observed that the equations of motion attached to the force models are not sufficient to determine the internal force field in a system, but for the particular case of a statically determinate one-dimensional element. In the general case, it is necessary to specify the mechanical constitutive equation of the constituent material. Given this physical data, such as the elastic, elastoplastic or viscoelastic constitutive laws, the principle of virtual work is the cornerstone of variational principles and numerical methods which are exposed extensively in numerous textbooks but do not fall within the scope of this book. Nevertheless, some examples related to the theory of yield design are briefly given as straightforward applications of the principle that yield relevant results in various practical circumstances.

Three appendices are devoted to the introduction and simple practice of mathematical concepts – Tensor Calculus, Differential Operators, Distributors and Wrenches – inasmuch as necessary for setting up the notations and making an autonomous reading possible.

Acknowledgments

I wish to express my gratitude to the many colleagues to whom I have been indebted, all along my teaching career, for helpful criticisms and suggestions.

Among them, I extend special thanks to Dr Jean-Michel Delbecq and Professors Michel Amestoy, Habibou Maitournam, Marcel Save and Pierre Suquet.

Finally, as a Senior Fellow, I wish to acknowledge the support of the Institute of Advanced Study at the City University of Hong Kong.

Jean SALENÇON
December 2017

1

The Emergence of
the Principle of Virtual Velocities

1.1. In brief

The historical path to Lagrange's statement of the principle of virtual velocities has been two-millennium long, a facet of what Benvenuto [BEN 91] calls *"The Enigma of Force and the Foundations of Mechanics"* and could *"be regarded as vague meandering, impotent struggles, foolish attempts at reduction, and justified doubt about the nature of force"* (Truesdell[1]). Many authors, after Lagrange himself [LAG 88a], have tracked the history and avatars of the concepts of virtual velocities and virtual work: a very comprehensive analysis appears in [DUH 05], [DUH 06] and we must also quote, among others, [DUG 50], [TRU 68], [BEN 81], [BEN 91] and [CAP 12]. The purpose of this chapter is just to present some milestones along this historical path, up to Lagrange's contribution.

1.2. Setting the principle as a cornerstone

In the first edition of the *Mécanique Analitique* [LAG 88a], Lagrange had some very extolling words about the principle of virtual velocities:

> *"But this principle is not only very simple and very general in itself; as an invaluable and unique advantage it can also be expressed in a general formula which encompasses all the problems that can be proposed regarding equilibrium. We will expose this formula in all its extent; we will even try to present it in a more general way than done usually up to now, and present new applications".*

---

1 Foreword to [BEN 91, p. IX].

But leafing over his *Complete Works* as published in [LAG 88b], we find that in the subsequent editions, without dimming his enthusiasm for this fundamental principle of statics, he would be somewhat cautious about the possibility of laying it as a first stone (*Méchanique Analitique*, Section 1, Part 1, section 18):

> *"Regarding the nature of the principle of virtual velocities, it must be recognized that it is not self-evident enough to be settled as a primitive principle".* [Quant à la nature du principe des vitesses virtuelles, il faut convenir qu'il n'est pas assez évident par lui-même pour pouvoir être érigé en principe primitif.]

The name of Lagrange shines at the top of the list of the professors of Mechanics at the École polytechnique in Paris where he taught from 1794, when the school was founded, until 1799 (his first successor was Fourier). Some 200 years later, Germain, his 27th successor in charge of teaching Mechanics at the École polytechnique from 1973 to 1985, did take up the challenge of setting the "Principe des Puissances Virtuelles"[2], as the cornerstone of his synthetic presentation of Mechanics [GER 86]. In the English version of our own textbook for the École polytechnique [SAL 01], where we followed the same track, we retained the wording *Principle of Virtual Work* for simplicity's sake with the corresponding method for the modeling of forces being called the *Virtual Work Method*, thus dropping the reference to "virtual rates", but we explicitly named the linear forms involved in the statements *"virtual **rates** of work"*.

1.3. The "simple machines"

When looking for the very roots of Mechanics, we inescapably encounter the study of the "simple machines" that provide mechanical advantage, or leverage, when applying a single active force to do work against a single load force, such as the weight of a body: *"early theoretical thinking about statics and mechanics took as its references particular objects, things like the lever, used since ancient times as necessary tools"*[3]. Aristotle's (384–322 BC) *Quaestionae Mechanicae* (Mechanical Problems)[4], as quoted by Benvenuto[5], defines Mechanics as an art:

> *"Miraculously some facts occur in physics whose causes are unknown; that is, those artifices that appear to transgress Nature in favour of man...Thus, when it is necessary to do something that goes*

---

2 i.e. "Principle of virtual powers" or "Principle of virtual rates of work".
3 [BEN 91, p. 4].
4 [ARI 36]; apocryphal?
5 [BEN 91, p. XVIII].

beyond Nature, the difficulties can be overcome with the assistance of art. Mechanics is the name of the art that helps us over these difficulties; as the poet Antiphon put it, "Art brings the victory that Nature impedes"".

Regarding the lever problem, he finds a marvelous explanation in the fact that the weight and "small force" describe their circular trajectories with different velocities:

"Among the problems included in this class are included those concerned with the lever. For it is strange that a great weight can be moved by a small force, and that, too, when a greater weight is involved. For the very same weight, which a man cannot move without a lever, he quickly moves by applying the weight of the lever. Now the original cause of all such phenomena is the circle; and this is natural, for it is in no way strange that something remarkable should result from something more remarkable, and the most remarkable fact is the combination of opposites with each other. The circle is made up of such opposites, for to begin with it is composed both of the moving and of the stationary, which are by nature opposite to each other ... 'Therefore, as has been said before, there is nothing strange in the circle being the first of all marvels'".

"...Again, no two points on one line drawn as a radius from the centre travel at the same pace, but that which is further from the fixed centre travels more rapidly; it is due to this that many of the remarkable properties in the movement of circles arise".

Aristotle's *Physicae Auscultationes* (Physics) [ARI 09] is usually referred to for the introduction of the concept of (motive) "Power" ($\delta \upsilon \nu \alpha \mu \iota \varsigma$ or $\iota \sigma \chi \upsilon \varsigma$) representing the product of the weight of the considered body by its velocity (the ratio of the displacement to the duration of the movement) in order to explain the principle of the rectilinear lever [DEG 08]. The equilibrium of the lever is just stated as the equality (equivalence) of the powers acting at each end, explaining the mechanical advantage by the comparison of the velocities of the active and load forces. The *"rule of proportion"* (Physics, vol. VII, Chapter V) clearly refers to motion, with the major ambiguity due to his reference to time that would be definitely ruled out by Descartes (section 1.5.1):

"Then, the movement A have moved B a distance G in a time D, then in the same time the same force A will move ½ B twice the distance G, and in ½ D it will move ½ B the whole distance for G: thus the rules of proportion will be observed".

Archimedes' (287–212 BC) approach to statics in *De Planorum Æquilibriis* (On the Equilibrium of Planes) [ARC] is completely different: "*While Aristotle relates mechanics to a physical theory, aiming for a universal synthesis, Archimedes thinks of statics as a rational and autonomous science, founded on almost self-evident postulates and built upon rigorous mathematical demonstrations*"[6].

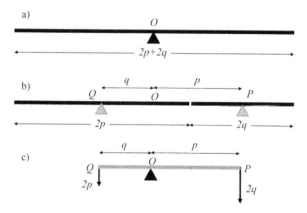

Figure 1.1. *Archimedes' proof of the law of the lever*

A very illustrative example is given by his proof of the law of the lever (or steelyard) that can be sketched as follows (Figure 1.1). The initial accepted demand is that a lever with arms of equal length $(p+q)$ over which the load – weight – is uniformly distributed is in equilibrium (a). Then, through a *thought experiment* (b), this lever is split into two parts of length $2p$ and $2q$ respectively (and the corresponding loads) which, anticipating the terminology we will use in the following chapters, we may call *subsystems* of the given physical *system* (the lever). Considering first the subsystem with length $2p$, we can state, from the same initial demand, that it is in equilibrium about its midpoint Q where it exerts the load $2p$; in the same way, the subsystem with length $2q$ is in equilibrium about its midpoint P where it exerts the load $2q$. These midpoints are respectively at a distance q and p from the midpoint O of the lever: thus, the equilibrium of the whole lever with equal arms is also the result of the equilibrium about O of the lever QP with unequal arms q and p, and loads $2p$ and $2q$ respectively (c).

---

6 [BEN 91, p. 43].

This proof based upon a statical thought experiment does not refer to motion and does not call upon any general principle either. It has been discussed by many authors and various "improvements" were put forward that are listed and analyzed in [BEN 91] and won't be discussed in this brief outline whose purpose is to introduce the two fundamental pathways that were to be followed all along the history of Mechanics. Schematically, we could say that Archimedes aimed at providing answers to given practical problems based upon a limited number of preliminary demands, while Aristotle would try to formulate a general principle, in the present case the equality of the powers of the active force and load force, to cope with any possible problem.

Making use of Descartes' own words[7] in his criticism of Galileo's analyses of the steelyard and lever similar to Archimedes' proof, Duhem[8] commented on these approaches: Archimedes "plainly explains *Quod ita sit* but not *Cur ita sit*" ("What" but not "Why") and, about Aristotle's analysis (see section 1.5.1),

> *"This insight is, indeed, the seed from which will come out, through a twenty century development, the powerful ramifications of the Principle of virtual velocities".*

A similar comment had been made by Fourier in his *Mémoire sur la Statique* (A Memoir on Statics)[9]:

> *"One may add that his writings offer the first insights on the Principle of virtual velocities".*

1.4. Leonardo, Stevin, Galileo

It is clear that we are still very far from a general statement of the principle. The concepts must be extracted as essences through a long lasting trial-and-error process that cannot be extensively presented here following the historical timeline and

---

7 [DES 68, *Correspondance 2*, p. 433]. As a matter of fact, Descartes was just following Aristotle's own distinction between "artists" and "men of experience" as it appears in Chapter 1 of the *Metaphysics*: *"But yet we think that knowledge and understanding belong to art rather than to experience, and we suppose artists to be wiser than men of experience (which implies that Wisdom depends in all cases rather on knowledge); and this because the former know the cause, but the latter do not"*, notwithstanding the fact that *"With a view to action, experience seems in no respect inferior to art, and men of experience succeed even better than those who have theory without experience"*.
8 [DUH 05, p. 332].
9 [FOU 98].

quoting all contributions that are reported in the analyses of Duhem, Dugas, Benvenuto, Capecchi, etc. The story will be made short.

Among the many topics he covered in his manuscripts, which are stored and preserved in the Library of the *Institut de France* in Paris[10], Leonardo da Vinci (1431–1519) detailed the properties of the simple machines (*Ms. A, E, F, I* and *M*) and tried to express them through a simple general law that turns out to be quite similar to Aristotle's statement (*Ms. F*):

> *"If a power* [puissance] *moves a given body along a given length of space during a given time span, it will move half this body during the same time span along twice the given length of space. Or the same power* [vertu] *will move half this body along the same length of space in half the same time span".*

Simon Stevin (1548–1620) also referred to the lever problem: he discarded Aristotle's argument about the velocities along the circular trajectories with the simple, hammer-like statement that [STE 05/08]:

> *"What is immobile does not describe circles, but two weights in equilibrium are immobile; thus two weights in equilibrium do not describe circles"*[11].

This actually underscores a true conceptual difficulty: why should the equilibrium of a system be studied by referring to motion?

Nevertheless, for the analysis of the inclined plane, Stevin derived the condition for the balance of forces using a diagram with a "wreath" or necklace containing evenly spaced round balls resting on a triangular wedge (Figure 1.2). He concluded that if the weights were not proportional to the lengths of the sides on which they rested, they would not be in equilibrium since the necklace would be in perpetual motion, which he considered obviously impossible. Incidentally, Stevin was so proud of his proof that the corresponding figure appears on the cover of his books *De Beghinselen der Weeghconst* [STE 86][12] and *Hypomnemata Mathematica* [STE 05/08] with his motto *"Wonder en is gheen wonder"* ("Magic is no magic" also translated by "Wonder, not miracle") as a refutation of Aristotle's "marvel". But we may wonder whether this proof (although questionable) was not, as a matter of fact, a kind of kinematical thought experiment.

---

10 [LEO 87/08] – *Les Manuscrits de Léonard de Vinci. Ms A-M.*
11 [BEN 91, p. 82], see also [DUH 05, p. 267].
12 The Elements of the Art of Weighing.

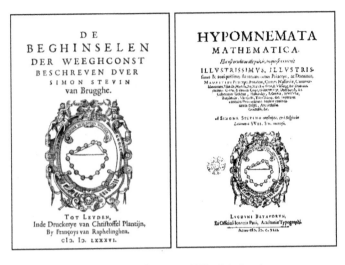

Figure 1.2. *Covers of Stevin's books*

Also, in *Hypomnemata Mathematica* [STE 05/08], when dealing with pulleys and pulley blocks, Stevin wrote the following remark:

"Ut spatium agentis, ad spatium patientis,
sic potentia patientis ad potentiam agentis"

that may be translated as

"As the space of the actor is to the space of the sufferer,
so is the power of the sufferer to the power of the actor"[13],

expressing that the displacement of the resistance is to the displacement of the power as the power to the resistance. We observe that there is no reference to time in this sentence that sounds like a *rule of proportion*. According to Benvenuto, Stevin would not give this statement the status of a principle, which he disliked, and considered it *"as a criterion, not an explanation of equilibrium"*[14].

In Galileo's (1564–1642) works [GAL 99, GAL 34, GAL 38], we encounter several occurrences of an implicit use of a concept close to what would be defined

---

13 [BEN 91, p. 81]: this principle is said to have been already stated by Guidobaldo dal Monte.
14 [BEN 91, p. 82].

later on as virtual work. A famous example is related to the analysis of the inclined plane in *Della Scienza Meccanica*[15]:

> "...*Thence the weight F moves downwards, drawing the body E on the sloped plane, this body will cover a distance along AC equal to the one described by the weight F in its fall. But this should be observed: it is true that the body E will have covered all the line AC in the time the weight F falls down an equal length; but during this time, the body E will not have moved away from the common centre of weights more than the vertical length BC, while the weight F, falling down according to the vertical, has dropped a length equal to all the line AC. Recall that weights only resist an oblique motion inasmuch as they move away from the centre of the Earth... We can thus say rightly that the travel [viaggio] of the force [forza] F is to the travel [viaggio] of the force [forza] E in the same ratio as the length AC to the length CB".*

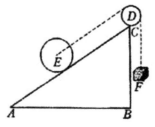

Figure 1.3. *The inclined plane [GAL 38]*

Despite the fact that this proof was only based upon the concomitant displacements or travels [*viaggi*] of the weight and the body with respect to the "common centre of weights", Galileo, as a foreword, still referred to time and velocity in the Aristotelian spirit:

> "*Finally, let us not overlook the following consideration: as a principle, we said that necessarily, in any mechanical instrument, as much the force was increased via this instrument, as much, on the other hand, one would lose time or velocity".*

---

15 [GAL 34].

1.5. Descartes and Bernoulli

1.5.1. *René Descartes (1596–1650)*

The correspondence of Descartes (1596–1650)[16], as published by Adam and Tannery shows, through many examples, that he had a much clearer vision of a virtual velocity principle than his predecessors or contemporaries, whom he would sometimes treat rather roughly as in a letter to Mersenne (November 15, 1638):

> *"Pour ce qu'a écrit Galilée touchant la balance et le levier, il explique fort bien quod ita sit, mais non pas cur ita sit, comme je fais par mon Principe. Et pour ceux qui disent que je devois considerer la vitesse, comme Galilée, plutot que l'espace, pour rendre raison des Machines, je croy, entre nous, que ce sont des gens qui n'en parlent que par fantaisie, sans entendre rien en cette matiere...."*[17]

> *"Regarding what Galileo wrote about the steelyard and the lever, he plainly explains what happens but not why it happens, as I do it myself through my Principle. And as for those who pretend that I should consider velocity, as Galileo does, instead of space, I believe, between us, that they are just people who talk without any understanding of the matter at hand".*

He had stated his principle plainly, answering a letter from Constantijn Huygens (Christian's father) on October 5, 1637 about the fundamental principle of the simple machines in its common form:

> *"The invention of all these machines is founded on one principle, which is that the same force which can lift a weight, for example of 100 pounds, up to two feet, can also lift a weight of 200 pounds up to one foot, or a weight of 400 pounds up to half a foot..."*

which expresses the conservation of the product of the load by its vertical displacement and looks like a reminder of Aristotle's rule of proportion *without any reference to time*. In the rest of the letter, he examined such simple machines as the pulley, inclined plane, wedge, etc. within this framework. Nevertheless, one point still needed to be corrected and later on, in a letter to

---

16 [DES 68].
17 [DES 68, *Correspondance* 2, p. 433].

Mersenne (July 13, 1638), Descartes insisted on the fact that the displacements to be considered were infinitesimal:

> *"From this it follows evidently that the gravity relative to a given body, or equivalently the force to be exerted to sustain it or prevent it from going down, when it is in a given position, should be measured by means of the beginning of the movement that would be done by the power which sustains it either for lifting it or following it if it went down".*

which is obviously a major step forward as regards the final formulation of the principle. In order to make himself clearer he added, a few lines below:

> *"Note that I say* begin to go down *and not simply go down, because it is only the beginning of the descent that must be taken into account".*

Besides insisting on the infinitesimal character of the displacements that must be considered, we should note that Descartes in the French wording makes use of the conditional or potential mode for the verb *"par le commencement du mouvement que devroit* [devrait] *faire la puissance qui le soustient* [soutient]*"*. This opens the way to the concept of virtuality of these displacements and also counters Stevin's objection about the contradiction between equilibrium and motion. Finally, let us quote a short sentence in a letter which is usually considered as having been sent to Boswell in 1646, where Descartes discarded actual velocities as the cause of the properties of such simple machines as the lever in the Aristotelian way:

> *"I do not deny the material truth of what Mechanicists usually say, namely that the higher the velocity of the longer arm of the lever compared with the shorter arm, the smaller the force necessary to move it; but I do deny that velocity or slowness be the cause of this effect".*

In other words, referring to time or velocities is not erroneous but just irrelevant.

Descartes' fundamental statement was generalized by Wallis (1616–1703)[18], dealing with any kind of forces with the proper definition of their forward or backward movements:

> *"And, as a general rule, the forward or backward movements of motor forces whatsoever* [virium motricium quarumcunque] *are obtained from the products of the forces by their forward or backward movements estimated along the directions of these forces".*

---

18 [WAL 70].

1.5.2. Johann Bernoulli (1667–1748)

Thanks to Pierre Varignon in his *Nouvelle Mécanique ou Statique*[19], we have the exact wording of the letter Johann Bernoulli sent him on January 26, 1717. In this letter, Bernoulli gave the first definitions of the concepts of *Energy* [Énergie] and *virtual velocities* [Vitesses virtuelles] in the case of a small rigid body motion. Defining virtual velocities, he considers small rigid body movements and the components of the corresponding small displacements of the forces along their lines of action:

> *"Imagine several different forces which act according to various trends or directions to maintain a point, a line, a surface or a body in equilibrium. Imagine that a small movement, either parallel to any direction or about any fixed point be imposed to all this system of forces. It will be easy for you to understand that in this movement each of these forces advances or moves back in its direction, unless one or more of these forces have their own tendencies [tendances] perpendicular to the direction of the small movement; in which case this force or these forces, would not advance nor move back; because these movements forward or backward, which are what I call virtual velocities, are just what the quantities in which each tendency line increase or decrease in the small movement".*

He then defines the *Energy* of each force as the product of the considered force by its virtual velocity in the movement, either "*affirmative*" or "*negative*" depending on whether the force moves forward or backward. As a matter of fact, this is just the definition of the work by the considered force in the small displacement of its point of application, a concept that had not been introduced before but by Salomon De Caus (1576–1630) with the French word "*Travail*" and present in his book *Les raisons des forces mouvantes* [DEC 15].

With these definitions, Bernoulli issues the general statement that

> *"For any equilibrated system of forces...the sum of the affirmative energies will be equal to the sum of negative energies counted positive".*

In the second volume of *Les Origines de la Statique*[20], Duhem could not help lamenting that Bernoulli adopted the terminology "*vitesses virtuelles*" [virtual velocities], instead of virtual displacements, since time and velocities have nothing

19 [VAR 25, II, ix, p. 176].
20 [DUH 06, footnote p. 268].

to do in that matter, and also that this terminology had been retained by many authors. As a response to that criticism, we may argue now that this terminology makes it impossible to forget about the infinitesimal character of the quantities involved. The word *virtual* qualifying those velocities may be considered sufficient to recall that they are no velocities at all but just test functions in the mathematical sense of functional analysis.

1.6. Lagrange (1736–1813)

1.6.1. *Lagrange's statement of the principle*

Up to this point, reading the statements we have reproduced, we implicitly assigned to the word "force", the meaning we give it today but it must be understood that the corresponding concept was still to receive a plain definition as in Lagrange's *Méchanique analitique*[21]:

> *"On entend, en général, par force ou puissance la cause, quelle qu'elle soit, qui imprime ou tend à imprimer du mouvement au corps auquel on la suppose appliquée ; et c'est aussi par la quantité du mouvement imprimé, ou prêt à imprimer, que la force ou puissance doit s'estimer. Dans l'état d'équilibre, la force n'a pas d'exercice actuel ; elle ne produit qu'une simple tendance au mouvement ; mais on doit toujours la mesurer par l'effet qu'elle produirait si elle n'était pas arrêtée".*

That, we may translate as follows:

> *"We generally mean by force or power the cause, whatever it is, which imparts or tends to impart a movement to the body to which it is supposed to be applied; and it is also by the quantity of the movement imparted, or ready to impart, that the force or power must be estimated. In the equilibrium state, the force does not have a current exercise; it only produces a simple tendency to movement; but one must always measure it by the effect it would produce if it were not stopped".*

Thence, echoing Galileo's and Newton's laws of inertia [NEW 87], the concept of force appears as the abstract cause of the alteration of motion it is actually imparting or would potentially tend to impart to a body. This last point is especially relevant when dealing with statics, where only "tendencies" can be considered.

---

21 [LAG 88a].

Lagrange then gave his consistent definition of virtual velocities:

"One must understand by virtual velocity, the velocity which a body in equilibrium would be ready to receive, in case this equilibrium should be upset; i.e. the velocity that this body would really take in the first instant of its movement"[22].

and his generalized statement of the principle of virtual velocities[23]:

"If a system of bodies or points, each of them being submitted to arbitrary powers, is in equilibrium, and if this system is given a small unspecified movement, in which each point moves along an infinitely small distance, which is its virtual velocity, the sum of the products of each power by the distance travelled by its point of application along the direction of that power, will always be equal to naught, with the small distances being counted positive when they are travelled in the direction of the power and negative in the opposite direction".

What is most important in this statement is that it explicitly deals with a *system* of bodies or points. It introduces "a small *unspecified* movement of the system", defined by independent small unspecified movements of each point of the system as illustrated in Lagrange's own proof of the principle given hereunder.

1.6.2. *Lagrange's proof of the principle*

A few years later, in a paper published in the *Journal de l'école polytechnique*[24], Lagrange expressed his dissatisfaction as to the principle of virtual velocities being usually derived from the principles of composition of forces and equilibrium of the lever which he considered not evident enough to be taken as a basis. He thus presented a new proof based upon the pulley block equilibrium principle.

---

22 [LAG 88]: "On doit entendre par *vitesse virtuelle* celle qu'un corps en équilibre est disposé à recevoir, en cas que l'équilibre vienne à être rompu, c'est-à-dire la vitesse que ce corps prendrait réellement dans le premier instant de son mouvement".
23 [LAG 88]: "Si un système quelconque de tant de corps ou points que l'on veut, tirés chacun par des puissances quelconques, est en équilibre, et qu'on donne à ce système un petit mouvement quelconque, en vertu duquel chaque point parcoure un espace infiniment petit qui exprimera sa vitesse virtuelle, la somme des puissances, multipliées chacune par l'espace que le point où elle est appliquée parcourt suivant la direction de cette même puissance, sera toujours égale à zéro, en regardant comme positifs les petits espaces parcourus dans le sens des puissances, et comme négatifs les espaces parcourus dans un sens opposé".
24 [LAG 97, pp. 115–118].

The main thrust of the reasoning is to consider that the forces applied to each body of a system (as in the preceding references, the term "body" refers to a material point) are exerted by means of a weight acting at one end of an ideally inextensible, flexible and weightless string through as many fixed and mobile pulley blocks and tackles as necessary, the other extremity of the string being fixed. It must be noted that no geometrical constraints, either internal or external, are imposed on the bodies of the system.

More precisely, using the notations Lagrange adopted in the subsequent editions of the *Méchanique analitique* where he reproduced this approach, we try to illustrate this description in Figure 1.4 (as a rule, Lagrange did not provide any figure: "*You will not find Figures in this Work. The methods I use require neither constructions nor geometrical or mechanical arguments, but only algebraic operations, in a regular and uniform course*"[25]). For simplicity, we consider the simple case of two bodies (points). One is connected to three pulley blocks where the numbers of pulleys are $P/2$, $Q/2$ and $T/2$ respectively with P, Q, T even integers. In the same way, the other body is connected to two pulley blocks with $R/2$ and $S/2$ pulleys (R and S even integers).

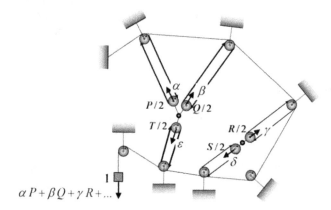

Figure 1.4. *Illustrating Lagrange's proof of the principle of virtual velocities*

As a consequence of the pulley block equilibrium principle, the forces acting on the bodies are commensurable and, taking the active weight at the extremity of the string as a unit, their values are P, Q, T along the corresponding directions of the string for the first body, and R and S for the second one.

---

25 "On ne trouvera point de Figures dans cet Ouvrage. Les méthodes que j'y emploie ne demandent ni constructions, ni raisonnements géométriques ou méchaniques, mais seulement des opérations algébriques, assujetties à une marche régulière et uniforme."

With this description of the system loading process, Lagrange states, as an obvious condition[26] defining the equilibrium state of the system, that any arbitrary infinitesimal displacement of each body (point) about its equilibrium position produces no downward movement of the weight at the free extremity of the string. This condition can be expressed explicitly as follows.

Infinitesimal arbitrary displacements of the bodies result in the distances between the mobile pulley blocks and the corresponding fixed ones being reduced (algebraically) by the quantities $\alpha, \beta, \gamma, ...$ respectively. As a result, taking in consideration the number of pulleys in each pulley block, the free extremity of the string will move downward along the infinitesimal distance $P\alpha + Q\beta + R\gamma + ...$. Writing down the equilibrium condition as stated here above, i.e. no downward motion of the active weight, results in:

$$P\alpha + Q\beta + R\gamma + ... = 0 \qquad [1.1]$$

"which is precisely the analytic expression of the general principle of virtual velocities".

Whatever its ingenuity, this proof obviously still suffers some shortcomings which do not appear in the proof given by Fourier in the same issue of the *Journal de l'école polytechnique*[27]. Anyhow, equation [1.1] is the starting point of the most important development in the subsequent editions of the *Méchanique analitique* regarding geometrical constraints.

1.6.3. *Lagrange's multipliers*

Substituting differential quantities for $\alpha, \beta, \gamma ...$, [1.1] may be written in the general differential form with respect to the coordinates (x_i, y_i, z_i) of each body (index i) of the system:

$$P\,dp + Q\,dq + R\,dr + ... = 0 \qquad [1.2]$$

where the differentials $dp, dq, dr, ...$ are typically written as

$$dp = \frac{\partial p}{\partial x_i}\,dx_i + \frac{\partial p}{\partial y_i}\,dy_i + \frac{\partial p}{\partial z_i}\,dz_i \qquad [1.3]$$

---

26 A "demand" as in Archimedes' rationales?
27 [FOU 98].

with the index i referring to the body concerned by the considered force. In the case of no geometrical constraints, equation [1.2] is valid $\forall dx_i, dy_i, dz_i, \forall i$. This leads to the equilibrium equations of the system by equating all the coefficients of the $dx_i, dy_i, dz_i, \forall i$ to zero.

Assuming the geometrical constraints that may be imposed on the evolution of the bodies are written as linear forms $dL, dM, dN...$ of the differentials $dx_i, dy_i, dz_i, \forall i$ assigned to be equal to zero:

$$dL = 0, \ dM = 0, \ dN = 0,...$$
[1.4]

Lagrange remarks that, from the theory of linear equations, writing [1.2] with [1.3] under the mathematical constraints [1.4] on $dx_i, dy_i, dz_i, \forall i$ is equivalent to writing:

$$P\,dp + Q\,dq + R\,dr + ... + \lambda\,dL + \mu\,dM + v\,dN + ... = 0, \ \forall dx_i, dy_i, dz_i, \forall i$$
[1.5]

where λ, μ, v are indeterminate (in other words, the linear form in [1.2] is a linear combination of the linear forms $dL, dM, dN ...$).

To this method Lagrange gave the name of *Méthode des multiplicateurs* (Multiplier Method) while referring to [1.5] as the general equation of equilibrium and explaining how to handle it in order to obtain the solution to the equilibrium problem.

But the most important step forward came out from his noting the mathematical similarity of the $\lambda\,dL, \mu\,dM...$ terms with the $P\,dp, Q\,dq...$ ones and thence giving a mechanical significance to the Lagrange multipliers. For instance, assuming the linear form dL to be the differential of a function L of the coordinates of the bodies in the system, the term $\lambda\,dL$ is written as

$$\lambda\,dL = \lambda\frac{\partial L}{\partial x_i}dx_i + \lambda\frac{\partial L}{\partial y_i}dy_i + \lambda\frac{\partial L}{\partial z_i}dz_i , \ i = 1, 2,... ,$$
[1.6]

which is quite similar to [1.3] but for the fact that the coordinates of more than one body may be involved. Thence, Lagrange's statement:

> "*It comes out then that each geometrical constraint equation is equivalent to one or several forces acting on the system, along given directions or, as a general rule, tending to vary the values*

of the given functions; so that the same state of equilibrium will be obtained for the system, either using these forces or the constraint equations. And here one encounters the metaphysical reason why introducing the terms $\lambda\,dL + \mu\,dM + ...$ in the general equilibrium equation makes it possible to treat this equation as if all bodies were completely free".

and further on:

"Conversely, these forces may be substituted for the geometrical constraint equations in such a way that, using these forces the constituent bodies of the system will be considered as completely free without any constraint... In proper words, these forces stand as the resistances that the bodies should meet for being linked to each other or due to the obstacles that may impede their motion; or rather, these forces are precisely the resistances, which must be equal and opposite to the pressures exerted by the bodies".

The scalars λ, μ, ν are now known as the *Lagrange multipliers* associated with the corresponding constraints.

These statements are crucial: we may say that they introduce and define binding and internal forces for the given geometrical constraints, either external or internal, through the concept of *duality*. Compared with the initial definition of forces by Lagrange (section 1.6.1), we observe that these forces are *defined through the movement they are supposed to impede*. It follows that, practically, resistances will not have a data status but be characterized by a limitation imposed on their magnitude.

Without getting into too many details, we must mention the contribution by Fossombroni[28] and the statement by Fourier[29]:

"Moreover, if one regards resistances as forces, which provides, as it is known, the means of estimating these resistances, the body can be considered free, and sum of the moments is nil for all possible displacements".

---

28 [FOS 96] which Fourier acknowledges in [FOU 98], *Œuvres publiées...*, p. 518.
29 [FOU 98], *Œuvres publiées...*, p. 488: "Au reste, si l'on considère les résistances comme des forces, ce qui fournit, comme on le sait, le moyen d'estimer ces résistances, le corps peut être regardé comme libre, et la somme des moments est nulle pour tous les déplacements possibles."

We may also remark that although Lagrange's proof assumes the geometrical constraints, either external or internal, to be written as linear forms of $dx_i, dy_i, dz_i, \forall i$ being equated to zero, the final interpretation of the scalars λ, μ, ν, related to the resistances associated with these geometrical constraints, yields the possibility of treating geometrical constraints that are expressed as inequalities such as $dL \geq 0$ or $dM \geq 0$ (as for unilateral support, for instance): the geometrical constraints are still considered as equalities, which we may call "bilateral", $dL = 0$, $dM = 0$, and inequalities $\lambda \leq 0, \mu \leq 0$ are imposed on λ, μ as conditions on the "resistances". This maintains the essential point that equation [1.5] is written as $\forall dx_i, dy_i, dz_i, \forall i$, as remarked by Fourier.

2

Dualization of Newton's Laws

2.1. In brief

The short historical survey in the preceding chapter has shown how the principle of virtual velocities was progressively stated in the form given by Lagrange, based upon "proofs" that started from simple accepted laws such as the law of the lever or the law of the pulleys. In the same way, quoting Lagrange's words "it is not self-evident enough", we will here establish the Principle of virtual work on a fully defined model – namely a system of material points – as the result of the dualization process applied to Newton's fundamental law of dynamics and law of action and reaction. The corresponding statement will then be given a more general formulation to fit any geometrical modeling of a material system and be "upgraded" to the status of a principle.

2.2. Newton's statements

2.2.1. *First law*

Figure 2.1. *Newton's Principia*[1]

---

1 Isaac Newton, 1642–1727.

Lex. I.

"Corpus omne perseverare in statu suo quiescendi vel movendi uniformiter in directum, nisi quatenus a viribus impressis cogitur statum illum mutare".

This is the *law of inertia*: "Every body persists in its state of being at rest or of moving uniformly straight forward, except insofar as it is compelled to change its state by force impressed".[2]

2.2.2. Second law

Lex. II.

"Mutationem motus proportionalem esse vi motrici impressæ, & fieri secundum lineam rectam qua vis illa imprimitur".

This is the *fundamental law of dynamics*: "The alteration of motion is ever proportional to the motive force impressed; and is made in the direction of the right line in which that force is impressed".[3]

2.2.3. Third law

Lex III.

"Actioni contrariam semper et aequalem esse reactionem: sive corporum duorum actiones in se mutuo semper esse aequales et in partes contrarias dirigi".

This is known as the *law of action and reaction*: "To every action there is always opposed an equal reaction: or the mutual actions of two bodies upon each other are always equal, and directed to contrary parts".[4]

Note that these three famous statements are preceded by eight definitions given as preliminaries where such concepts as the inertia force (*vis insita* or *vis inertiae*), absolute space (*spatium absolutum*) and relative space, and accelerative force (*vis acceleratrix*) are introduced and commented on.

---

2 [NEW 87], *The Principia*, transl. by Cohen and Whitman [COH 99].
3 [NEW 87], *The Principia*, transl. by Andrew Motte, 1729 [MOT 29].
4 [NEW 87], *The Principia*, transl. by Andrew Motte, 1729 [MOT 29].

2.2.4. Material points

Using our present mathematical terminology, we may say that any "body" whatsoever (an apple or even a planet) is modeled as a material point with a mass m, defined by its position in a reference frame and whose evolution is thus characterized by its velocity $\underline{U}$ in the reference frame, a vector in $\mathbb{R}^3$; besides this geometrical modeling, the mechanical modeling is completed by the force modeling: a concentrated force $\underline{F}$ is a vector in $\mathbb{R}^3$.

For a material point with mass m, the fundamental law of dynamics postulates the existence of Galilean reference frames $\mathcal{R}$ in which the law of inertia is valid in the form

$$\text{in a Galilean frame } \mathcal{R}, \ \underline{F} = m\underline{a}, \tag{2.1}$$

where $\underline{a}$ is the acceleration of the point mass in the reference frame $\mathcal{R}$. The product $m\underline{a}$ is the quantity of acceleration (or rate of change of the momentum $m\underline{U}$) of the mass m.

The law of action and reaction governing interaction forces for a pair of material points M_1 and M_2 is written as

$$\begin{cases} \underline{F}_{12} + \underline{F}_{21} = 0 \\ \underline{M_1 M_2} \wedge \underline{F}_{12} = 0 \end{cases} \tag{2.2}$$

where $\underline{F}_{12}$ denotes the force exerted by the material point M_1 on the material point M_2, (respectively $\underline{F}_{21}$ exerted by M_2 on M_1).

2.3. System of material points

2.3.1. System of material points

We now consider a set of n material points in its configuration κ at time t. For each material point (j) of mass m_j, the force to be entered on the left-hand side of the fundamental law [2.1] is the resultant by *vector addition* of the external force $\underline{F}_j$ exerted on (j) by influences originating outside the set of n material points, and forces exerted on (j) by the other material points (i) making up the set $(i \neq j, i = 1, 2, ...n)$. The forces exerted by a material point of the set on the other ones are assumed to be decoupled pairwise interactions, meaning that the influence of point (i) on point (j) does not depend on the presence of point (k). As in [2.2], we denote by $\underline{F}_{ij}$ the force exerted by the material point (i) on the material point (j), $(i \neq j, i = 1, 2, ... n)$.

The set of n material points constitutes the considered mechanical system, which we denote by S. Our model thus distinguishes *external forces* $\underline{F}_j$ from *internal forces* $\underline{F}_{ij}$, as shown in Figure 2.2.

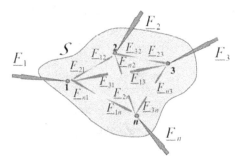

Figure 2.2. *System of point masses: internal and external forces*

For this system, the laws of dynamics are just Newton's laws, expressed as follows.

To begin with, we write down the fundamental law of dynamics [2.1] for each material point of the system S [5]:

$$\begin{cases} \text{in a Galilean frame,} \\ \forall(j) \in S, \ \underline{F}_j + \sum_{\substack{(i) \in S \\ i \neq j}} \underline{F}_{ij} = m_j \, \underline{a}_j, \end{cases} \qquad [2.3]$$

which exhibits the structure

$$\boxed{\text{External force on } (j)} + \boxed{\text{Internal force on } (j)} = \boxed{\text{Quantity of acceleration of } (j)} \qquad [2.4]$$

5 The summation convention for repeated indices *will not be used here*: all summations will be indicated explicitly.

In the next step, we formulate the law of action and reaction [2.2] governing interaction forces for each pair of material points:

$$
\begin{cases}
\forall (i) \in S, \forall (j) \neq (i) \in S, \\
\underline{F}_{ij} + \underline{F}_{ji} = 0 \\
M_i M_j \wedge \underline{F}_{ij} = 0
\end{cases} \tag{2.5}
$$

2.3.2. Subsystem

In a *thought experiment*, we consider a part of the previous system, thus defining a *subsystem* S' of S. In this subsystem, the distinction between internal and external forces no longer refers to S but to S'.

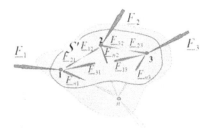

Figure 2.3. *Subsystem S': internal and external forces*

From the description of interaction forces between material points given previously, we derive the external force $\underline{F}'_j$ on a material point (j) of S' straightforwardly in terms of the forces $\underline{F}_j$ and $\underline{F}_{ij}$:

$$
\begin{cases}
\forall S' \subset S, \forall (j) \in S', \\
\underline{F}'_j = \underline{F}_j + \sum_{(i) \in (S - S')} \underline{F}_{ij}.
\end{cases} \tag{2.6}
$$

The internal forces on the material point (j) of S' reduce to those exerted by the other material points of S': $\underline{F}_{ij}$, $(i) \in S'$, $i \neq j$ (Figure 2.3).

As a result, the fundamental law of dynamics for each material point of S' [2.3] can also be expressed in the form:

$$\left\{ \begin{array}{l} \text{in a Galilean frame,} \\ \forall (j) \in S' \subset S, \\ \underline{F}'_j + \sum_{\substack{(i) \in S' \\ i \neq j}} \underline{F}_{ij} = m_j \, \underline{a}_j \, , \end{array} \right. \qquad [2.7]$$

which exhibits the same structure with respect to S' as [2.3] to S :

External force $/S'$ on (j)	+	Internal force $/S'$ on (j)	=	Quantity of acceleration of (j)	[2.8]

Regarding the law of action and reaction for the interaction forces between each pair of material points in S', it is just the restriction of [2.5] to each pair of material points in S'.

2.3.3. *Law of mutual actions*

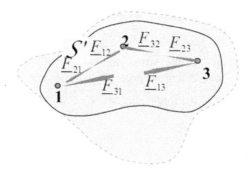

Figure 2.4. *System of point masses: law of mutual actions*

It is worth noting that the law of action and reaction expressed in [2.5] is equivalent to statement [2.9] which considers, rather than just a pair of material points, the whole set of material points in any arbitrary subsystem S' of S (including S itself). We will call this the *law of mutual actions*. It states that the internal forces in *any arbitrary subsystem* form a system of forces equivalent to zero, i.e. with zero resultant and zero resultant moment (Figure 2.4):

$$
\left\{
\begin{array}{l}
\forall S' \subseteq S, \\[2mm]
\displaystyle\sum_{(j)\in S'}\sum_{\substack{(i)\in S' \\ i\neq j}} \underline{F}_{ij} = 0 \\[6mm]
\displaystyle\sum_{(j)\in S'}\sum_{\substack{(i)\in S' \\ i\neq j}} \underline{OM}_j \wedge \underline{F}_{ij} = 0
\end{array}
\right.
\qquad [2.9]
$$

where O denotes an arbitrary geometrical point in κ.

2.3.4. *Summing up...*

For a system S of material points, we have seen that the description of internal and external forces can be used to determine the forces corresponding to any subsystem S' of S. The laws of mechanics are the fundamental law of dynamics and the law of mutual actions, expressed in the same form for the system as a whole S and for any subsystem S' of S.

2.4. Dualization and virtual work for a system of material points

2.4.1. *System comprising a single material point*

The fundamental law for a single point system can be dualized easily. Let $\hat{\underline{U}}$ denote an arbitrary vector in $\mathbb{R}^3$. Then, [2.3] is equivalent to the statement

$$
\left\{
\begin{array}{l}
\text{in a Galilean frame,} \\[2mm]
\forall \hat{\underline{U}} \in \mathbb{R}^3 \\[2mm]
\underline{F}.\hat{\underline{U}} = m\,\underline{a}.\hat{\underline{U}}.
\end{array}
\right.
\qquad [2.10]
$$

Defining linear forms $\mathcal{P}$ and $\mathcal{A}$ on $\mathbb{R}^3$ by

$$
\left\{
\begin{array}{l}
\mathcal{P}(\hat{\underline{U}}) = \underline{F}.\hat{\underline{U}} \\[2mm]
\mathcal{A}(\hat{\underline{U}}) = m\,\underline{a}.\hat{\underline{U}}
\end{array}
\right.
\qquad [2.11]
$$

we have the dual version of [2.3]

$$\begin{cases} \text{in a Galilean frame,} \\ \forall \underline{\hat{U}} \in \mathbb{R}^3, \ \mathcal{P}(\underline{\hat{U}}) = \mathcal{A}(\underline{\hat{U}}). \end{cases} \qquad [2.12]$$

2.4.2. System comprising several material points

Consider again the system S made up of n point masses m_j associated with material points (j) located at M_j at time t in the configuration κ, as shown in Figure 2.1.

2.4.2.1. Dualizing the fundamental law of dynamics

2.4.2.1.1. System S

For each material point (j) of S, we introduce an arbitrary vector $\underline{\hat{U}}_j$ of $\mathbb{R}^3$ in order to dualize the fundamental law of dynamics [2.3]. The dual statement, equivalent to [2.3] is:

$$\begin{cases} \text{in a Galilean frame,} \\ \forall (j) \in S, \forall \underline{\hat{U}}_j \in \mathbb{R}^3, \\ \underline{F}_j \cdot \underline{\hat{U}}_j + \sum_{\substack{(i) \in S \\ i \neq j}} \underline{F}_{ij} \cdot \underline{\hat{U}}_j = m_j \, \underline{a}_j \cdot \underline{\hat{U}}_j. \end{cases} \qquad [2.13]$$

Clearly, since the n vectors $\underline{\hat{U}}_j$ are arbitrary, the above statement is equivalent to the global statement

$$\begin{cases} \text{in a Galilean frame,} \\ \forall \underline{\hat{U}}_1, \dots \underline{\hat{U}}_n \in \mathbb{R}^3 \times \dots \mathbb{R}^3, \\ \sum_{(j) \in S} \underline{F}_j \cdot \underline{\hat{U}}_j + \sum_{(j) \in S} \sum_{\substack{(i) \in S \\ i \neq j}} \underline{F}_{ij} \cdot \underline{\hat{U}}_j = \sum_{(j) \in S} m_j \, \underline{a}_j \cdot \underline{\hat{U}}_j. \end{cases} \qquad [2.14]$$

On $\mathbb{R}^3 \times \dots \mathbb{R}^3$, we now define three linear forms

$$\left\{ \begin{aligned} \mathcal{P}_e(\hat{\underline{U}}_1,...\hat{\underline{U}}_n) &= \sum_{(j)\in S} \underline{F}_j \cdot \hat{\underline{U}}_j \\ \mathcal{P}_i(\hat{\underline{U}}_1,...\hat{\underline{U}}_n) &= \sum_{(j)\in S}\sum_{\substack{(i)\in S \\ i\neq j}} \underline{F}_{ij} \cdot \hat{\underline{U}}_j \\ \mathcal{A}(\hat{\underline{U}}_1,...\hat{\underline{U}}_n) &= \sum_{(j)\in S} m_j\,\underline{a}_j \cdot \hat{\underline{U}}_j \end{aligned} \right. \qquad [2.15]$$

and we can state the dual formulation of the fundamental law of dynamics for the system S as

$$\left\{ \begin{aligned} &\text{in a Galilean frame,} \\ &\forall \hat{\underline{U}}_1,...\hat{\underline{U}}_n \in \mathbb{R}^3 \times ...\mathbb{R}^3, \\ &\mathcal{P}_e(\hat{\underline{U}}_1,...\hat{\underline{U}}_n) + \mathcal{P}_i(\hat{\underline{U}}_1,...\hat{\underline{U}}_n) = \mathcal{A}(\hat{\underline{U}}_1,...\hat{\underline{U}}_n). \end{aligned} \right. \qquad [2.16]$$

2.4.2.1.2. Subsystem S'

For an arbitrary subsystem S' of S, dualization is performed in exactly the same way, starting from [2.7], which only involves the material points of S'. An arbitrary vector $\hat{\underline{U}}_j$ of $\mathbb{R}^3$ is associated with each of these points and the separation into internal and external forces relative to S', stated in [2.6] and [2.8], leads us to define the following linear forms on $\mathbb{R}^3 \times ...\mathbb{R}^3$:

$$\left\{ \begin{aligned} &S' = (\ell),...(p), \\ &\forall \hat{\underline{U}}_\ell,...\hat{\underline{U}}_p \in \mathbb{R}^3 \times ...\mathbb{R}^3 \\ &\mathcal{P}'_e(\hat{\underline{U}}_\ell,...\hat{\underline{U}}_p) = \sum_{(j)\in S'} \underline{F}'_j \cdot \hat{\underline{U}}_j \\ &\mathcal{P}'_i(\hat{\underline{U}}_\ell,...\hat{\underline{U}}_p) = \sum_{(j)\in S'}\sum_{\substack{(i)\in S' \\ i\neq j}} \underline{F}_{ij} \cdot \hat{\underline{U}}_j \\ &\mathcal{A}'(\hat{\underline{U}}_\ell,...\hat{\underline{U}}_p) = \sum_{(j)\in S'} m_j\,\underline{a}_j \cdot \hat{\underline{U}}_j. \end{aligned} \right. \qquad [2.17]$$

The dualized statement of the fundamental law of dynamics for the subsystem S' is then given by [2.18], the obvious counterpart of [2.16]:

$$\begin{cases} \text{in a Galilean frame,} \\ S' = (\ell), ...(p) \subset S, \\ \forall \underline{\hat{U}}_\ell, ...\underline{\hat{U}}_p \in \mathbb{R}^3 \times ...\mathbb{R}^3, \\ \mathcal{P}'_e(\underline{\hat{U}}_\ell, ...\underline{\hat{U}}_p) + \mathcal{P}'_i(\underline{\hat{U}}_\ell, ...\underline{\hat{U}}_p) = \mathcal{A}'(\underline{\hat{U}}_\ell, ...\underline{\hat{U}}_p). \end{cases}$$ [2.18]

2.4.2.2. *Dualizing the law of mutual actions*

The law of mutual actions, as expressed in [2.9], constitutes a system of two vector equations for each subsystem S' of S, including S itself. Each of these systems of two equations is dualized by introducing two arbitrary vectors of $\mathbb{R}^3$: $\underline{\hat{U}}_0$ and $\underline{\hat{\omega}}_0$.

Then, [2.9] is equivalent to

$$\begin{cases} \forall S' \subseteq S, \\ \forall \underline{\hat{U}}_0 \in \mathbb{R}^3, \forall \underline{\hat{\omega}}_0 \in \mathbb{R}^3, \\ \left(\sum_{(j) \in S'} \sum_{\substack{(i) \in S' \\ i \neq j}} \underline{F}_{ij} \right) . \underline{\hat{U}}_0 + \left(\sum_{(j) \in S'} \sum_{\substack{(i) \in S' \\ i \neq j}} \underline{OM}_j \wedge \underline{F}_{ij} \right) . \underline{\hat{\omega}}_0 = 0. \end{cases}$$ [2.19]

Putting

$$\begin{cases} \forall \underline{\hat{U}}_0 \in \mathbb{R}^3, \forall \underline{\hat{\omega}}_0 \in \mathbb{R}^3, \\ \forall (j) \in S', \\ \underline{\hat{U}}_j = \underline{\hat{U}}_0 + \underline{\hat{\omega}}_0 \wedge \underline{OM}_j, \end{cases}$$ [2.20]

statement [2.19] can be reformulated as

$$\begin{cases} \forall S' \subseteq S, \\ \forall \underline{\hat{U}}_0 \in \mathbb{R}^3, \forall \underline{\hat{\omega}}_0 \in \mathbb{R}^3, \\ \left(\sum_{(j) \in S'} \sum_{\substack{(i) \in S' \\ i \neq j}} \underline{F}_{ij} \right) . \underline{\hat{U}}_j = 0 \text{ if [2.20].} \end{cases}$$ [2.21]

Here, we recognize the linear form $\mathcal{P}'_i$ defined in [2.17]. The dual formulation of the law of mutual actions [2.9] then becomes

$$\begin{cases} \forall S' = (\ell),...(p) \subseteq S, \\ \forall \hat{\underline{U}}_0 \in \mathbb{R}^3, \forall \hat{\underline{\omega}}_0 \in \mathbb{R}^3, \\ \mathcal{P}'_i(\hat{\underline{U}}_\ell,...\hat{\underline{U}}_p) = 0 \text{ if } [2.20]. \end{cases} \qquad [2.22]$$

Note that the vectors $\hat{\underline{U}}_j$ constrained to satisfy [2.20] are such that

$$\begin{cases} \forall (i) \in S', \forall (j) \in S', \\ \underline{M_i M_j}.(\hat{\underline{U}}_j - \hat{\underline{U}}_i) = 0 \text{ if } [2.20], \end{cases} \qquad [2.23]$$

which means that, if the geometrical points M_j, $(j) \in S'$, in κ had velocities $\hat{\underline{U}}_j$ defined from $\hat{\underline{U}}_0$ and $\hat{\underline{\omega}}_0$ by [2.20], their mutual separations $\left| M_i M_j \right|$ would be conserved. The corresponding instantaneous motion is just the *rigid body motion* defined by arbitrary vectors $\hat{\underline{U}}_0$ (velocity of point O) and $\hat{\underline{\omega}}_0$ (angular velocity).

2.4.3. *Virtual velocity, virtual motion and virtual work*

2.4.3.1. *Material point*

For a material point, where a force $\underline{F}$ is applied at a point M in κ at time t, the scalar product $\underline{F}.\hat{\underline{U}}$ occurring in [2.10] associates with each vector $\hat{\underline{U}}$ in $\mathbb{R}^3$ the *rate* at which the force $\underline{F}$ would do *work* if the material point were moving with velocity $\hat{\underline{U}}$.

We shall choose our terminology carefully, in order to emphasize the arbitrary nature of the vector $\hat{\underline{U}}$ in $\mathbb{R}^3$ as mentioned in [2.10]. In particular, it implies that $\hat{\underline{U}}$ is fully independent of any restrictions which might be imposed on the actual velocity of the *material point* at M in its real (or actual) motion.

$\hat{\underline{U}}$ is the *virtual velocity* of the material point M. It defines a virtual motion of this material point.

$P(\hat{\underline{U}}) = \underline{F}.\hat{\underline{U}}$ is the *virtual rate of work by the force* $\underline{F}$ in this virtual motion.

$\mathcal{A}(\hat{\underline{U}}) = m\underline{a}.\hat{\underline{U}}$ is therefore the *virtual rate of work by the quantity of acceleration* $m\underline{a}$ of the material point in the virtual motion.

2.4.3.2. *System of material points*

The same arguments apply to a system S comprising several material points.

The whole set of arbitrary vectors $\hat{\underline{U}}_1, ... \hat{\underline{U}}_n$ in $\mathbb{R}^3$ introduced in [2.13] is a distribution of virtual velocities in κ for the material points (j) of S. It defines a *virtual motion* (v.m.) of the system S. Then

$$P_e(\hat{\underline{U}}_1, ... \hat{\underline{U}}_n) = \sum_{(j) \in S} \underline{F}_j.\hat{\underline{U}}_j \quad \text{is, for } S \text{, the virtual rate of work by external forces}$$

in this virtual motion;

$$P_i(\hat{\underline{U}}_1, ... \hat{\underline{U}}_n) = \sum_{(j) \in S} \sum_{\substack{(i) \in S \\ i \neq j}} \underline{F}_{ij}.\hat{\underline{U}}_j \quad \text{is, for } S \text{, the virtual rate of work by}$$

internal forces;

$$\mathcal{A}(\hat{\underline{U}}_1, ... \hat{\underline{U}}_n) = \sum_{(j) \in S} m_j\, \underline{a}_j .\hat{\underline{U}}_j \quad \text{is the virtual rate of work by quantities}$$

of acceleration.

Similarly, for a subsystem $S' = (\ell), ... (p)$, the *virtual motion* of S' is defined by $\hat{\underline{U}}_\ell, ... \hat{\underline{U}}_p$ and the virtual rates of work corresponding to those specified above are $P'_e(\hat{\underline{U}}_\ell, ... \hat{\underline{U}}_p)$, $P'_i(\hat{\underline{U}}_\ell, ... \hat{\underline{U}}_p)$, $\mathcal{A}'(\hat{\underline{U}}_\ell, ... \hat{\underline{U}}_p)$ defined by [2.17].

For the system S or an arbitrary subsystem S', the set of all possible virtual motions can clearly be given a *vector space structure*.

Among these virtual motions, the dualized form of the law of mutual actions only refers to those distributions of virtual velocities $\hat{\underline{U}}_j$ that satisfy the condition [2.20] on S or on the considered subsystem S'. Bearing in mind the comment at the end of section 2.4.2 as regards statement [2.23], it is natural to refer to this class of virtual motions as *rigid body virtual motions* (r.b.v.m.) of the system S or considered subsystem S'.

2.4.4. *Statement of the principle of virtual work (P.V.W.)*

Let us sum up the above discussion. For a system of material points and the model of forces presented in section 2.3, we have shown equivalence between the laws of mechanics (Newton's laws) and the following statement of the principle of *virtual work*, using the definitions given earlier:

$$\left\{ \begin{array}{l} \text{in a Galilean frame,} \\ \forall \hat{\underline{U}}_1, ... \hat{\underline{U}}_n \text{ v.m. of } S, \\ \mathcal{P}_{\mathrm{e}}(\hat{\underline{U}}_1, ... \hat{\underline{U}}_n) + \mathcal{P}_{\mathrm{i}}(\hat{\underline{U}}_1, ... \hat{\underline{U}}_n) = \mathcal{A}(\hat{\underline{U}}_1, ... \hat{\underline{U}}_n) \end{array} \right. \qquad [2.24]$$

$$\left\{ \begin{array}{l} \forall S' = (\ell), ... (p) \subset S, \\ \text{in a Galilean frame ,} \\ \forall \hat{\underline{U}}_\ell, ... \hat{\underline{U}}_p \text{ v.m. of } S', \\ \mathcal{P}'_{\mathrm{e}}(\hat{\underline{U}}_\ell, ... \hat{\underline{U}}_p) + \mathcal{P}'_{\mathrm{i}}(\hat{\underline{U}}_\ell, ... \hat{\underline{U}}_p) = \mathcal{A}'(\hat{\underline{U}}_\ell, ... \hat{\underline{U}}_p) \end{array} \right. \qquad [2.25]$$

$$\left\{ \begin{array}{l} \forall\ S' = (\ell), ... (p) \subseteq S \\ \forall \hat{\underline{U}}_\ell, ... \hat{\underline{U}}_p \text{ r.b.v.m. of } S' \\ \mathcal{P}'_{\mathrm{i}}(\hat{\underline{U}}_\ell, ... \hat{\underline{U}}_p) = 0. \end{array} \right. \qquad [2.26]$$

The first propositions [2.24] and [2.25] dualize the fundamental law of dynamics, and the last one, [2.26], the law of mutual actions.

Considering *rigid body virtual motions*, we observe that, as a consequence of [2.24] and [2.26], we obtain the following statement for the system:

$$\left\{ \begin{array}{l} \text{in a Galilean frame,} \\ \forall \hat{\underline{U}}_1, ... \hat{\underline{U}}_n \text{ r.b.v.m. of } S, \\ \mathcal{P}_{\mathrm{e}}(\hat{\underline{U}}_1, ... \hat{\underline{U}}_n) = \mathcal{A}(\hat{\underline{U}}_1, ... \hat{\underline{U}}_n) \end{array} \right. \qquad [2.27]$$

and, in the same way, for S'

$$\begin{cases} \forall S' = (\ell),...(p) \subset S, \\ \text{in a Galilean frame ,} \\ \forall \hat{\underline{U}}_\ell,...\hat{\underline{U}}_p \text{ r.b.v.m. of } S', \\ \mathcal{P}'_e(\hat{\underline{U}}_\ell,...\hat{\underline{U}}_p) = \mathcal{A}'(\hat{\underline{U}}_\ell,...\hat{\underline{U}}_p). \end{cases} \qquad [2.28]$$

These statements express the dualized form of the fundamental law of dynamics for the system S (respectively subsystem S') *considered as a whole*.

2.4.5. *Virtual motions in relation to the modeling of forces*

We said at the outset that our aim was to identify the principles of dualization and also those statements that could be laid down as fundamental principles, starting from a known model of forces and corresponding statements of the laws of mechanics.

In this process, it is clear that the spaces on which the dualization is carried out for a system or its subsystems are directly determined by our prior knowledge of the model: since forces are modeled by vectors in $\mathbb{R}^3$, the duality established via the Euclidean scalar product introduces arbitrary vectors $\hat{\underline{U}}_j$ in $\mathbb{R}^3$.

We might attempt to start from another force model, in which the idea of point masses would be abandoned, apart from the concentrated forces $\underline{F}_j$ and $\underline{F}_{ij}$ already proposed in the previous case, with the introduction of couples $\underline{\Gamma}_j$ and $\underline{\Gamma}_{ij}$ at M_j. Dualization would then introduce, in addition to the vectors $\hat{\underline{U}}_j$, a further set of arbitrary vectors $\hat{\underline{r}}_j$ in $\mathbb{R}^3$. These $\hat{\underline{r}}_j$, associated with the couples in the scalar products, would be interpreted as virtual angular velocities at each point M_j. Products of type $\underline{\Gamma}_j.\hat{\underline{r}}_j$ would complete the previous expressions of the linear forms into $\mathcal{P}_e(\hat{\underline{U}}_1,...\hat{\underline{U}}_n,\hat{\underline{r}}_1,...\hat{\underline{r}}_n)$, $\mathcal{P}'_e(\hat{\underline{U}}_\ell,...\hat{\underline{U}}_p,\hat{\underline{r}}_\ell,...\hat{\underline{r}}_p)$, $\mathcal{P}_i(\hat{\underline{U}}_1,...\hat{\underline{U}}_n,\hat{\underline{r}}_1,...\hat{\underline{r}}_n)$ and $\mathcal{P}'_i(\hat{\underline{U}}_\ell,...\hat{\underline{U}}_p,\hat{\underline{r}}_\ell,...\hat{\underline{r}}_p)$. A *virtual motion* of S would then be defined by $(\hat{\underline{U}}_1,...\hat{\underline{U}}_n,\hat{\underline{r}}_1,...\hat{\underline{r}}_n)$. The *rigid body motions* of S', in which $\mathcal{P}'_i$ vanishes, would be defined by [2.20] for $\hat{\underline{U}}_\ell,...\hat{\underline{U}}_p$ and, in addition,

$$\hat{\underline{r}}_\ell = ..\hat{\underline{r}}_p = \hat{\underline{\omega}}_0. \qquad [2.29]$$

Such a geometrical modeling is a way to take into account an underlying oriented microstructure in what may appear as a micro-macro modeling process – micropolar media, curvilinear media (see Chapter 7), plates (see Chapter 8), etc. – which stresses the link that exists between the virtual motions considered in the dualization and the force model we end up with.

2.5. Virtual work method for a system of material points

2.5.1. *Presentation of the virtual work method*

The guiding idea behind the virtual work method is to dualize the problem of modeling forces by taking the principle of virtual work [2.24]–[2.26] as the fundamental starting point.

We first provide the geometrical description of the system, which involves its actual motions, and then define the *virtual motions* which are to be considered for the system and its subsystems and which must constitute vector spaces.

On these vector spaces, we write down the linear forms $\mathcal{P}_e$, $\mathcal{P}_e'$, $\mathcal{P}_i$, $\mathcal{P}_i'$, $\mathcal{A}$, $\mathcal{A}'$, which are to represent the virtual rates of work by external forces, internal forces and quantities of acceleration for the system and its subsystems. The choice of these linear forms, through the cofactors it introduces, sketches out the force model that we wish to build: the cofactors are a first representation of internal and external forces that will be specified in the following steps of the method.

Using the principle of virtual work, we will be able to complete the model by:

– specifying the linear forms $\mathcal{P}_e$, $\mathcal{P}_e'$, $\mathcal{P}_i$, $\mathcal{P}_i'$, if necessary, in order to comply with the statements of the principle;

– deriving the dynamical equations for the system and its subsystems.

Obviously, the choice of the vector spaces of virtual motions, which are the mathematical tools for the dualization construction, is of key importance. It must be clear that the virtual motions are in no way restricted by conditions that may be imposed on the actual motions of the system in whatever mechanical evolution it may be following. But, they must encompass all the actual motions of the system since, otherwise, the construction would have no practical utility.

2.5.2. *Example of an application*

The aim here is to show how the force model which we started with in section 2.3 can be set up using the virtual work method.

2.5.2.1. *Geometrical model and actual motions*

The system S is described as being made up of n material points (j) of mass m_j occupying geometrical positions M_j in the configuration κ. A subsystem S' is a part of S, *delimited in a thought experiment*, comprising material points $(\ell),...(p)$. Since we are dealing with material points, the geometrical configuration is simply specified by giving the geometrical positions of the points M_j at time t. The actual motions of the system S or a subsystem S' are defined in κ by velocities $\underline{U}_j(t)$ of the relevant points.

2.5.2.2. *Vector spaces of virtual motions*

The vector space of virtual motions of the system S is generated by n vectors $\hat{\underline{U}}_j$ in $\mathbb{R}^3$, the virtual velocities associated with points M_j. The virtual motions of a subsystem S' are defined in the same way. These vector spaces clearly contain the actual motions of the system S (or subsystem S') and the rigid body motions of S (or S').

2.5.2.3. *Writing down the linear forms $\mathcal{P}_e$ and $\mathcal{P}'_e$*

Once these choices have been made, the most general expression for the linear form $\mathcal{P}_e$ on the vector space generated by the $\hat{\underline{U}}_j$ is

$$\mathcal{P}_e(\hat{\underline{U}}_1,...\hat{\underline{U}}_n) = \sum_{(j)\in S} \underline{F}_j \cdot \hat{\underline{U}}_j \qquad [2.30]$$

where the cofactors $\underline{F}_j$ are arbitrary vectors.

Similarly for $\mathcal{P}'_e$ relative to S', the most general expression is

$$\mathcal{P}'_e(\hat{\underline{U}}_\ell,...\hat{\underline{U}}_p) = \sum_{(j)\in S'} \underline{F}^{S'}_j \cdot \hat{\underline{U}}_j \qquad [2.31]$$

where the cofactors $\underline{F}_j^{S'}$ associated with M_j are arbitrary vectors which *depend on the subsystem S'* to which the material point (j) belongs.

External forces relative to S and S' respectively are thus modeled by the vector sets $\underline{F}_j$ ($j = 1...n$) and $\underline{F}_j^{S'}$ ($j = \ell...p$), respectively.

2.5.2.4. *Writing down the linear forms $\mathcal{A}$ and $\mathcal{A}'$*

The actual motions of the system define accelerations $\underline{a}_j$ of the material points, and hence the quantities of acceleration $m_j \underline{a}_j$. The virtual rate of work by quantities of acceleration is thus written as

$$\mathcal{A}(\hat{\underline{U}}_1,...\hat{\underline{U}}_n) = \sum_{(j) \in S} m_j \underline{a}_j . \hat{\underline{U}}_j \quad \text{for } S \tag{2.32}$$

$$\mathcal{A}'(\hat{\underline{U}}_\ell,...\hat{\underline{U}}_p) = \sum_{(j) \in S'} m_j \underline{a}_j . \hat{\underline{U}}_j \quad \text{for } S' \tag{2.33}$$

2.5.2.5. *Writing down the linear forms $\mathcal{P}_i$ and $\mathcal{P}_i'$*

In the virtual work statements, the distinction between external and internal forces is unambiguous. As a result, for the system S, the most general form for $\mathcal{P}_i$ is

$$\mathcal{P}_i(\hat{\underline{U}}_1,...\hat{\underline{U}}_n) = \sum_{(j) \in S} \underline{\Phi}_j . \hat{\underline{U}}_j \tag{2.34}$$

where each vector $\underline{\Phi}_j$ models the internal forces exerted by the material points of $(S - (j))$ on (j).

Similarly, for a subsystem S', the most general form for $\mathcal{P}_i'$ is written as

$$\mathcal{P}_i'(\hat{\underline{U}}_\ell,...\hat{\underline{U}}_p) = \sum_{(j) \in S'} \underline{\Phi}_j^{S'} . \hat{\underline{U}}_j \tag{2.35}$$

where the $\underline{\Phi}_j^{S'}$ model the forces exerted by the material points of $(S' - (j))$ on (j).

As a particular case, consider a subsystem S' comprising only two elements (i), (j); equation [2.35] for the virtual rate of work $\mathcal{P}_i'(\hat{\underline{U}}_i, \hat{\underline{U}}_j)$ will be written as

$$\mathcal{P}_i'(\hat{\underline{U}}_i, \hat{\underline{U}}_j) = \underline{F}_{ji} . \hat{\underline{U}}_i + \underline{F}_{ij} . \hat{\underline{U}}_j \qquad [2.36]$$

where $\underline{F}_{ji} = \underline{\Phi}_i^{S'}$ (respectively $\underline{F}_{ij} = \underline{\Phi}_j^{S'}$) is the force exerted by (i) on (j) (respectively (j) on (i)).

Recalling that the concept of subsystem is just the result of a *thought experiment* which does not influence the interactions between the elements of the system, we derive for any subsystem S', S included, the most general form for $\mathcal{P}_i'$

$$\begin{cases} \forall S' = (\ell), \dots (p) \subseteq S \\ \forall \hat{\underline{U}}_\ell, \dots \hat{\underline{U}}_p \text{ v.m. of } S' \\ \mathcal{P}_i'(\hat{\underline{U}}_\ell, \dots \hat{\underline{U}}_p) = \sum_{(j) \in S'} \sum_{\substack{(i) \in S' \\ i \neq j}} \underline{F}_{ij} . \hat{\underline{U}}_j \end{cases} \qquad [2.37]$$

where the cofactors $\underline{F}_{ij}$ are arbitrary vectors, defined on the subsystems which comprise only two elements (binary subsystems).

Thus, the virtual work done by internal forces for the system S (respectively for a subsystem S') in an arbitrary virtual motion $(\hat{\underline{U}}_1, \dots \hat{\underline{U}}_n)$ is the *sum* of the virtual work for each binary subsystem within S (respectively S') in this virtual motion of S (respectively S').

2.5.2.6. Implementing the principle of virtual work

2.5.2.6.1. Complying with the dual statement of the law of mutual actions

Consider any subsystem comprising two arbitrary material points (i) and (j).

Statement [2.26] of the principle of virtual work results in

$$\begin{cases} \forall (i), \forall (j) \in S \\ \forall \underline{\hat{U}}_0 \in \mathbb{R}^3, \ \forall \underline{\hat{\omega}}_0 \in \mathbb{R}^3, \\ \underline{\hat{U}}_j = \underline{\hat{U}}_0 + \underline{\hat{\omega}}_0 \wedge \underline{OM}_j, \\ \underline{F}_{ij} \cdot \underline{\hat{U}}_j + \underline{F}_{ji} \cdot \underline{\hat{U}}_i = 0, \end{cases} \qquad [2.38]$$

from which we derive

$$\begin{cases} \underline{F}_{ij} + \underline{F}_{ji} = 0 \\ \underline{M_i M_j} \wedge \underline{F}_{ij} = 0. \end{cases} \qquad [2.39]$$

Hence

$$\mathcal{P}'_i(\underline{\hat{U}}_i, \underline{\hat{U}}_j) = \underline{F}_{ij} \cdot (\underline{\hat{U}}_j - \underline{\hat{U}}_i) \qquad [2.40]$$

and also, introducing the unit vector $\underline{e}_{ij}$ collinear with $\underline{M_i M_j}$

$$\underline{e}_{ij} = \underline{M_i M_j} \big/ \left| \underline{M_i M_j} \right|, \qquad [2.41]$$

it follows that $\underline{F}_{ij}$ takes the form

$$\underline{F}_{ij} = -F_{ij}\, \underline{e}_{ij} \qquad [2.42]$$

where we note the choice of the minus sign which will be commented later on.

Denoting by $\hat{\delta}_{ij}$ the virtual rate of extension of the length $\left| \underline{M_i M_j} \right|$ due to the virtual velocities $\underline{\hat{U}}_i$ and $\underline{\hat{U}}_j$ of the geometrical points M_i and M_j (the terminology reminds us that the relation [2.43] is purely geometrical and identical to the equation giving the actual rate of extension $\dot{\delta}_{ij}$ of $\left| \underline{M_i M_j} \right|$ for actual velocities $\underline{U}_i$ and $\underline{U}_j$)

$$\hat{\delta}_{ij} = (\underline{\hat{U}}_j - \underline{\hat{U}}_i) \cdot \underline{e}_{ij}, \qquad [2.43]$$

equation [2.40] becomes

$$\mathcal{P}_i'(\hat{\underline{U}}_i,\hat{\underline{U}}_j)=-F_{ij}\,\hat{\delta}_{ij}\,. \qquad\qquad [2.44]$$

For an arbitrary subsystem S', including S itself, we then have the transformed statement of [2.37]:

$$\begin{cases} \forall\ S'=(\ell),...(p)\subseteq S \\ \forall \hat{\underline{U}}_\ell,...\hat{\underline{U}}_p \ \text{v.m. of } S' \\ \mathcal{P}_i'(\hat{\underline{U}}_\ell,...\hat{\underline{U}}_p)=-\sum_{\substack{(j)\in S'\\}}\sum_{\substack{(i)\in S'\\ i>j}}F_{ij}\,\hat{\delta}_{ij}\,. \end{cases} \qquad [2.45]$$

Note the constraint $i>j$ under the Σ sign of [2.45] instead of $i\neq j$ in [2.37] to avoid that one binary subsystem being counted twice!

2.5.2.6.2. Virtual rate of work by internal forces

Thus, for any subsystem $S'\subseteq S$, the linear form expressing the virtual rate of work by internal forces, whose most general form was written in [2.37], must take the form specified by [2.45] in order to comply with the principle of virtual work (as anticipated in section 2.5.1): $\mathcal{P}_i'$ is calculated by summing a kind of discrete density of virtual rate of work by the internal forces defined from the binary subsystems within $S'\subseteq S$. The cofactors in [2.45] stand for the internal force modeling: a set of $n(n-1)/2$ independent scalars F_{ij}. These scalars are the intensities of the concentrated binary interaction forces between the material points of the system. These forces are collinear with the lines joining the concerned material points and only do work in the virtual rate of extension of the distances between the material points. With the sign convention adopted in [2.42], $F_{ij}>0$ corresponds to a negative value of the virtual rate of work when $\hat{\delta}_{ij}$, the virtual rate of extension of the distance between M_i and M_j, is positive: it means that $F_{ij}>0$ indicates an *attractive force* exerted by (i) on (j) – conversely by (j) on (i).

2.5.2.6.3. Fundamental law of dynamics

With [2.30], [2.32] and [2.37] for the linear forms involved, statement [2.24] of the principle of virtual work to the system S becomes

in a Galilean frame,

$\forall \hat{\underline{U}}_1, ... \hat{\underline{U}}_n$ v.m. of S, [2.46]

$$\sum_{(j) \in S} \underline{F}_j \cdot \hat{\underline{U}}_j + \sum_{(j) \in S} \sum_{\substack{(i) \in S \\ i \neq j}} \underline{F}_{ij} \cdot \hat{\underline{U}}_j = \sum_{(j) \in S} m_j \underline{a}_j \cdot \hat{\underline{U}}_j.$$

Taking advantage of $\hat{\underline{U}}_1, \hat{\underline{U}}_2, ... \hat{\underline{U}}_n$ being arbitrary we obtain the fundamental law in the form:

in a Galilean frame,

$\forall (j) \in S, \underline{F}_j + \sum_{\substack{(i) \in S \\ i \neq j}} \underline{F}_{ij} = m_j \underline{a}_j$ [2.47]

where $\underline{F}_{ij} = -F_{ij} \underline{e}_{ij}$.

In the same way, for an arbitrary subsystem $S' = (\ell), ... (p)$, we obtain from [2.25]:

$\forall S' \subset S$

in a Galilean frame,

$\forall (j) \in S', \underline{F}_j^{S'} + \sum_{\substack{(i) \in S' \\ i \neq j}} \underline{F}_{ij} = m_j \underline{a}_j.$ [2.48]

2.5.2.6.4. External forces relative to a subsystem

Writing that for the material point $(j) \in S'$ equations [2.47] and [2.48] must be identical, we obtain [2.49] which yields the expression of the forces external to S' exerted on $(j) \in S'$:

$$\forall (j) \in S', \underline{F}_j^{S'} = \underline{F}_j + \sum_{(i) \in (S-S')} \underline{F}_{ij} = \underline{F}_j - \sum_{(i) \in (S-S')} F_{ij} \underline{e}_{ij}.$$ [2.49]

2.5.2.6.5. Fundamental law of dynamics for the system

Writing [2.27] explicitly, we have

$$\begin{cases} \text{in a Galilean frame,} \\ \forall \underline{\hat{U}}_0 \in \mathbb{R}^3, \ \forall \underline{\hat{\omega}}_0 \in \mathbb{R}^3, \\ (\sum_{(j) \in S} \underline{F}_j).\underline{\hat{U}}_0 + (\sum_{(j) \in S} \underline{OM}_j \wedge \underline{F}_j).\underline{\hat{\omega}}_0 \\ \qquad = (\sum_{(j) \in S} m_j \underline{a}_j).\underline{\hat{U}}_0 + (\sum_{(j) \in S} \underline{OM}_j \wedge m_j \underline{a}_j).\underline{\hat{\omega}}_0 \end{cases} \qquad [2.50]$$

from which we derive the fundamental law of dynamics for the system:

$$\begin{cases} \text{in a Galilean frame,} \\ \sum_{(j) \in S} \underline{F}_j = \sum_{(j) \in S} m_j \underline{a}_j \\ \sum_{(j) \in S} \underline{OM}_j \wedge \underline{F}_j = \sum_{(j) \in S} \underline{OM}_j \wedge m_j \underline{a}_j. \end{cases} \qquad [2.51]$$

Introducing the concept of "wrench of forces" (see Appendix III), with $[\mathcal{F}_e]$ denoting the resultant wrench of all external forces $\underline{F}_j$ exerted on S and $[\mathcal{M}a]$ the resultant wrench of quantities of acceleration $m\underline{a}_j$ of the elements of S, [2.51] takes the simple form [2.52], the counterpart of [2.1]:

$$\begin{cases} \text{in a Galilean frame,} \\ [\mathcal{F}_e] = [\mathcal{M}a]. \end{cases} \qquad [2.52]$$

In the same way, for S':

$$\begin{cases} \text{in a Galilean frame,} \\ [\mathcal{F}'_e] = [\mathcal{M}a']. \end{cases} \qquad [2.53]$$

2.5.3. Comments

As was pointed out earlier, the first and crucial step of this application of the virtual work method was the geometrical modeling, providing the mathematical description of the system under consideration, defining its constituent elements and the geometric parameters that measure their evolution. It opened the way to the definition of the virtual motion vector spaces for the system and subsystems, then the general expressions of the linear forms for the virtual rates of work.

Applying the principle of virtual work first led us to specify the internal force modeling through [2.42] in order that the virtual rate of work by the internal forces complies with the dualized law of mutual action.

Then, it yielded the fundamental law of dynamics for the material points of the system [2.47] or subsystems [2.48], and the fundamental law for the system or subsystems considered as a whole, [2.52] and [2.53].

Finally, it gave the expression of the external forces for any subsystem $S' \neq S$. It must be underscored here that, despite the similarity of notation, there is a key difference between the way [2.24] is written for the system S and the way [2.25] is written for an arbitrary subsystem S' other than S. In the first case, the external forces $\underline{F}_j$ are known, being given as part of the statement of the problem, while in the second case, the external forces $\underline{F}_j^{S'}$ belong to the set of *unknowns* and are only explicitly formulated when the model has been constructed and they can be read off from [2.49].

Letting aside the specificities of the considered model, these general features will be encountered when using the general formulation of the method of virtual work for force modeling that will be presented in the next chapter.

2.6. Practicing

2.6.1. *Law of mutual actions*

For a system S of point masses $m_1, m_2, ..m_n$ submitted to concentrated forces $\underline{F}_j$ exerted on (j) by influences originating outside the set of material points $m_1, m_2, ..m_n$ and $\underline{F}_{ij}$ $(i \neq j, i = 1, 2, ...n)$ exerted on (j) by the other material points (i) of the system S, prove the equivalence between the *law of action and reaction*

$$\begin{cases} \forall i \neq j, \\ i = 1, 2, ...n, \; j = 1, 2, ...n \\ \underline{F}_{ij} + \underline{F}_{ji} = 0 \\ \underline{F}_{ij} \wedge \underline{M_i M_j} = 0 \end{cases}$$

and the *law of mutual actions*

$$
\left\{
\begin{array}{l}
\text{for } S \text{ and } \forall S' \subset S \\[2ex]
\displaystyle\sum_{\substack{(j) \in S' \\ (i) \in S'}} \sum_{i \neq j} \underline{F}_{ij} = 0 \\[3ex]
\displaystyle\sum_{\substack{(j) \in S' \\ (i) \in S'}} \sum_{i \neq j} \underline{OM_j \wedge \underline{F}_{ij}} = 0.
\end{array}
\right.
$$

Solution

The equivalence is self-evident for any subsystem of two elements. It follows that the *law of mutual actions* implies the *law of action and reaction*. Then, the general proof that the *law of action and reaction* implies the *law of mutual actions* can be exposed on a subsystem of three elements and generalized easily.

The *law of action and reaction* is written as

$$
\left\{
\begin{array}{l}
\underline{F}_{12} + \underline{F}_{21} = 0, \ \underline{F}_{13} + \underline{F}_{31} = 0, \ \underline{F}_{23} + \underline{F}_{32} = 0 \\[1ex]
\underline{F}_{12} \wedge \underline{M_1 M_2} = 0, \ \underline{F}_{13} \wedge \underline{M_1 M_3} = 0, \ \underline{F}_{23} \wedge \underline{M_2 M_3} = 0.
\end{array}
\right.
$$

From its first line, we easily derive the first line of the law of mutual actions:

$$
\underline{F}_{12} + \underline{F}_{13} + \underline{F}_{21} + \underline{F}_{23} + \underline{F}_{31} + \underline{F}_{32} = 0 .
$$

In the same way, the second line of the *law of action and reaction* yields

$$
\underline{F}_{12} \wedge \underline{M_1 M_2} + \underline{F}_{13} \wedge \underline{M_1 M_3} + \underline{F}_{23} \wedge \underline{M_2 M_3} = 0
$$

which can be rearranged, taking the first line into account, as

$$
\underline{OM_1 \wedge \underline{F}_{21}} + \underline{OM_1 \wedge \underline{F}_{31}} + \underline{OM_2 \wedge \underline{F}_{12}} + \underline{OM_2 \wedge \underline{F}_{32}} + \underline{OM_3 \wedge \underline{F}_{13}} + \underline{OM_3 \wedge \underline{F}_{23}} = 0
$$

which is actually the second line of the *law of mutual action* for the system.

Principle and Method of Virtual Work

3.1. Why and what for?

It may be observed that, in a sense, the rationale developed in the preceding chapter followed the same path as those we briefly exposed previously, such as Lagrange's proof. The goal was, starting from the fully described model of a system with internal forces, to establish a dualized statement equivalent to the primal formulations of the fundamental law of dynamics and law of mutual actions. The equivalence was proven through what was named the *virtual work method* which, moreover, appeared as a powerful tool for conversely building the mechanical model. Although it had only been established in this particular case, the dualized statement was named a *principle* and the simple mathematical concepts it involves make its generalization easy.

A question may now be asked: why and what for? Referring to the first chapter, we could say that the principle of virtual work comes out as a good candidate to answering the "*Cur*-question" as it may be seen that, in its generalized form, it can be established for any available mechanical model[1].

Together with the principle, the virtual work method will now be generalized into an operative method for mechanical modeling in the many and various geometrical circumstances encountered in practice. Once again the necessity of such a method may be questioned as opposed to a purely heuristic rationale. An answer can be gained from the "virtuous circle of modeling" shown in Figure 3.1.

---

1 Although the principle itself might also be considered just as an "*Ita*-answer".

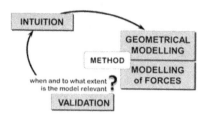

Figure 3.1. *The virtuous circle of mechanical modeling*

As a physical science, mechanics proceeds from experimentally-based intuition for its modeling processes that aim at providing the mathematical frameworks making further analyses possible, adapted to the expected applications. The first step of such a process is the geometrical modeling of the considered system issued from what may be called its general demeanor: more precisely, it is not surprising that a thin slender body would be geometrically described as curvilinear along a line or a thin plate as bi-dimensional on a plane. Then, forces must be modeled on this geometry. Intuition is again at the very roots of that modeling and the virtual work method provides the guidelines for ensuring the consistency of the model and deriving the corresponding mechanical laws.

3.2. General presentation of the virtual work method

We now take the *principle of virtual work,* as established in section 3.4.4 of the preceding chapter in the case of a discrete system made up of material points, as the fundamental principle for a force modeling method. We will thereby generalize the concepts of virtual velocities, virtual motions and virtual rate of work introduced in section 3.4.3.

3.2.1. *Geometrical modeling*

The geometrical description of the system S under consideration and its subsystems is given by all the parameters that define the state of the system and subsystems at a given instant of time.

A *subsystem* S' of S is the result of a *thought experiment* where only the elements in a part of S are considered.

Actual motions of the system or its subsystems are described by the evolution of these parameters with respect to time. They may be subject to *geometrical constraints*.

We must emphasize the meaning of the term *actual motions*. Once the geometrical model of the system has been set up, these are the motions taken into account in the actual evolutions of the system within the context of this model. This does not mean that they represent the physical motions actually observed[2].

3.2.2. *Virtual motions*

The *virtual motions* of the system and its subsystems are defined through the virtual rates of variation of the geometrical (or non-geometrical, see below) parameters.

Virtual means that these rates of variation *do not have to comply* with any constraint that may apply to actual motions. Thence, the *virtual motions* of the system generate a vector space. So do the virtual motions of any subsystem.

The *vector spaces* of the *virtual motions* of the system or its subsystems encompass the actual motions and the rigid body virtual motions of the system or subsystems.

Besides classical virtual velocity fields, the definition of virtual motions may involve the rates of variation of other parameters, *geometrical*[3] or *non-geometrical*[4]. To take account of this possibility, *virtual motions* will be generically denoted by $\hat{\mathbf{U}}$.

3.2.3. *Virtual (rates of) work*

The virtual rates of work by external forces, internal forces and quantities of acceleration in the system are expressed by means of *continuous linear forms* on the vector space of virtual motions. We proceed in the same manner for subsystems, whose virtual motions are defined on the corresponding geometry. These linear forms are driven by *intuition* derived from *experimental observations* and *practical issues* at the considered scale.

$$\mathcal{P}_{e}(\hat{\mathbf{U}}), \mathcal{P}_{i}(\hat{\mathbf{U}}), \mathcal{A}(\hat{\mathbf{U}}) \text{ for the system } S \qquad\qquad [3.1]$$

and

---

2 The remark applies in particular to the case of curvilinear or bi-dimensional media.
3 e.g. beams, plates and micropolar media.
4 e.g. composite materials and shape memory alloys.

$$\mathcal{P}'_e(\hat{\mathbf{U}}), \mathcal{P}'_i(\hat{\mathbf{U}}), \mathcal{A}'(\hat{\mathbf{U}}) \text{ for any subsystem } S'. \tag{3.2}$$

The *cofactors* introduced in $\mathcal{P}_e(\hat{\mathbf{U}}), \mathcal{P}_i(\hat{\mathbf{U}}), \mathcal{P}'_e(\hat{\mathbf{U}}), \mathcal{P}'_i(\hat{\mathbf{U}})$ constitute the *proposed modeling* of the *external* and *internal forces*. Thus, the representations we build for the forces are defined via the virtual motions, which play the role of "test functions". They thereby mark out the range of validity.

3.2.3.1. Remarks

– It is usually a simple matter to write down the continuous linear forms, $\mathcal{A}(\hat{\mathbf{U}})$ and $\mathcal{A}'(\hat{\mathbf{U}})$ since the actual motions of the system and its subsystems are known from the geometrical model. Quantities of acceleration, the cofactors in $\mathcal{A}(\hat{\mathbf{U}})$ and $\mathcal{A}'(\hat{\mathbf{U}})$ are therefore also known.

– The similarity of notation used for the linear forms relative to the system itself and its subsystems should not obscure the key difference between the system, which we can understand in a concrete way and on which we can carry out experiments, and its subsystems which are merely an intellectual construction, *delimited through a thought experiment*. As remarked earlier, this difference is manifested in particular when we write down the linear forms $\mathcal{P}_e(\hat{\mathbf{U}})$ and $\mathcal{P}'_e(\hat{\mathbf{U}})$. Experience is a guide when establishing $\mathcal{P}_e(\hat{\mathbf{U}})$: the corresponding external forces, cofactors in $\mathcal{P}_e(\hat{\mathbf{U}})$, can then be *considered as given* or known. In contrast, for a subsystem S', external forces – at least those exerted by $(S - S')$ on S' – are not supported by such an experimental base and the way $\mathcal{P}'_e(\hat{\mathbf{U}})$ is expressed depends on hypotheses and intuitive judgment. These external forces, cofactors in $\mathcal{P}'_e(\hat{\mathbf{U}})$, therefore have the status of *unknowns*.

– Expressions for the continuous linear forms $\mathcal{P}_i(\hat{\mathbf{U}})$ and $\mathcal{P}'_i(\hat{\mathbf{U}})$, constrained to satisfy [3.3], are also the result of hypotheses which must be consistent with those made for $\mathcal{P}_e(\hat{\mathbf{U}})$ and $\mathcal{P}'_e(\hat{\mathbf{U}})$. The cofactors introduced will end up providing the corresponding model for internal forces.

3.2.4. Principle of virtual (rates of) work

The principle of virtual work is now stated in a generalized form by substituting the generalized virtual motions $\hat{\mathbf{U}}$ for $\hat{\underline{U}}$ introduced in the preceding chapter:

$$\begin{cases} \forall S' \subseteq S \\ \forall \hat{\mathbf{U}} \text{ r.b.v.m. of } S', \ \ \mathcal{P}_i'(\hat{\mathbf{U}}) = 0 \end{cases} \qquad [3.3]$$

$$\begin{cases} \text{in a Galilean frame,} \\ \text{for } S, \\ \forall \hat{\mathbf{U}} \text{ v.m. }, \ \ \mathcal{P}_e(\hat{\mathbf{U}}) + \mathcal{P}_i(\hat{\mathbf{U}}) = \mathcal{A}(\hat{\mathbf{U}}) \\ \forall S' \subset S, \\ \forall \hat{\mathbf{U}} \text{ v.m. }, \ \ \mathcal{P}_e'(\hat{\mathbf{U}}) + \mathcal{P}_i'(\hat{\mathbf{U}}) = \mathcal{A}'(\hat{\mathbf{U}}). \end{cases} \qquad [3.4]$$

At a first glance, the statement of the principle of virtual work given here above does not appear as straightforwardly linked to Lagrange's statement of the principle of virtual velocities

$$P \, dp + Q \, dq + R \, dr + \ldots + \lambda \, dL + \mu \, dM + \nu \, dN + \ldots = 0, \ \forall dx_i, dy_i, dz_i, \forall i \qquad [3.5]$$

where λ, μ, ν are indeterminate. But comparing this equation with [3.4] written for the system S, we note that:

– [3.4] is written $\forall \hat{\mathbf{U}}$ virtual motion of S, which means that, using Lagrange's own words, the constituent particles of the system are *"considered as completely free without any constraint"*.

– The virtual rate of work by external forces $\mathcal{P}_e(\hat{\mathbf{U}})$ covers all external forces, including the reactions associated with the geometrical constraints due to *"the obstacles that may impede"* the motion of the system. It stands for the $P \, dp + Q \, dq + R \, dr + \ldots + \lambda \, dL$ terms in [3.5].

– The rate of work by internal forces $\mathcal{P}_i(\hat{\mathbf{U}})$ is the counterpart of the $\mu \, dM + \nu \, dN + \ldots$ terms, standing for the *"resistances"* that the particles *"should meet for being linked to each other"*.

– The presence of the rate of work by quantities of acceleration $\mathcal{A}(\hat{\mathbf{U}})$ may also be interpreted as the result of d'Alembert's *principle*[5] to shift from the Statics viewpoint to the Dynamics one.

---

5 [ALE 43] Jean Le Rond d'Alembert (1717–1783).

3.2.5. *Implementing the principle of virtual work*

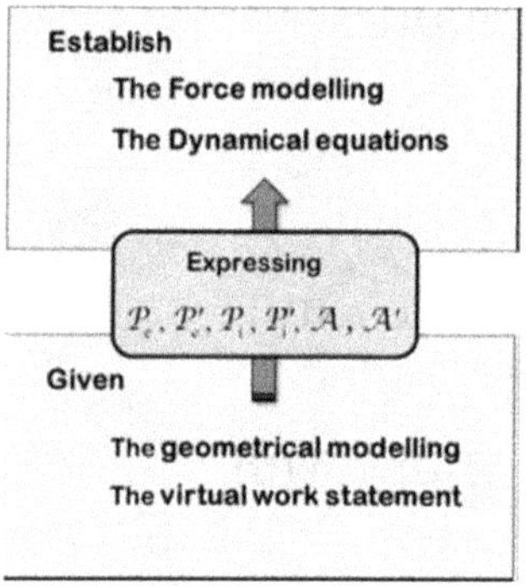

Figure 3.2. *Principle of the virtual work modeling method*

– To begin with, using the dual statement of the law of mutual actions [3.3], we check the conformity of the proposed expressions $\mathcal{P}_i(\hat{\mathbf{U}})$ and $\mathcal{P}'_i(\hat{\mathbf{U}})$ for the virtual rates of work by internal forces for the system and its subsystems or, if need be, we specify these expressions in such a way that [3.3] be satisfied.

– Using the dual statement of the fundamental law [3.4] for the system, provided that the expressions proposed for the rates of work $\mathcal{P}_e(\hat{\mathbf{U}}), \mathcal{P}_i(\hat{\mathbf{U}}), \mathcal{A}(\hat{\mathbf{U}})$ are mutually consistent from a mathematical standpoint (consistency of underlying physical hypotheses), we obtain the dynamical equations for the external and internal forces of the system within the framework of the model we have set up.

– From the same equation applied to a subsystem *stricto sensu*, under the same condition of consistency for $\mathcal{P}'_e(\hat{\mathbf{U}}), \mathcal{P}'_i(\hat{\mathbf{U}}), \mathcal{A}'(\hat{\mathbf{U}})$, we obtain the dynamical equations for this subsystem and derive the expression for the external forces on this subsystem based upon the preceding determination of the field of internal forces within the system.

– As a consequence of [3.3] and [3.4], the *fundamental law of dynamics* for the system or any subsystem is written in the dualized form:

$$\begin{cases} \text{in a Galilean frame,} \\ \forall S' \subseteq S, \\ \forall \hat{\mathbf{U}} \text{ r.b.v.m. of } S', \\ \mathcal{P}'_e(\hat{\mathbf{U}}) = \mathcal{A}'(\hat{\mathbf{U}}) \end{cases} \qquad [3.6]$$

Written for the system S, [3.6] comes out as a *necessary condition* to be satisfied by the data on the external forces and acceleration fields (see Chapter 6, section 6.3.5).

– Let $\hat{\mathbf{U}}_1$ and $\hat{\mathbf{U}}_2$ be two virtual motions of a subsystem $S' \subseteq S$ that differ one from the other by the addition of an arbitrary rigid body virtual motion:

$$\hat{\mathbf{U}}_1 = \hat{\mathbf{U}}_2 + \text{r.b.v.m.} \tag{3.7}$$

It follows from the linearity of $\mathcal{P}'_i(\hat{\mathbf{U}})$ and [3.3] that

$$\mathcal{P}'_i(\hat{\mathbf{U}}_1) = \mathcal{P}'_i(\hat{\mathbf{U}}_2). \tag{3.8}$$

This result is often referred to as the *objectivity* of the virtual rate of work by internal forces. It implies, as a particular case, that when the actual motion of the considered subsystem is implemented as a virtual motion, the rate of work by internal forces is independent of the reference frame where that motion is defined (the very definition of objectivity).

3.2.6. *Comments*

This rather systematic and structured presentation of the method should not overshadow the fact that it is *in no way axiomatic*! The various choices involved, first when defining the geometrical modeling, then when writing down the continuous linear forms for $\mathcal{P}_e(\hat{\mathbf{U}}), \mathcal{P}_i(\hat{\mathbf{U}}), \mathcal{A}(\hat{\mathbf{U}})$ and $\mathcal{P}'_e(\hat{\mathbf{U}}), \mathcal{P}'_i(\hat{\mathbf{U}}), \mathcal{A}'(\hat{\mathbf{U}})$, clearly underscore the many hypotheses that must be made in the light of experience, intuition and also the type of mathematical model of "physical reality" we wish to produce. Moreover, the above-mentioned choices and hypotheses are not independent from each other: this is part of the consistency we have to and can check through the method. In the end, the models obtained through such a process must be validated: this is the following step in the virtuous circle in Figure 3.1 where the domain of relevance of the results it makes possible to obtain is delimited. Completing the cycle may point out that this domain should be enlarged, modified or shifted, thus calling for a new modeling process based on new intuitive hypotheses.

We should add that the virtual work method is not the only way of model building. Its main advantage lies in its unified and systematic approach, which makes it possible and easier to construct more original models in a consistent way.

3.3. General results

From equation [3.6], we can already derive some immediately accessible results, for which we do not need to specify the vector spaces of the motions considered in the model, or the linear forms constructed on these spaces. Those results are then valid for any geometrical modeling and associated force modeling.

3.3.1. *System, subsystems, actual and virtual motions*

S and S' denote the system and an arbitrary subsystem as schematically represented in Figure 3.3.

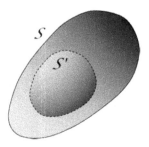

Figure 3.3. *Schematic representation of the system S and a subsystem S'*

Whatever the geometrical modeling (section 3.2.1) the vector spaces of virtual motions $\hat{U}$ for S (respectively for S') contain the vector space of rigid body motions and the vector space of actual motions of S (respectively S') (section 3.2.2).

3.3.2. *Virtual (rates of) work in rigid body virtual motions*

Referring to section 3.2.3, the virtual rates of work by external or internal forces and quantities of acceleration are linear forms $\mathcal{P}_e(\hat{U}), \mathcal{P}_i(\hat{U}), \mathcal{A}(\hat{U})$ on the vector space of virtual motions of S. Similarly for $\mathcal{P}'_e(\hat{U}), \mathcal{P}'_i(\hat{U}), \mathcal{A}'(\hat{U})$ on the vector space of virtual motions of S'. Independently of the precise definition of the vector spaces of virtual motions, the restrictions of these forms on the vector subspace of rigid body virtual motions of S define the wrenches of external forces $[\mathcal{F}_e]$, internal forces $[\mathcal{F}_i]$ and quantities of acceleration $[\mathcal{M}a]$ through the equations (Appendix 3):

$$\forall\{\mathcal{D}\}, \quad \begin{cases} [\mathcal{F}_e].\{\mathcal{D}\} = \mathcal{P}_e(\hat{\underline{U}}) \\ [\mathcal{F}_i].\{\mathcal{D}\} = \mathcal{P}_i(\hat{\underline{U}}) \\ [\mathcal{M}a].\{\mathcal{D}\} = \mathcal{A}(\hat{\underline{U}}) \end{cases} \qquad [3.9]$$

where $\hat{\underline{U}}$ denotes the rigid body velocity field defined by the velocity distributor $\{\mathcal{D}\}$. The wrenches $[\mathcal{F}_e']$, $[\mathcal{F}_i']$ and $[\mathcal{M}a']$ are defined in the same way for any subsystem S'.

3.3.3. *Fundamental law of dynamics*

With these definitions equation [3.6] is equivalent to:

$$\begin{cases} \text{in a Galilean frame,} \\ \text{for } S, \\ \forall\{\mathcal{D}\}, \ [\mathcal{F}_e].\{\mathcal{D}\} = [\mathcal{M}a].\{\mathcal{D}\}, \end{cases} \qquad [3.10]$$

thence

$$\begin{cases} \text{in a Galilean frame,} \\ \text{for } S, \\ [\mathcal{F}_e] = [\mathcal{M}a]. \end{cases} \qquad [3.11]$$

Equation [3.11] expresses the primal statement of the fundamental law of dynamics for the system as a whole:

In a Galilean frame, the wrench of external forces is equal to the wrench of quantities of acceleration, whatever the model.

The same statement is obviously valid for any subsystem. Note that equations [2.52] and [2.53] obtained in the preceding chapter are particular examples of this general result.

3.3.4. *Wrench of internal forces*

Regarding equation [3.3], we obtain,

$$\begin{cases} \forall S' \subseteq S \\ \forall \{\mathcal{D}\}, \ [\mathcal{F}_i'].\{\mathcal{D}\} = 0 \end{cases} \qquad [3.12]$$

thence

$$\begin{cases} \forall S' \subseteq S \\ [\mathcal{F}_i'] = 0. \end{cases} \qquad [3.13]$$

Whatever the force modeling, the wrench of internal forces defined through [3.9] for the system or any subsystem is equal to zero.

3.3.5. *Law of mutual actions*

Let S_1 and S_2 be two disjoint systems and define S as $(S_1 + S_2)$. Define $[\mathcal{F}_{21}]$ as the wrench of the forces external to S_1 due to S_2, and $[\mathcal{F}_{12}]$ in the same way. Then, applying [3.13] to S, we obtain

$$[\mathcal{F}_i]_1 + [\mathcal{F}_i]_2 + [\mathcal{F}_{21}] + [\mathcal{F}_{12}] = 0 \qquad [3.14]$$

where $[\mathcal{F}_i]_1$ and $[\mathcal{F}_i]_2$ the wrenches of internal forces for S_1 and S_2 are equal to zero; it follows that:

$$[\mathcal{F}_{21}] + [\mathcal{F}_{12}] = 0. \qquad [3.15]$$

This is the *law of mutual actions* for the two systems S_1 and S_2.

3.3.6. *Remarks*

At this stage, it seems essential to insist on the fact that it would be a serious mistake to think that these results are only valid for undeformable systems or for systems that are not actually deformed in the actual motion. This would amount to confusing the concepts of virtual and actual motions. The fact that we exploit the principle of virtual work on the subspace of rigid body virtual motions bears no relation either to the mechanical properties of the system (constitutive equation of its constituent material) or to its actual motion under the effect of the forces under study.

As announced initially, statements [3.11], [3.13] and [3.15] are the most general results that can be obtained whatever the geometrical and force modeling. These global statements expressed in terms of wrenches must and will be satisfied by any mechanically consistent force model.

Also, we find them at the roots of the classical heuristic presentation of various force models in continuum mechanics where the equations of dynamics for the model are established by writing down the fundamental law of dynamics and the law of mutual actions in terms of wrenches, [3.11], [3.13] and [3.15], for "small elements" (e.g. small tetrahedron and small parallelepiped in three-dimensional continuum mechanics, small triangle and small rectangle for plates).

3.4. Particular results

By getting into some specification of the geometrical modeling of the system and subsystems, we will now be able, on the basis of the general results established in the preceding paragraphs, to derive particular results that cover a wide range of practical applications.

3.4.1. *System, subsystems, actual and virtual motions*

In the three-dimensional Euclidean space, the system, generically denoted by S, consists of point masses m_i characterized by their position vectors $\underline{OM}_i = \underline{x}_i$, and distributed masses defined by mass densities at the geometric point considered as the position of the "diluted" mass element, namely:

– A volume density, mass per unit volume, $\rho_\Omega(\underline{x})$ such that the mass dm_Ω of the volume element $d\Omega$ at point $\underline{x}$ is

$$dm_\Omega = \rho_\Omega(\underline{x})\,d\Omega \qquad\qquad [3.16]$$

– A surface density, mass per unit area, $\rho_\Sigma(\underline{x})$ such that the mass dm_Σ of the surface element with area da at point $\underline{x}$ on a surface Σ is

$$dm_\Sigma = \rho_\Sigma(\underline{x})\,da \qquad\qquad [3.17]$$

– A line density, mass per unit length, $\rho_L(\underline{x})$ such that the mass dm_L of the line element with length ds at point $\underline{x}$ on a curve L is

$$dm_L = \rho_L(\underline{x})\,ds. \tag{3.18}$$

The total mass of the system is then written in the general form:

$$\mathcal{M} = \sum_i m_i + \int_\Omega dm_\Omega + \int_\Sigma dm_\Sigma + \int_L dm_L = \int_S dm. \tag{3.19}$$

Virtual rates of variation $\hat{\underline{U}}(\underline{x})$ of the position vectors $\underline{x}$ over S generate the virtual velocity fields that define the virtual motions $\hat{\underline{U}}$ of the system.

Actual motions are defined by the actual rates of variation of $\underline{x}$:

$$\underline{U}(\underline{x},t) = \frac{d\underline{x}}{dt} \tag{3.20}$$

and the acceleration is

$$\underline{a}(\underline{x},t) = \frac{d\underline{U}}{dt}(\underline{x},t). \tag{3.21}$$

Thence, the wrench of quantities of acceleration for the system is written as

$$[\,\mathcal{M}a\,] = \left[\, O, \int_S \frac{d\underline{U}}{dt}(\underline{x},t)\,dm, \int_S \underline{OM} \wedge \frac{d\underline{U}}{dt}(\underline{x},t)\,dm \,\right]^{\,6} \tag{3.22}$$

3.4.2. Momentum theorem

The *momentum* of a mass element m_i or dm is defined with respect to the velocity in the same way as the quantity of acceleration with respect to the acceleration, and the wrench of momenta is

$$[\,\mathcal{M}U\,] = \left[\, O, \int_S \underline{U}(\underline{x},t)\,dm, \int_S \underline{OM} \wedge \underline{U}(\underline{x},t)\,dm \,\right]. \tag{3.23}$$

Recalling that the *conservation of the mass* of a material element is a fundamental principle of mechanics, the derivation of equation [3.23] with respect to time yields straightforwardly

---

6 With the same definition of the integrals over the system S as in [3.19].

$$\frac{d}{dt}[\,\mathcal{MU}\,]=\left[\,O,\int_{S}\frac{d}{dt}\underline{U}(\underline{x},t)\,dm,\int_{S}\underline{U}(\underline{x},t)\wedge\underline{U}(\underline{x},t)\,dm\right.$$
$$\left.+\int_{S}\underline{OM}\wedge\frac{d\underline{U}}{dt}(\underline{x},t)\,dm\,\right]$$

[3.24]

that is

$$\frac{d}{dt}[\,\mathcal{MU}\,]=[\,\mathcal{Ma}\,].$$

[3.25]

This identity states that the *material time derivative*[7] of the momentum wrench of the system is equal to the wrench of quantities of acceleration. The same result is obviously valid for any subsystem S'.

Then, taking [3.25] into account in expression [3.11] of the fundamental law of dynamics, we obtain the new result, valid for the system described in section 3.4.1:

$$\left\{\begin{array}{l}\text{in a Galilean frame,}\\[4pt]\text{for }S,\\[4pt][\,\mathcal{F_e}\,]=\dfrac{d}{dt}[\,\mathcal{MU}\,]\end{array}\right.$$

[3.26]

that is also valid for any subsystem S'.

In particular, if the system S is *isolated*, meaning that it is subject to no external effects, we have $[\,\mathcal{F_e}\,]=0$ and thus:

in a Galilean frame, the momentum of an isolated system is conserved.

3.4.3. Center of mass theorem

The center of mass of the system at a given instant of time is the *geometrical* point G defined through the equation

$$\underline{OG}=\underline{x}_{G}=\frac{1}{\mathcal{M}}\left(\sum_{i}\underline{x}_{i}\,m_{i}+\int_{\Omega}\underline{x}\,dm_{\Omega}+\int_{\Sigma}\underline{x}\,dm_{\Sigma}+\int_{L}\underline{x}\,dm_{L}\right)=\frac{1}{\mathcal{M}}\int_{S}\underline{x}\,dm.\quad[3.27]$$

---

7 The term "material time derivative" underscores the fact that the derivative is taken following the material element through the principle of mass conservation.

In the actual motion of the system defined by the velocity field $\underline{U}(\underline{x},t) = \dfrac{d\underline{x}}{dt}$, this geometrical point moves with the velocity $\underline{U}_G(t)$ obtained from the derivation of [3.27] with respect to time; thence

$$\mathcal{M}\underline{U}_G(t) = \int_S \underline{U}(\underline{x},t)\,dm. \qquad [3.28]$$

It follows that the resultant of the momentum wrench [3.23], commonly called *the momentum of S*, is equal to the momentum of the total mass $\mathcal{M}$ of the system concentrated at the center of mass of the system with the velocity $\underline{U}_G(t)$ of that geometrical point.

Time derivation of [3.28] yields the acceleration of the geometrical point G

$$\mathcal{M}\underline{a}_G(t) = \int_S \frac{d\underline{U}}{dt}(\underline{x},t)\,dm. \qquad [3.29]$$

Comparing this result with [3.22], we find that $\mathcal{M}\underline{a}_G(t)$, which is just the resultant of $[\,\mathcal{M}\underline{a}\,]$, commonly called the quantity of acceleration of the system, is equal to the quantity of acceleration of the total mass $\mathcal{M}$ of the system concentrated at the center of mass of the system if it has the motion of that geometrical point.

Let $\underline{F}_e$ denote the resultant of the wrench of external forces $[\,\mathcal{F}_e\,]$, then the fundamental law of dynamics yields the vector equation

$$\left\{ \begin{array}{l} \text{in a Galilean frame,} \\ \text{for } S, \\ \underline{F}_e = \mathcal{M}\underline{a}_G(t). \end{array} \right. \qquad [3.30]$$

Together with [3.28], this equation expresses the *theorem of the center of mass*. It shows that the center of mass of the system, a geometrical point, moves as if the total mass of the system were concentrated at that point and submitted to the resultant of all external forces that act on the system.

In other words, it means that: *if the system is simply modeled as a point mass with the total mass of the system located at its center of mass at a given instant of time, submitted to all external forces acting on it, this point mass, according to* *the fundamental law of dynamics where the external force is the resultant $\underline{F}_e$ of all external forces, will move as the center of mass of the system during all its evolution[8].*

3.4.4. *Kinetic energy theorem*

As stated in section 3.2.2, the vector space of virtual motions of the system includes the real (or actual) motions. We may therefore make use of the actual motion $\underline{U}$ in the principle of virtual work [3.4]. We obtain for the system:

$$
\left\{
\begin{array}{l}
\text{in a Galilean frame,} \\
\underline{U} \text{ real motion of } S, \\
\mathcal{P}_e(\underline{U}) + \mathcal{P}_i(\underline{U}) = \mathcal{A}(\underline{U}).
\end{array}
\right.
\qquad [3.31]
$$

In this equation, $\mathcal{P}_e(\underline{U}) + \mathcal{P}_i(\underline{U})$ represent the total rate of work by external and internal forces in the actual motion. As for $\mathcal{A}(\underline{U})$, it can be written as

$$
\mathcal{A}(\underline{U}) = \int_S \frac{d\underline{U}(\underline{x},t)}{dt} \cdot \underline{U}(\underline{x},t)\, dm.
\qquad [3.32]
$$

Denoting by $K(\underline{U})$ the *kinetic energy* of the system in the actual motion $\underline{U}$

$$
K(\underline{U}) = \frac{1}{2}\int_S \underline{U}^2(\underline{x},t)\, dm,
\qquad [3.33]
$$

$\mathcal{A}(\underline{U})$, as expressed in [3.32], comes out as the material time derivative of $K(\underline{U})$, the kinetic energy of the system S:

$$
\mathcal{A}(\underline{U}) = \frac{d}{dt} K(\underline{U}).
\qquad [3.34]
$$

Thence, [3.31] becomes

---

8 Once again, we insist onto the fact that these results *do not assume* the system to be undeformable or undeformed in its actual motion.

$$
\left\{
\begin{array}{l}
\text{in a Galilean frame,} \\[4pt]
\underline{U}\ \text{real motion of } S, \\[4pt]
\mathcal{P}_e(\underline{U}) + \mathcal{P}_i(\underline{U}) = \dfrac{\mathrm{d}}{\mathrm{d}t} K(\underline{U}).
\end{array}
\right.
\qquad\qquad [3.35]
$$

This equation expresses the kinetic energy theorem:

In a Galilean frame, the time derivative of the kinetic energy of the system S in the actual motion is equal to the rate of work by all forces external and internal to S in that motion (analogous statement for any subsystem S′).

3.4.5. Comments

The concept of *material time derivative*, which has been introduced in sections 3.4.2 and 3.4.3, will be thoroughly dealt with in Chapter 5. It means essentially that the derivative with respect to time is taken following the material element. As an example, in [3.21], the derivative $\dfrac{\mathrm{d}U}{\mathrm{d}t}(\underline{x},t)$ is the total derivative of $\underline{U}(\underline{x},t)$ with respect to time taking into account that the position $\underline{x}$ of the material element is a function of time. With the results concerning the material time derivative of a volume integral, it will be possible, on the basis of [3.24], to establish the important "particular result" known as *Euler's theorem*.

No particular differentiability conditions have been specified here as to the actual velocity field, implicitly assuming it to be continuously differentiable. Piecewise continuous and continuously differentiable actual velocity fields are sometimes encountered (shock waves), which make it necessary to modify the definition of the kinetic energy (see Chapter 5, section 5.3.5). More comments on this topic are given in classical textbooks such as [GER 86, SAL 01].

3.5. About equilibrium

3.5.1. Equilibrium

We consider the case of a system S, as described in section 3.3.1, which is in the state of equilibrium in a Galilean reference frame characterized by

$$\begin{cases} \text{in a Galilean frame,} \\ \forall \hat{\mathbf{U}} \text{ v.m. of } S, \\ \mathcal{A}(\hat{\mathbf{U}}) = 0, \end{cases} \qquad [3.36]$$

which implies that quantities of acceleration are identically equal to zero over S and, as an obvious consequence,

$$\begin{cases} \text{in a Galilean frame,} \\ \forall S' \subset S \\ \forall \hat{\mathbf{U}} \text{ v.m. of } S', \\ \mathcal{A}'(\hat{\mathbf{U}}) = 0. \end{cases} \qquad [3.37]$$

It then follows from [3.4] that

$$\begin{cases} \text{for } S, \\ \forall \hat{\mathbf{U}} \text{ v.m. of } S \ , \ \ \mathcal{P}_{\mathrm{e}}(\hat{\mathbf{U}}) + \mathcal{P}_{\mathrm{i}}(\hat{\mathbf{U}}) = 0 \\ \forall S' \subset S, \\ \forall \hat{\mathbf{U}} \text{ v.m. of } S' \ , \ \ \mathcal{P}'_{\mathrm{e}}(\hat{\mathbf{U}}) + \mathcal{P}'_{\mathrm{i}}(\hat{\mathbf{U}}) = 0, \end{cases} \qquad [3.38]$$

which is at the origin of the terminology "*virtual resisting rate of work*" that is sometimes used for the quantity $-\mathcal{P}'_{\mathrm{i}}(\hat{\mathbf{U}})$.

It obviously follows from the definition of the equilibrium state through [3.36] that $\forall S' \subseteq S, [\mathcal{M}a'] = 0$ and with [3.11], we obtain

$$\begin{cases} \text{for } S, \\ [\mathcal{F}_{\mathrm{e}}] = 0, \\ \forall S' \subset S \\ [\mathcal{F}'_{\mathrm{e}}] = 0, \end{cases} \qquad [3.39]$$

with the same difference of status between $[\mathcal{F}_{\mathrm{e}}]$ and $[\mathcal{F}'_{\mathrm{e}}]$ as already mentioned in section 3.2.3 for $\mathcal{P}_{\mathrm{e}}(\hat{\mathbf{U}})$ and $\mathcal{P}'_{\mathrm{e}}(\hat{\mathbf{U}})$.

3.5.2. *Self-equilibrating fields of internal forces*

We now assume that the system S, in the state of equilibrium defined by [3.36], is submitted to no external force, which implies

$$\forall \hat{\mathbf{U}} \text{ v.m. of } S, \ \mathcal{P}_e(\hat{\mathbf{U}}) = 0 \tag{3.40}$$

It follows from [3.38] that the virtual rate of work by internal forces in any virtual motion is equal to zero:

$$\forall \hat{\mathbf{U}} \text{ v.m. of } S, \ \mathcal{P}_i(\hat{\mathbf{U}}) = 0 \tag{3.41}$$

The fields of internal forces in this case are said to be *self-equilibrating* and [3.41] shows that they generate a vector space orthogonal to the vector space of virtual motions of S through the linear form $\mathcal{P}_i(\hat{\mathbf{U}})$.

It is important to note here that equations [3.40] and [3.41] are concerned with S only. The counterpart of [3.41] for the rate of work by the forces internal to an arbitrary subsystem $S' \subset S$ *is not satisfied* as a rule.

As an example, we consider a system S of n mass points as described in Chapter 2 (section 2.5). The vector space of internal forces for S is generated by $n(n-1)/2$ scalars F_{ij} (see Chapter 2, equation [2.45]). It is, by construction, orthogonal to the vector space of rigid body virtual motions of S, with dimension 6. We now assume that all external forces $\underline{F}_j$ are equal to zero. As a consequence of [3.41], the vector space of internal forces is orthogonal to the vector space of all virtual motions of S, with dimension $3n$, which implies $(3n-6)$ additional orthogonality conditions. Thence, the dimension of the vector space of self-equilibrating internal forces for the system S is equal to

$$k = \frac{n(n-1)}{2} - 3(n-2).$$

Geometrical Modeling of the Three-dimensional Continuum

4.1. The concept of a continuum

4.1.1. *Geometrical modeling*

The first step of the virtuous circle (Figure 4.1) starts from the intuition gained through experiments and practice, and proposes a geometrical modeling of the considered system that needs to be validated afterwards. Schematically, we may say that the concept of a continuum, as opposed to that of a system made of countable discrete point masses, aims at representing Newton's celebrated apple, not as a single point mass (see Chapter 3, section 3.4.3) or as a system of point masses (like a pomegranate!), but as a volume made of "diluted particles". Obviously, this intuitive concept of a mechanical continuum needs to be mathematically formulated and its relationship with the very definition of continuity in mathematics should be investigated and given a precise meaning.

Figure 4.1. *The first step of the virtuous circle*

4.1.2. *Experiments*

Figure 4.2 presents the result of an extrusion process performed on a thick metal tube which was first sawn along a meridian plane and marked with a regular square grid, and then pushed and extruded through a die.

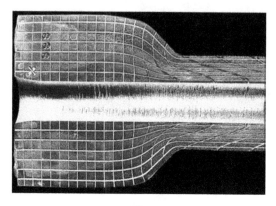

Figure 4.2. *Extrusion of a thick tube (document kindly supplied by M. Sauve, CEA)*

Similarly, Figure 4.3 shows the result of a die stamping process performed on an aluminum alloy rectangular specimen initially marked with a rectangular square grid.

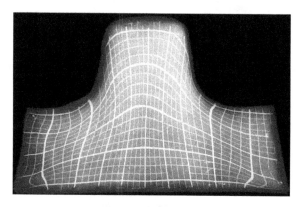

Figure 4.3. *Die stamping of an aluminum alloy specimen (document kindly supplied by A. Le Douaron)*

Both experiments show that the solid constituent material of the considered system actually undergoes deformation during the process and that particles of matter that were initially close to each other remain neighbors.

The same conclusion arises when observing the flow of a liquid as in Figure 4.4, which shows two frames of a film presenting the steady flow of a fluid, initially marked with a regular square grid, in a converging channel (excerpt from *Flow visualization*, [KLI 63]).

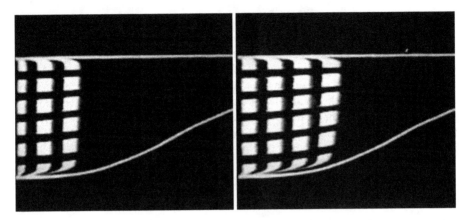

Figure 4.4. *Excerpt from Flow visualization [KLI 63]*

Figure 4.5 relates to the two-dimensional biaxial compression test performed on a sample of sand. A regular square grid was initially drawn on the plane surface of the specimen with a touch of paint being lightly sprayed to make the granular texture more visible. The figure presents the result of a first step of the compression test where it appears that, at the "macroscopic" level where the grains of the material are not considered individually, we can still draw a conclusion similar to what was already stated from Figures 4.2–4.4 regarding neighboring macroscopic particles.

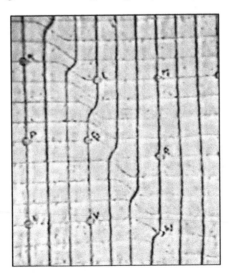

Figure 4.5. *Excerpt from Strain localization [DES 86]*

4.1.3. *Comments*

This last example enhances an important point about the relevance of the concept of a continuum as a mechanical model, which will be studied in greater detail later on: validation, the third step of the virtuous circle, shall not ask whether matter is or is not continuous but when and to what extent a continuum model is relevant for the practical applications that will be considered. This implies that a proper physical meaning should be given to the very concept of a "particle" at the considered macroscopic level.

4.2. System and subsystems

4.2.1. *Particles and system*

Within the framework of the classical three-dimensional continuum mechanics modeling, as it is currently used for the preceding examples, a system S is geometrically described, at a given instant of time t, by a volume Ω_t with boundary $\partial\Omega_t$. The constituent particles may be called "diluted" as opposed to point masses, meaning that they are geometrically represented by an infinitesimal volume element $d\Omega_t$ with mass dm, and geometrically characterized by their position vector $\underline{OM} = \underline{x}$ with respect to a reference frame.

The mass per unit volume is $\rho(\underline{x},t)$ such that:

$$dm = \rho(\underline{x},t)\,d\Omega_t. \tag{4.1}$$

At a given instant of time t, the state of the system in the reference frame defined that way is called its (spatial) configuration and denoted by κ_t.

The definition of a subsystem $S' \subset S$ on a volume Ω'_t is similar but for the fact that a subsystem is just the result of a thought experiment.

4.2.2. *Actual motions. Eulerian and Lagrangian descriptions*

At time t, the geometrical evolution of the system S is given by the velocity field $\underline{U}$ over Ω_t

$$\underline{U}(\underline{x},t) = \frac{d\underline{x}}{dt} \tag{4.2}$$

which defines the motion of the particles and conveys the infinitesimal volume elements $d\Omega_t$.

This description is commonly named the *Eulerian description* of the continuum. At each instant of time, it refers to the present configuration κ_t. Any physical quantity associated with a particle is defined in κ_t as a function of the geometrical position of that particle at that instant of time in the form:

$$\mathcal{B} = b(\underline{x}, t). \tag{4.3}$$

This implies that in order to follow the evolution of such a quantity attached to a particle, for instance when taking the derivative of the quantity with respect to time, we have to take into account the motion of the particle in the considered reference frame as defined by [4.2]. This yields the concept of *material* or *convective time derivative* already introduced in Chapter 3 and which will be dealt with in detail in the following chapter.

Considering the configuration κ_0 of the system at a fixed instant of time t_0, let us denote by $\underline{X}$ the position vector of the particle with mass dm and volume $d\Omega_0$ at point M_0 in that configuration: $\underline{OM_0} = \underline{X}$. The volume and boundary of the system S in that configurations are denoted by Ω_0 and $\partial\Omega_0$ respectively and the mass per unit volume is $\rho_0(\underline{X}, t_0)$ such that

$$dm = \rho_0(\underline{X}, t_0) \, d\Omega_0. \tag{4.4}$$

The governing principle of the *Lagrangian description* is to use the coordinate vector $\underline{X}$ to label the particle located at that position at t_0 in the same way as, in the experiments referred to earlier, particles were marked in order to follow their evolution. With this description, the position vector of the particle $\underline{X}$ at time t is written as

$$\underline{x} = \underline{\phi}(\underline{X}, t) \tag{4.5}$$

with the obvious equation

$$\underline{X} = \underline{\phi}(\underline{X}, t_0). \tag{4.6}$$

where $\underline{X}$ identifies the particle.

The vector field $\underline{\phi}$ on Ω_0 defines the transport between time t_0 and time t of the particle $\underline{X}$ with the infinitesimal volume element $d\Omega_0$ to its location $\underline{x}$ in Ω_t with volume $d\Omega_t$.

In the same way, the value at time t of any physical quantity associated with a particle $\underline{X}$ is written as:

$$B = B(\underline{X},t). \tag{4.7}$$

This applies to the mass per unit volume of the particle at time t in the form

$$dm = \rho_0(\underline{X},t)d\Omega_0. \tag{4.8}$$

The velocity of the particle $\underline{X}$ at time t is obtained straightforwardly from [4.5]

$$\underline{U}_0(\underline{X},t) = \frac{\partial \underline{\phi}}{\partial t}(\underline{X},t). \tag{4.9}$$

Putting together equations [4.2], [4.5], [4.6] and [4.9], we establish the equivalence between the Eulerian and Lagrangian descriptions through the differential equation for the vector function $\underline{\phi}$

$$\begin{cases} \underline{U}_0(\underline{X},t) = \frac{\partial \underline{\phi}}{\partial t}(\underline{X},t) = \underline{U}(\underline{\phi}(\underline{X},t),t) \\ \underline{\phi}(\underline{X},t_0) = \underline{X} \end{cases} \tag{4.10}$$

or equivalently

$$\begin{cases} d\underline{x} = \underline{U}(\underline{x},t) \\ \underline{x}|_{t_0} = \underline{X}, \end{cases} \tag{4.11}$$

a system of three differential equations in three unknown scalar functions (the coordinates of $\underline{x}$) of the time variable t.

The solution $\underline{x} = \underline{\phi}(\underline{X},t)$ to this system is just the equation of the pathline of the particle $\underline{X}$ for $t \geq t_0$ (Figure 4.6).

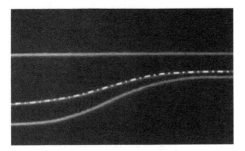

Figure 4.6. *Pathline of a particle*

4.3. Continuity hypotheses

4.3.1. *Lagrangian description*

We must now examine the mathematical conditions to be imposed on the vector field $\underline{\phi}$ in the Lagrangian description to provide a suitable formulation of the intuitive concept of continuity introduced in section 4.1.

– At any time t, the vector function $\underline{\phi}$ must be a *one-to-one mapping* (bijection) of Ω_0 onto Ω_t with inverse bijection denoted by $\underline{\psi} = \underline{\phi}^{-1}$.

This hypothesis expresses that, at time t, a particle $\underline{X}$ cannot be located in two different positions and, conversely, that two different particles $\underline{X}$ and $\underline{X'}$ cannot be located at the same place.

– The vector functions $\underline{\phi}$ and $\underline{\psi}$ are assumed to be *continuous* with respect to all space variables (homeomorphism) and time.

This hypothesis implies that two particles whose positions differ by an infinitesimal separation in Ω_0 will maintain this relationship in Ω_t at any time. Also, particles occupying a connected region in Ω_0 occupy a connected region of the same dimension in Ω_t. In the same way, particles within some closed surface in Ω_0 remain within the transported surface in Ω_t at any time.

As a result, the particles at the boundary $\partial\Omega_0$ of Ω_0 remain at the boundary $\partial\Omega_t$ of Ω_t at any time. This conclusion is consistent with everyday practical observations, as shown in Figure 4.7 that presents the result of a punch indentation test performed on a multilayer plasticine block.

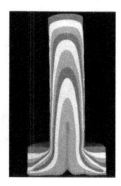

Figure 4.7. *Punch indentation of a plasticine block (CEMEF, Sophia Antipolis)*

– The vector functions $\underline{\phi}$ and $\underline{\psi}$ are assumed to be *continuously differentiable*, C^1 *or even* C^2, with respect to space variables (diffeomorphisms) and time.

The gradient (Appendix 2) of $\underline{\phi}$ with respect to space variables is defined at point $\underline{X}$ and time t by the differential relationship

$$\begin{cases} \mathbb{R}^3 \xrightarrow{\;\text{grad}\,\underline{\phi}\;} \mathbb{R}^3 \\ \underline{dX} \in \mathbb{R}^3 \mapsto \underline{\underline{\text{grad}}}\,\underline{\phi}(\underline{X},t).\underline{dX} \in \mathbb{R}^3, \end{cases} \qquad [4.12]$$

which describes the tangent linear mapping to the homeomorphism at point $\underline{X}$ and time t. The gradient $\underline{\underline{\text{grad}}}\,\underline{\psi}(\underline{x},t)$ is defined in the same way.

In an orthonormal basis $\{\underline{e}_i\}$ defining orthonormal Cartesian coordinates:

$$\underline{\underline{\text{grad}}}\,\underline{\phi}(\underline{X},t) = \frac{\partial \phi_j(\underline{X},t)}{\partial X_i}\underline{e}_j \otimes \underline{e}_i. \qquad [4.13]$$

The matrix of the components of $\underline{\underline{\text{grad}}}\,\underline{\phi}(\underline{X},t)$ is the Jacobian matrix of the homeomorphism $\underline{\phi}$ at point $\underline{X}$ and time t and its determinant is the *Jacobian determinant*:

$$J(\underline{X},t) = \frac{D(x_1,x_2,x_3)}{D(X_1,X_2,X_3)}(\underline{X},t). \qquad [4.14]$$

The Jacobian matrix of $\underline{\psi}$ at point $\underline{x}$ and time t is the inverse of the preceding one and its Jacobian determinant is equal to $J^{-1}(\underline{\psi}(\underline{x},t),t)$.

Both $J(\underline{X},t)$ and $J^{-1}(\underline{\psi}(\underline{x},t),t)$ cannot be zero since ϕ is a homeomorphism:

$$\forall t, J(\underline{X},t) \neq 0 \text{ and } J^{-1}(\underline{X},t) \neq 0 \text{ that implies } J(\underline{X},t) \neq \infty. \qquad [4.15]$$

Moreover, continuous differentiability of ϕ being assumed, $J(\underline{X},t)$ is continuous with respect to space and time variables. Its sign is therefore constant over Ω_0 and throughout the motion from κ_0 to κ_t. Obviously, in κ_0, we have $J(\underline{X},t_0) = 1$, thence

$$\forall \underline{X}, \forall t, \ 0 < J(\underline{X},t) < +\infty. \qquad [4.16]$$

Geometrically, $J(\underline{X},t)$ yields the volume ratio between the infinitesimal volume element $d\Omega_0$ and its image $d\Omega_t$ through the homeomorphism

$$d\Omega_t = J(\underline{X},t)d\Omega_0. \qquad [4.17]$$

The result in [4.16] can thus be pictured in the following way: the volume of a particle conserves its sign and can become neither zero nor infinite, which sounds quite sensible from the physical viewpoint!

– Mathematical continuity hypotheses are also imposed to physical quantities associated with the particles: as a general rule, they are supposed of class C^1 or even C^2 with respect to space and time variables.

4.3.2. Eulerian description

Compared with the Lagrangian description, the Eulerian description can be labeled as *incremental*, like jumping from one frame of a film at time t to the following one at time $(t+dt)$. In comparison, the Lagrangian description follows the whole sequence of the evolution of the system from a reference frame at time t_0 to the present one at time t and is therefore often called a *pathline description*.

From equations [4.10] or [4.11], it follows that the mathematical continuity hypotheses to be imposed on $\underline{U}(\underline{x},t)$ and on any physical quantity $b(\underline{x},t)$ in the

Eulerian description consistently with section 4.3.1 are *continuity and continuous differentiability* with respect to space and time variables.

4.3.3. *Conservation of mass. Equation of continuity*

As recalled in Chapter 3 (section 3.4.2), the conservation of mass of a mechanical system is a fundamental principle of mechanics. Presently, it means that for the system S and any subsystem $S' \subset S$

$$\forall \Omega_0' \subseteq \Omega_0, \forall t, \ \mathcal{M}' = \int_{S'} \mathrm{d}m = \int_{\Omega_0'} \rho_0(\underline{X},t)\,\mathrm{d}\Omega_0 \text{ is constant.} \qquad [4.18]$$

It follows that $\rho_0(\underline{X},t)$ is independent of t and may be written unambiguously as

$$\rho_0(\underline{X},t) = \rho_0(\underline{X},t_0) = \rho_0(\underline{X}). \qquad [4.19]$$

Then, for any particle,

$$\rho_0(\underline{X})\,\mathrm{d}\Omega_0 = \rho(\underline{x},t)\,\mathrm{d}\Omega_t = \mathrm{d}m. \qquad [4.20]$$

Putting together [4.17] and [4.20], we obtain

$$\begin{cases} \underline{x} = \underline{\phi}(\underline{X},t) \\ \rho(\underline{x},t) = J^{-1}(\underline{X},t)\rho_0(\underline{X}) \end{cases} \qquad [4.21]$$

These equations are the Lagrangian expression of the *equation of continuity*, meaning that the geometrical transport of the particles defined by the vector field $\underline{\phi}$ is actually a *material* transport. The Eulerian form of the equation of continuity will be given in the next chapter.

4.4. Validation of the model

4.4.1. *Weakening of continuity hypotheses*

Although it may appear from the preceding section that the continuity hypotheses which we introduced seem to conform to our intuitive idea of continuity, we will now investigate whether a slight change to these hypotheses would make it easier to treat certain phenomena within the three-dimensional continuum model.

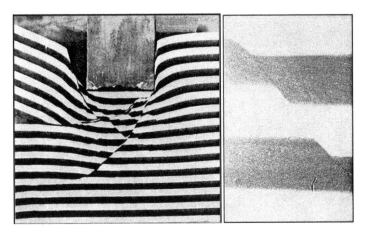

Figure 4.8. *Asymmetrical punch indentation test on a plasticine block (Application to tectonics in East Asia) [PEL 87]*

Figure 4.8 presents the result of a punch indentation test performed on a multilayer plasticine block. It shows that our intuitive idea of continuity referring to the conservation of proximity for neighboring points throughout the evolution is valid but for some "slip surfaces" where discontinuities of the vector function $\underline{\phi}$ should be allowed. The same phenomenon is observed in Figure 4.9 with the "slip line" pattern under a strip footing acting on a purely cohesive material in a medium-scale experiment.

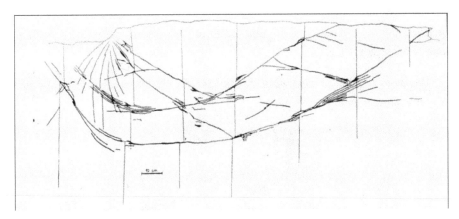

Figure 4.9. *"Slip line" pattern under a strip footing on a purely cohesive soil [HAB 84]*

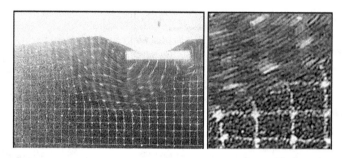

Figure 4.10. *Foundation on a two-dimensional layer (reduced scale experiment): short exposure visualizing the velocity field [BON 72]*

In Figure 4.10, we see the velocity field induced in a punch indentation test of a two-dimensional granular layer (made of Schneebeli rolls) performed as a reduced scale analysis of the bearing capacity of a strip footing. It clearly exhibits a jump surface in the velocity field, which delimitates the mass of soil set in motion from the rest that stays quite motionless. The same phenomenon is also observed in fluid mechanics as shown in Figure 4.11, presenting the two-dimensional flow around a flat plate with a velocity jump surface that is usually called a "jet surface".

These examples suggest that the condition which we initially formulated about neighboring particles is too restrictive. Therefore, the continuity hypotheses will be weakened by requiring only *piecewise continuity and continuous differentiability* with respect to space and time variables: $\underline{\phi}(\underline{X},t)$ and its derivatives of the first and second orders in the Lagrangian description, and $\underline{U}(\underline{x},t)$ and its derivatives of the first order in the Eulerian one are allowed to be *discontinuous across a countable infinity* of surfaces of $\mathbb{R}^3$.

The same weakened continuity conditions will be imposed on $B(\underline{X},t)$ and $b(\underline{x},t)$ representing physical quantities.

Figure 4.11. *Flow around a flat plate (ONERA, France)*

4.4.2. *Physical validation*

As anticipated in section 4.1.3, physical validation of the three-dimensional continuum model refers to the very "heart" and "art" of modeling: the question is not to decide whether matter is or is not continuous – provided a proper physical meaning is given to that term – but to investigate when and how far a three-dimensional continuum model can be referred to and provide relevant results.

The example of granular materials addressed in Figure 4.9 enhances the importance of the scale of the considered problem compared with the scale of the microstructure of the material in order that the model be asked good questions it can provide relevant answers to. Thus, slope stability problems, foundation analysis, seismic stability of earth structures, silos, etc. can be addressed within the three-dimensional continuum model with the weakened continuity hypotheses stated in section 4.4.1, and would not be solved at the granular level from a practical viewpoint. However, this model in its classical form would not be suitable for the in-depth analysis of slip bands, for instance. Similar examples are encountered in fluid mechanics with complex fluids.

This being said, the domain of physical relevance of the model comes out astonishingly wide: from geophysics to biological fluids, through geotectonics, metal forming, fluid mechanics, acoustics, soil mechanics, mechanics of materials, polymers, composite materials, rheology, tribology, etc.

The case of composite materials deserves a special interest and brings us back to the question of the physical meaning or relevance of the notion of particles that we made the basis of our model.

Composite materials are often made of a matrix that is reinforced by oriented fiber layers, which confer an oriented microstructure to the physical representative volume element to be considered as a material particle. It follows that restricting the kinematic description to the velocity field $\underline{U}(\underline{x},t)$ in the Eulerian description does not provide a suitable description of the microstructure associated with the particle. Introducing additional geometric parameters in the form of a rotational vector field independent of $\underline{U}(\underline{x},t)$ to account for the orientation of the microstructure yields a non-classical three-dimensional continuum that offers a possibility of modeling such materials.

To conclude with this chapter where three-dimensional continuum modeling was introduced, let us refer to Saint-Venant who described our initial intuitive approach quite clearly:

> *"But one does not usually need to consider real or individual displacements of molecules... It is enough to consider average displacements which are just the displacements of the centres of gravity of groups of a certain number of molecules. Each point of the space occupied by the body can always be regarded as the centre of gravity of a similar group, contained in an unperceivable element, but finite, of this space. And so, despite the allotted small dimensions of the element, it can always be supposed to contain a rather considerable number of molecules to compensate for the irregular inequalities of their individual displacements one by the other, average displacements generally vary with continuity and in a simple way from one point to another of the body..."[1].*

4.5. Practicing

4.5.1. *Homogeneous transformation*

In orthonormal Cartesian coordinates, consider the motion defined by $\underline{x} = \underline{\phi}(\underline{X}, t) = \underline{a}(t) + \underline{\underline{\alpha}}(t) . \underline{X}$ with $\underline{a}(t)$ continuous and continuously differentiable such that $\underline{a}(0) = 0$ and $\underline{\underline{\alpha}}(t)$ continuous and continuously differentiable such that $\underline{\underline{\alpha}}(0) = \underline{\underline{1}}$. Consider the set of particles occupying a straight line segment $M_0 M_0'$ in the reference configuration. Prove that these particles occupy a straight line segment MM' in the present configuration at time t. Defining this set of particles in κ_0 as

---

1 "Mais on n'a pas besoin, ordinairement, de considérer les déplacements réels ou individuels des molécules... Il suffit de considérer des déplacements moyens qui ne sont autre chose que les déplacements des centres de gravité de groupes d'un certain nombre de molécules. Chaque point de l'espace occupé par le corps peut toujours être considéré comme le centre de gravité d'un pareil groupe, contenu dans un élément imperceptible, mais fini, de cet espace. Et comme, malgré la petitesse attribuée aux dimensions de l'élément, il peut toujours être supposé contenir un nombre assez considérable de molécule pour compenser, l'une par l'autre, les inégalités irrégulières de leurs déplacements individuels, les déplacements moyens varient généralement, d'un point à l'autre du corps, avec continuité et d'une manière simple." [BAR 55].

a material vector $\underline{V} = M_0 M_0'$ gives the expression $\underline{v} = \underline{MM'}$ of this material vector in the present configuration κ_t .

Solution

The particles in M_0 and M_0' with position vectors $\underline{X}$ and $\underline{X}'$ respectively are transported onto M and M' with position vectors $\underline{x}$ and $\underline{x}'$. Let P_0 denote a particle on $M_0 M_0'$ with position vector $\underline{\chi} = \lambda \underline{X}' + (1-\lambda)\underline{X}$, $0 \le \lambda \le 1$. It is transported by the motion onto P with position vector $\underline{\xi}(t)$

$$\underline{\xi}(t) = \underline{a}(t) + \underline{\underline{\alpha}}(t).(\lambda \underline{X}' + (1-\lambda)\underline{X}) \ , \ 0 \le \lambda \le 1,$$

which can be written as

$$\underline{\xi}(t) = \lambda(\underline{a}(t) + \underline{\underline{\alpha}}(t).\underline{X}') + (1-\lambda)(\underline{a}(t) + \underline{\underline{\alpha}}(t).\underline{X}) = \lambda \underline{x}' + (1-\lambda)\underline{x} \ , \ 0 \le \lambda \le 1$$

That is actually the position vector of a particle on the straight line segment MM' at time t . Thus, the material vector $\underline{V} = M_0 M_0'$ remains a vector when transported by the motion into the present configuration through the considered transformation and we have: $\underline{v} = \underline{\underline{\alpha}}(t).\underline{V}$.

This transformation is called a *homogeneous transformation*.

4.5.2. Simple shear

In orthonormal Cartesian coordinates, consider the motion defined by $\underline{x} = \underline{\phi}(\underline{X}, t)$ with $x_1 = X_1 + \alpha t X_2, x_2 = X_2, x_3 = X_3$. Calculate the volume expansion $J(\underline{X}, t)$. Determine the pathlines. Give the Eulerian representation of this motion.

Solution

Volume expansion: $J(\underline{X}, t) = 1$. No volume change.

The pathlines are straight lines parallel to $\underline{e}_1$.

The Eulerian description of the motion is: $\underline{U}(\underline{x}, t) = \alpha x_2 \underline{e}_1 = \alpha(\underline{e}_1 \otimes \underline{e}_2).\underline{x}$.

4.5.3. *"Lagrangian" double shear*

In orthonormal Cartesian coordinates, consider the motion defined by $\underline{x} = \underline{\phi}(\underline{X},t)$ with $x_1 = X_1 + \alpha t X_2, x_2 = X_2 + \alpha t X_1, x_3 = X_3$. Calculate the volume expansion $J(\underline{X},t)$.

Solution

The volume expansion is $J(\underline{X},t) = (1 - \alpha^2 t^2)$. We observe that although the motion looks like the addition of two simple shear motions with no volume change, the volume expansion is non-zero and the duration of the motion must be limited to $0 \le t < 1/\alpha$.

4.5.4. *Rigid body motion*

In orthonormal Cartesian coordinates, consider the motion defined by $\underline{x} = \underline{\phi}(\underline{X},t) = \underline{a}(t) + \underline{\underline{\alpha}}(t).\underline{X}$ with $\underline{a}(t)$ continuous and continuously differentiable such that $\underline{a}(0) = \underline{0}$ and $\underline{\underline{\alpha}}(t)$ continuous and continuously differentiable such that $\underline{\underline{\alpha}}(0) = \underline{\underline{1}}$ and $\underline{\underline{\alpha}}(t).^t\underline{\underline{\alpha}}(t) = \underline{\underline{1}}$. Calculate the volume expansion $J(\underline{X},t)$.

Let $\underline{V}$ denote a material vector in κ_0 that is transported onto $\underline{v}$ in κ_t. Calculate $\underline{v}.\underline{v}$ as a function of $\underline{V}$.

With $\underline{V}$ and $\underline{W}$ two material vectors in κ_0 transported onto $\underline{v}$ and $\underline{w}$ in κ_t, calculate $\underline{v}.\underline{w}$ as a function of $\underline{V}$ and $\underline{W}$.

Solution

$\underline{x} = \underline{\phi}(\underline{X},t) = \underline{a}(t) + \underline{\underline{\alpha}}(t).\underline{X}$ can be written explicitly as $x_i = a_i(t) + \alpha_{ij}(t)X_j$.

Thence, $\dfrac{\partial x_i}{\partial X_j} = \alpha_{ij}(t)$ and $J(\underline{X},t) = \det[\alpha_{ij}(t)] = \det[\underline{\underline{\alpha}}(t)]$. As $\det[^t\underline{\underline{\alpha}}(t)] = \det\underline{\underline{\alpha}}(t)$, we derive from $\underline{\underline{\alpha}}(t).^t\underline{\underline{\alpha}}(t) = \underline{\underline{1}}$ that $(\det\underline{\underline{\alpha}}(t))^2 = 1$ and then $\det\underline{\underline{\alpha}}(t) = 1$ from $\det\underline{\underline{\alpha}}(0) = 1$ and $\underline{\underline{\alpha}}(t)$ being continuous with respect to time. Hence, $J(\underline{X},t) = 1$: no volume change.

The transformation is homogeneous, thence $\underline{v} = \underline{\underline{\alpha}}(t).\underline{V}$ and

$$\underline{v}.\underline{v} = (\underline{\underline{\alpha}}(t).\underline{V}).(\underline{\underline{\alpha}}(t).\underline{V}) = \underline{V}.{}^{t}\underline{\underline{\alpha}}(t).\underline{\underline{\alpha}}(t).\underline{V};$$

with $\underline{\underline{\alpha}}(t).{}^{t}\underline{\underline{\alpha}}(t) = {}^{t}\underline{\underline{\alpha}}(t).\underline{\underline{\alpha}}(t) = \underline{\underline{1}}$, this yields $\underline{v}.\underline{v} = \underline{V}.\underline{V}$ that is $|\underline{v}| = |\underline{V}|$.

In plain words, the considered motion is a *direct isometry function of time*.

Similarly, $\underline{v}.\underline{w} = \underline{V}.\underline{\underline{1}}.\underline{W} = \underline{V}.\underline{W}$. From $|\underline{v}| = |\underline{V}|$ and $|\underline{w}| = |\underline{W}|$, it follows that the angle made by $\underline{v}$ with $\underline{w}$ is constant with respect to time, equal to the angle made by by $\underline{V}$ with $\underline{W}$ (as expected from any direct isometry).

Kinematics of the
Three-dimensional Continuum

5.1. Kinematics

5.1.1. *The issues*

In contrast with the Lagrangian *pathline description*, the Eulerian *incremental* description defines the motion of the system shifting from κ_t to κ_{t+dt} through the velocity field given at time t. During this infinitesimal evolution, any particle of S moves from its location $\underline{x}$ in Ω_t to $\underline{x}+\underline{U}(\underline{x},t)dt$ in Ω_{t+dt}. Geometrically, the material system is transported by the motion. If this motion is not a *rigid body motion*, the system undergoes incremental geometrical changes that can be determined from the velocity field $\underline{U}$ following the material elements: this will be the first issue addressed in this chapter. Then, the concept of convective time derivative will be dealt with in more details including the case of piecewise continuous and continuously differentiable fields.

5.1.2. *Material time derivative of a vector*

The concept of material time derivative was incidentally introduced in Chapters 3 (section 3.4.2) and 4 (section 4.2.2) with the meaning of the derivative with respect to time of a physical quantity attached to a particle or a system, following that particle or system conveyed by the motion. It will be treated more extensively in the next section, while, in this section, we will concentrate on the geometrical evolution of an infinitesimal linear material element in the infinitesimal motion from the time t to the time $(t+dt)$, thus defining the *convective derivative of a material vector*.

In κ_t, let $\underline{dM}$ denote an infinitesimal material vector at the point $\underline{x}$ (the notation "d" to denote an infinitesimal element has the same meaning as in $d\Omega_t$). At time $(t+dt)$, this material vector is located at the point $\underline{x}+\underline{U}(\underline{x},t)\,dt$ and becomes

$$\underline{dM} + \frac{d}{dt}(\underline{dM})\,dt, \qquad\qquad [5.1]$$

which defines the material derivative $\dfrac{d}{dt}(\underline{dM})$ of the vector $\underline{dM}$ at point $\underline{x}$ and time t.

As sketched in Figure 5.1, we may write

$$\underline{dM} + \frac{d}{dt}(\underline{dM})\,dt = -\underline{U}(\underline{x},t)\,dt + \underline{dM} + \underline{U}(\underline{x}+\underline{dM},t)\,dt \qquad [5.2]$$

where, assuming the vector field to be continuously differentiable,

$$\underline{U}(\underline{x}+\underline{dM},t) - \underline{U}(\underline{x},t) = \underline{\underline{\operatorname{grad}}}\,\underline{U}(\underline{x},t).\underline{dM}. \qquad [5.3]$$

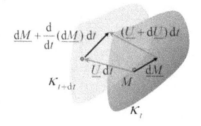

Figure 5.1. *Material derivative of a vector*

We then obtain the pivotal formula for the convective or material derivative of a vector in the Eulerian description:

$$\frac{d}{dt}(\underline{dM}) = \underline{\underline{\operatorname{grad}}}\,\underline{U}(\underline{x},t).\underline{dM}. \qquad\qquad [5.4]$$

Note that this equation defines the linear mapping of $\mathbb{R}^3$ onto $\mathbb{R}^3$ that associates with any material vector its convective derivative:

$$\underline{dM} \in \mathbb{R}^3 \mapsto \frac{d}{dt}(\underline{dM}) = \underline{\underline{grad\,U(\underline{x},t)}}.\underline{dM} \in \mathbb{R}^3. \qquad [5.5]$$

5.1.3. *Strain rate*

Let $\underline{dM}_1$ and $\underline{dM}_2$ be two infinitesimal material vectors at the field point $\underline{x}$ in κ_t. At time $(t+dt)$, these vectors are transported by the motion in κ_{t+dt} and become respectively

$$\underline{dM}_1 + \frac{d}{dt}(\underline{dM}_1)\,dt = \underline{dM}_1 + \left(\underline{\underline{grad\,U(\underline{x},t)}}.\underline{dM}_1\right)dt \qquad [5.6]$$

and

$$\underline{dM}_2 + \frac{d}{dt}(\underline{dM}_2)\,dt = \underline{dM}_2 + \left(\underline{\underline{grad\,U(\underline{x},t)}}.\underline{dM}_2\right)dt. \qquad [5.7]$$

Thence, the convective derivative of the scalar product $\underline{dM}_1.\underline{dM}_2$ is written as

$$\frac{d}{dt}(\underline{dM}_1.\underline{dM}_2) = \left(\underline{\underline{grad\,U(\underline{x},t)}}.\underline{dM}_1\right).\underline{dM}_2 + \underline{dM}_1.\left(\underline{\underline{grad\,U(\underline{x},t)}}.\underline{dM}_2\right). \qquad [5.8]$$

Taking advantage of the definition of $^t\underline{\underline{grad\,U(\underline{x},t)}}$, we obtain the symmetric formula

$$\frac{d}{dt}(\underline{dM}_1.\underline{dM}_2) = \underline{dM}_1.\left(\underline{\underline{grad\,U(\underline{x},t)}} + {}^t\underline{\underline{grad\,U(\underline{x},t)}}\right).\underline{dM}_2, \qquad [5.9]$$

defining the *Eulerian strain rate tensor* $\underline{\underline{d}}(\underline{x},t)$, a symmetric bilinear form on $\mathbb{R}^3 \times \mathbb{R}^3$

$$\underline{\underline{d}}(\underline{x},t) = \frac{1}{2}\left(\underline{\underline{grad\,U(\underline{x},t)}} + {}^t\underline{\underline{grad\,U(\underline{x},t)}}\right) \qquad [5.10]$$

so that

$$\frac{d}{dt}(\underline{dM}_1.\underline{dM}_2) = 2\,\underline{dM}_1.\underline{\underline{d}}(\underline{x},t).\underline{dM}_2. \qquad [5.11]$$

5.1.4. Extension rate

Equation [5.11] defining the evolution of the scalar product of any pair of material vectors associated with the particle at point $\underline{x}$ in κ_t in the infinite motion between t and $(t + dt)$ makes it possible to quantify the deformation of the material system at that point in that motion.

Considering first the case when $d\underline{M}_1 = d\underline{M}_2 = ds\,\underline{e}_1$, where $\underline{e}_1$ is a unit vector, we have obviously

$$\frac{d}{dt}(d\underline{M}_1 \cdot d\underline{M}_1) = \frac{d}{dt}(ds^2) = 2\,ds\,\frac{d}{dt}(ds), \qquad [5.12]$$

while we derive from [5.11]

$$\frac{d}{dt}(d\underline{M}_1 \cdot d\underline{M}_1) = 2\,ds^2\,\underline{e}_1 \cdot \underline{\underline{d}}(\underline{x},t) \cdot \underline{e}_1. \qquad [5.13]$$

Putting these two results together, we obtain the *extension rate* of a material vector along the $\underline{e}_1$ direction:

$$\frac{1}{ds}\frac{d}{dt}(ds) = \underline{e}_1 \cdot \underline{\underline{d}}(\underline{x},t) \cdot \underline{e}_1. \qquad [5.14]$$

In an orthonormal basis built upon $\underline{e}_1$, we write

$$\underline{\underline{d}}(\underline{x},t) = d_{ij}(\underline{x},t)\underline{e}_i \otimes \underline{e}_j = \frac{1}{2}(\frac{\partial U_i}{\partial x_j} + \frac{\partial U_j}{\partial x_i})\underline{e}_i \otimes \underline{e}_j \qquad [5.15]$$

and the *extension rate* along $\underline{e}_1$ is given by

$$\frac{1}{ds}\frac{d}{dt}(ds) = d_{11}(\underline{x},t). \qquad [5.16]$$

5.1.5. Distortion rate

In an orthonormal basis as described here above, we consider two infinitesimal material vectors along $\underline{e}_1$ and $\underline{e}_2$: $d\underline{M}_1 = ds_1\,\underline{e}_1$ and $d\underline{M}_2 = ds_2\,\underline{e}_2$. After the infinitesimal motion, they are transported onto $\underline{dM}_1 + \dfrac{d}{dt}(\underline{dM}_1)dt$ and

$\underline{dM}_2 + \dfrac{d}{dt}(\underline{dM}_2)dt$ in κ_{t+dt} that make the angle $(\dfrac{\pi}{2} - \dot{\theta}\,dt)$ counted positive about

$\underline{e}_3$ (Figure 5.2).

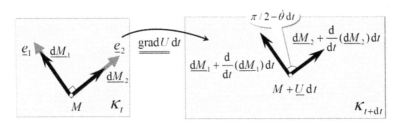

Figure 5.2. *Distortion rate*

The scalar product $\underline{dM}_1 . \underline{dM}_2$ initially equal to zero in κ_t becomes equal to $ds_1 \, ds_2 \dot{\theta}\, dt$ in κ_{t+dt}. This variation can also be derived from [5.11]

$$\frac{d}{dt}(\underline{dM}_1 . \underline{dM}_2) = 2\, ds_1 \, ds_2 \, \underline{e}_1 . \underline{\underline{d}}(\underline{x},t) . \underline{e}_2 \qquad [5.17]$$

and we obtain the final result

$$\dot{\theta}(\underline{x},t) = 2\, d_{12}(\underline{x},t) \qquad [5.18]$$

which is called the *rate of distortion* of two orthogonal material vectors in κ_t at point $\underline{x}$.

5.1.6. *Principal axes of the strain rate tensor*

The strain rate tensor $\underline{\underline{d}}(\underline{x},t)$ is a symmetric bilinear form. It follows that a triad of *principal axes* can be determined for $\underline{\underline{d}}(\underline{x},t)$ that is orthogonal and, in an orthonormal basis $\{\underline{e}_i\}$ built upon these axes, $\underline{\underline{d}}(\underline{x},t)$ takes the form

$$\underline{\underline{d}}(\underline{x},t) = d_{11}(\underline{x},t)\underline{e}_1 \otimes \underline{e}_1 + d_{22}(\underline{x},t)\underline{e}_2 \otimes \underline{e}_2 + d_{33}(\underline{x},t)\underline{e}_3 \otimes \underline{e}_3 \qquad [5.19]$$

where, as a consequence of [5.16], $d_{11}(\underline{x},t)$, $d_{22}(\underline{x},t)$ and $d_{33}(\underline{x},t)$ are the extension rates along the three principal axes and denoted by $d_1(\underline{x},t)$, $d_2(\underline{x},t)$ and $d_3(\underline{x},t)$. It follows immediately from [5.18] that for each pair of principal axes at point $\underline{x}$, we have $\dot{\theta}(\underline{x},t) = 0$. Stated in plain words, this means that a triad of infinitesimal vectors lying along the principal axes of $\underline{\underline{d}}(\underline{x},t)$ *remains orthogonal* in the infinitesimal motion between κ_t and κ_{t+dt} where it conserves its orientation (Figure 5.3). Conversely, we see easily from equation [5.18] that this is a characteristic property of the principal axes.

5.1.7. *Volume dilatation rate*

This latter result makes it easy to compute the rate of volume dilatation in the infinitesimal motion from κ_t to κ_{t+dt}. From now on, the orthonormal basis $\{\underline{e}_i\}$ is assumed to be right handed and we consider three infinitesimal material vectors along $\underline{e}_1$, $\underline{e}_2$ and $\underline{e}_3$. The volume of the parallelepiped defined by $\underline{dM}_1$, $\underline{dM}_2$ and $\underline{dM}_3$ in κ_t is

$$d\Omega_t = (\underline{dM}_1, \underline{dM}_2, \underline{dM}_3) = ds_1 \, ds_2 \, ds_3, \qquad [5.20]$$

and becomes

$$d\Omega_t + \frac{d}{dt}(d\Omega_t)\,dt = ds_1 \, ds_2 \, ds_3 \left(1 + (d_1 + d_2 + d_3)\,dt\right) \text{ in } \kappa_{t+dt}. \qquad [5.21]$$

We derive the expression for the *rate of volume dilatation*

$$\frac{1}{d\Omega_t}\frac{d}{dt}(d\Omega_t) = d_1(\underline{x},t) + d_2(\underline{x},t) + d_3(\underline{x},t) = \text{tr}\,\underline{\underline{d}}(\underline{x},t) \qquad [5.22]$$

also equal to

$$\frac{1}{d\Omega_t}\frac{d}{dt}(d\Omega_t) = \text{div}\,\underline{U}(\underline{x},t)^{\,1} \qquad [5.23]$$

---

1 Note incidentally that this result is at the origin of the terminology "divergence".

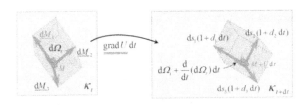

Figure 5.3. *Volume dilatation rate*

As an important consequence, we derive that any motion in which there is no volume change at time t at the point $\underline{x}$ is characterized by the condition $\operatorname{div}\underline{U}(\underline{x},t) = 0$. It follows that any *isochoric* evolution of a system (solid or fluid) shall satisfy the condition

$$\forall t,\ \forall \underline{x} \in \Omega_t,\ \operatorname{div}\underline{U}(\underline{x},t) = 0. \tag{5.24}$$

5.1.8. *Spin tensor*

Noting that the strain rate tensor, which is the symmetric part of the velocity gradient, appears to be quite sufficient to quantify the deformation of the material system in the infinitesimal motion, we may wonder about the antisymmetric part of $\operatorname{grad}U(\underline{x},t)$ that is named the *spin tensor* and denoted by

$$\underline{\underline{\Omega}}(\underline{x},t) = \frac{1}{2}\left(\operatorname{grad}U(\underline{x},t) - {}^{t}\operatorname{grad}U(\underline{x},t)\right). \tag{5.25}$$

In a right-handed orthonormal basis, where $\underline{\underline{\Omega}}(\underline{x},t)$ is written as

$$\underline{\underline{\Omega}}(\underline{x},t) = \frac{1}{2}\left(\frac{\partial U_i}{\partial x_j} - \frac{\partial U_j}{\partial x_i}\right)\underline{e}_i \otimes \underline{e}_j \tag{5.26}$$

we can associate $\underline{\underline{\Omega}}(\underline{x},t)$ with the *spin vector* $\underline{\Omega}(\underline{x},t)$ through the canonical formula

$$\forall \underline{V} \in \mathbb{R}^3,\ \underline{\underline{\Omega}}(\underline{x},t).\underline{V} = \underline{\Omega}(\underline{x},t) \wedge \underline{V} \tag{5.27}$$

from which it comes out that

$$\underline{\Omega}(\underline{x},t) = \frac{1}{2}\operatorname{curl}\underline{U}(\underline{x},t). \tag{5.28}$$

The linear mapping [5.5] can thus be written as

$$\underline{dM} \in \mathbb{R}^3 \mapsto \frac{d}{dt}(\underline{dM}) = \underline{\underline{d}}(\underline{x},t).\underline{dM} + \underline{\Omega}(\underline{x},t) \wedge \underline{dM} \in \mathbb{R}^3 \qquad [5.29]$$

and, being applied to a triad of infinitesimal material vectors along the principal axes of $\underline{\underline{d}}(\underline{x},t)$, it yields:

$$\begin{cases} \dfrac{d}{dt}(\underline{dM}_1) = d_1(\underline{x},t)\underline{dM}_1 + \underline{\Omega}(\underline{x},t) \wedge \underline{dM}_1 \\[2mm] \dfrac{d}{dt}(\underline{dM}_2) = d_2(\underline{x},t)\underline{dM}_2 + \underline{\Omega}(\underline{x},t) \wedge \underline{dM}_2 \\[2mm] \dfrac{d}{dt}(\underline{dM}_3) = d_3(\underline{x},t)\underline{dM}_3 + \underline{\Omega}(\underline{x},t) \wedge \underline{dM}_3 . \end{cases} \qquad [5.30]$$

These three equations show that, in the infinitesimal motion from κ_t to κ_{t+dt}, each infinitesimal material vector $\underline{dM}_i$ is subject to the elongation $d_i(\underline{x},t)\,dt$ while the whole triad is subject to the rigid body infinitesimal rotation defined by the vector $\underline{\Omega}(\underline{x},t)\,dt$.

The spin vector $\underline{\Omega}(\underline{x},t)$ is the instantaneous angular velocity vector of the *material triad of principal axes* of $\underline{\underline{d}}(\underline{x},t)$ at time t in the transport due to the motion. More significantly from a physical viewpoint, we may introduce the notion of a mean motion of the matter at the point $\underline{x}$ and time t as in [SAL 01]: the spin vector appears then as the instantaneous mean angular velocity of a small spherical material region centered at the point $\underline{x}$[2]. The spin vector can thus be followed experimentally by means of a small device called a vorticity meter that floats on the surface of a fluid and rotates showing the vorticity as in [KLI 63], or more sophisticated devices.

In a similar way, as in section 5.1.7, an *irrotational* motion is characterized by the condition

$$\forall t, \ \forall \underline{x} \in \Omega_t, \ \underline{\mathrm{curl}}\,\underline{U}(\underline{x},t) = 0 \Leftrightarrow \underline{\Omega}(\underline{x},t) = 0. \qquad [5.31]$$

---

[2] This result is at the origin of the terminology "rotationnel" in French and similar terms in Latin languages for the curl. The curl of $\underline{U}$ is also called the *vorticity vector*, while the spin tensor is sometimes called the vorticity tensor, which may cause confusion as they are not associated through the canonical formula!

5.1.9. *Rigid body motion*

As already stated in Chapter 2, any velocity field defined in the form

$$\underline{U}(\underline{x},t) = \underline{U}(t) + \underline{\omega}(t) \wedge \underline{OM}, \qquad [5.32]$$

where $\underline{U}(t)$ and $\underline{\omega}(t)$ are arbitrary vectors in $\mathbb{R}^3$ and O an arbitrary point, is such that the distance between any pair of geometrical points in κ_t is conserved in κ_{t+dt}. With the introduction of $\underline{\underline{\omega}}(t)$, the antisymmetric tensor associated with $\underline{\omega}(t)$ through the canonical formula, [5.32] may also be written as

$$\underline{U}(\underline{x},t) = \underline{U}(t) + \underline{\underline{\omega}}(t).\underline{OM}, \qquad [5.33]$$

which yields

$$\underline{\underline{\text{grad}\, U}}(\underline{x},t) = \underline{\underline{\omega}}(t) \qquad [5.34]$$

and thus

$$\underline{\underline{d}}(\underline{x},t) = 0, \;\; \underline{\underline{\Omega}}(\underline{x},t) = \underline{\underline{\omega}}(t), \;\; \underline{\Omega}(\underline{x},t) = \underline{\omega}(t). \qquad [5.35]$$

These last equations express the fact that the system undergoes no deformation and is subject, as a whole, to a rigid body motion (translation and rigid body rotation) between t and $(t+dt)$. These two results stated in plain words may appear somewhat redundant and we may wonder whether assuming $\underline{\underline{d}}(\underline{x},t) = 0, \forall \underline{x}$ would be sufficient to imply that $\underline{\Omega}(\underline{x},t)$ is constant over Ω_t. The answer to this question is positive, as a particular case of the general issue of the geometrical compatibility of a strain rate field.

5.1.10. *Geometrical compatibility of a strain rate field*

5.1.10.1. *Physical context*

The question of the geometrical compatibility of a strain rate field arises from many practical problems, especially in the mechanics of deformable solids. As an example, we may consider the case when a temperature change rate field $\tau(\underline{x},t)$, imposed to a system, generates a strain rate field $\underline{\underline{d}}_\tau(\underline{x},t)$.

– If the field $\underset{=t}{d}(\underline{x},t)$ is *geometrically compatible*, meaning that it actually derives from a velocity field $\underline{U}(\underline{x},t)$ over $\varOmega_t$ through [5.10], the system will deform freely and continuously.

– If *no continuous and continuously differentiable* velocity field can be found from which $\underset{=t}{d}(\underline{x},t)$ would be derived, internal forces (that will be modeled in the following chapter) must necessarily come into action. Through the constitutive equation of the material, they induce a complementary strain rate field $\underset{=c}{d}(\underline{x},t)$ such that the complete strain rate field $\underset{=t}{d}(\underline{x},t)+\underset{=c}{d}(\underline{x},t)$ is geometrically compatible and defines a continuous evolution of the system.

– In this latter case, it may happen that the internal forces necessary to maintain the continuity of the system cannot be sustained by the material. In such a case, the solution that takes place is "piecewise geometrically compatible" with localized fracture in the system, a classical problem with fragile materials.

5.1.10.2. Compatibility conditions

Clearly, from a mathematical viewpoint, it is quite natural to find that *compatibility conditions* should be satisfied by a symmetric tensor field $\underset{=}{d}$, which depends on six independent scalar fields, in order that it be derived through [5.10] from a three-component vector field. We will establish these compatibility conditions by a direct method that will also provide a practical integration procedure; more extensive presentations and comments are available in classical textbooks (e.g. [SAL 01]).

Orthogonal Cartesian coordinates will be used and spatial derivatives denoted by a comma in order to simplify the equations and compatibility conditions to be written (also, the arguments $\underline{x}$, t are dropped for simplicity sake).

5.1.10.2.1. Necessary conditions

Let $\underset{=}{d}$ be a strain rate field derived from a velocity field $\underline{U}$ and $\underset{=}{\varOmega}$ the associated spin tensor field:

$$d_{ij} = \frac{1}{2}(U_{i,j} + U_{j,i}) \, , \, \varOmega_{ij} = \frac{1}{2}(U_{i,j} - U_{j,i}).$$ [5.36]

Taking the spatial derivatives of these components, we obtain the relation

$$\varOmega_{ij,k} = d_{ki,j} - d_{jk,i}$$ [5.37]

and the second order derivatives

$$\begin{cases} \Omega_{ij,k\ell} = d_{ki,j\ell} - d_{jk,i\ell} \\ \Omega_{ij,\ell k} = d_{\ell i,jk} - d_{j\ell,ik}. \end{cases} \tag{5.38}$$

Writing that $\Omega_{ij,k\ell} = \Omega_{ij,\ell k}$, we obtain six necessary compatibility conditions

$$d_{ki,j\ell} + d_{j\ell,ik} - d_{jk,i\ell} - d_{\ell i,jk} = 0 \tag{5.39}$$

where i, j, k, ℓ take arbitrary values amongst 1, 2, 3. Explicitly:

$$\begin{aligned} 2d_{23,23} &= d_{33,22} + d_{22,33} \quad + \text{ cyclic permutation} \\ d_{13,23} - d_{12,33} - d_{33,21} + d_{32,31} &= 0 \quad + \text{ cyclic permutation}. \end{aligned} \tag{5.40}$$

These conditions were established by Saint-Venant[3].

5.1.10.2.2. Sufficient conditions

Conversely, given $\underset{=}{d}$ a symmetric tensor field complying with [5.40], the integration procedure starts with the computation of the quantities $\Omega_{ij,k}$ through [5.37], which are antisymmetric with respect to i and j. The conditions [5.40] guarantee the integrability of the differential forms $d\Omega_{ij} = \Omega_{ij,k} \, dx_k$ into an antisymmetric tensor field, provided the region we are working in is *simply connected*. Then, it comes out, as a consequence of [5.37], that the integrability conditions of the differential forms $(d_{ij,k} + \Omega_{ij,k}) \, dx_k$ are identically satisfied. A vector field $\underline{U}$ can thus be constructed, provided the region is simply connected, such that $\underset{=}{d}$ is the symmetric part of its gradient field.

It is worth noting that each step of this integration procedure introduces an arbitrary constant field over the considered region: first, $\underline{\underline{\Omega}}_0$ an arbitrary antisymmetric constant tensor field with $\underline{\Omega}_0$ its associated vector field; then, $\underline{U}_0$ an arbitrary constant vector field. This results in the presence of an arbitrary rigid body velocity field $\underline{U}_0 + \underline{\Omega}_0 \wedge \underline{OM}$ in the vector field $\underline{U}$.

---

3 [BAR 64].

The proof that the conditions [5.40] are sufficient provided the region is simply connected was given by Beltrami[4].

An obvious consequence, already announced in section 5.1.9, is that $\underline{d}(\underline{x},t) = 0$, $\forall \underline{x}$ actually implies that the system under consideration is subject to a rigid body motion.

5.2. Convective derivatives

5.2.1. *General comments*

As recalled earlier, the very definition of the convective derivative of a quantity attached to a material system implies that the value of that quantity is computed at the times t and $(t+dt)$ following the particles of the system, whatever it is, as they are transported by the motion defined by the velocity field $\underline{U}$. Before going into the specific important cases that will be treated in the following section, it is worth trying to sketch out the general demeanor of the expressions we expect to obtain for these derivatives.

As a typical notation, let us denote by $\mathcal{B}(S,t)$ a quantity (scalar, vector, tensor, etc.) attached to a material system S (particle, volume, surface, line, etc.). Its definition in the Eulerian description refers to the geometrical position occupied by the considered system in κ_t and may be written symbolically as

$$\mathcal{B}(S,t) = b(S,t) \tag{5.41}$$

where S stands for the geometrical location of the system at the time t. In view of its definition, the material derivative of $\mathcal{B}(S,t)$ must cumulate the partial derivative of $b(S,t)$ with respect to time, as if the system were motionless or, so to speak, "frozen", and the contribution due to the very motion of the material system. This contribution is called the *convective term* and it is expected to be linear with respect to the velocity field as it is linearly derived from the infinitesimal motion of the system defined by $\underline{U}$:

---

4 [BEL 86, BEL 89]. It may seem surprising that we end up with six compatibility conditions to ensure that a six-component tensor field derives from a three-component vector field. In fact, the six equations [5.40] are not independent and they will hold in a region provided three of them are satisfied as field equations and the three others as boundary conditions.

$$\frac{d}{dt}\mathcal{B}(S,t) = \frac{d}{dt}b(S,t) = \frac{\partial}{\partial t}b(S,t) + \boxed{\text{convective term}}.$$

[5.42]

5.2.2. Convective derivative of a "point function"

In this first case, the considered quantity is defined as a function of a particle with the meaning that, in the Eulerian description, its depends on $\underline{x}$ and t :

$$\mathcal{B} = b(\underline{x},t).$$

[5.43]

Following the rationale presented here above, the convective term is easily identified as the result of the variation of $\mathcal{B}$ when the particle located in the point $\underline{x}$ in κ_t moves to the point $\underline{x} + \underline{U}(\underline{x},t)\,dt$ in κ_{t+dt}. Assuming $b(\underline{x},t)$ to be continuous and continuously differentiable, we may write

$$\begin{aligned}
\frac{d}{dt}b(\underline{x},t)\,dt &= b(\underline{x} + \underline{U}(\underline{x},t)\,dt, t+dt) - b(\underline{x},t) \\
&= \frac{\partial}{\partial t}b(\underline{x},t)\,dt + \underline{\text{grad }b.\underline{U}(\underline{x},t)}\,dt,
\end{aligned}$$

[5.44]

thence

$$\frac{d}{dt}b(\underline{x},t) = \frac{\partial}{\partial t}b(\underline{x},t) + \underline{(\text{grad }b(\underline{x},t)).\underline{U}(\underline{x},t)}.$$

[5.45]

Equation [5.45] written here in the case of a scalar function, is valid whatever the rank of a tensor $\mathcal{B} = b(\underline{x},t)$. As an example, for a vector $\underline{b}(\underline{x},t)$, we have

$$\frac{d}{dt}\underline{b}(\underline{x},t) = \frac{\partial}{\partial t}\underline{b}(\underline{x},t) + \underline{\underline{(\text{grad }b(\underline{x},t)).\underline{U}(\underline{x},t)}}.$$

[5.46]

5.2.3. Equation of continuity

Applying [5.45] to the mass per unit volume in the Eulerian description $\rho(\underline{x},t)$, we obtain the material derivative

$$\frac{d}{dt}\rho(\underline{x},t) = \frac{\partial}{\partial t}\rho(\underline{x},t) + \underline{(\text{grad }\rho(\underline{x},t)).\underline{U}(\underline{x},t)}.$$

[5.47]

This makes it possible to write the Eulerian expression of the equation of continuity. Recalling that

$$dm = \rho(\underline{x},t)\,d\Omega_t \tag{5.48}$$

the conservation of mass of the material element is written as

$$\frac{d}{dt}(dm) = \frac{d}{dt}(\rho(\underline{x},t)\,d\Omega_t) = (\frac{d}{dt}\rho(\underline{x},t))\,d\Omega_t + \rho(\underline{x},t)\frac{d}{dt}(d\Omega_t) = 0 \tag{5.49}$$

thence, with [5.23],

$$\frac{d}{dt}\rho(\underline{x},t) + \rho(\underline{x},t)\operatorname{div}\underline{U}(\underline{x},t) = 0. \tag{5.50}$$

This is the *Eulerian form of the equation of continuity* when $\underline{U}(\underline{x},t)$ is assumed to be continuous and continuously differentiable. Taking into account the explicit form of $\frac{d}{dt}\rho(\underline{x},t)$, [5.50] becomes

$$\frac{\partial}{\partial t}\rho(\underline{x},t) + (\operatorname{grad}\rho(\underline{x},t)).\underline{U}(\underline{x},t) + \rho(\underline{x},t)\operatorname{div}\underline{U}(\underline{x},t) = 0 \tag{5.51}$$

and, taking advantage of the identity

$$\operatorname{grad}\rho(\underline{x},t)).\underline{U}(\underline{x},t) + \rho(\underline{x},t)\operatorname{div}\underline{U}(\underline{x},t) = \operatorname{div}(\rho(\underline{x},t)\underline{U}(\underline{x},t)), \tag{5.52}$$

we have the equivalent Eulerian form of the equation of continuity

$$\frac{\partial}{\partial t}\rho(\underline{x},t) + \operatorname{div}(\rho(\underline{x},t).\underline{U}(\underline{x},t)) = 0. \tag{5.53}$$

5.2.4. Acceleration field

Applied to the velocity field itself, [5.46] yields the expression of the acceleration field

$$\underline{a}(\underline{x},t) = \frac{d}{dt}\underline{U}(\underline{x},t) = \frac{\partial}{\partial t}\underline{U}(\underline{x},t) + (\operatorname{grad}\underline{U}(\underline{x},t)).\underline{U}(\underline{x},t) \tag{5.54}$$

where it is recalled that the velocity field has been assumed to be continuous and continuously differentiable since section 5.1.2.

5.2.5. Convective derivative of a volume integral

The quantity $I(S,t)$ which we are now considering is defined in the Eulerian description as the integral of a volume density $b(\underline{x},t)$ assumed to be continuous and continuously differentiable

$$I(S,t) = i(S,t) = \int_{\Omega_t} b(\underline{x},t)\,\mathrm{d}\Omega_t.$$ [5.55]

Its convective derivative is thus obtained as

$$\frac{\mathrm{d}}{\mathrm{d}t}I(S,t) = \frac{\mathrm{d}}{\mathrm{d}t}i(S,t)$$
$$= \lim\left(\int_{\Omega_{t+\mathrm{d}t}} b(\underline{x}+\underline{U}\,\mathrm{d}t,t+\mathrm{d}t)\,\mathrm{d}\Omega_{t+\mathrm{d}t} - \int_{\Omega_t} b(\underline{x})\,\mathrm{d}\Omega_t\right) \text{ as } \mathrm{d}t \to 0.$$ [5.56]

A simple way of computing this derivative is to take advantage of the equation of continuity by introducing the *density per unit mass* $\dfrac{b(\underline{x},t)}{\rho(\underline{x},t)}$ so that

$$I(S,t) = i(S,t) = \int_{\Omega_t} \frac{b(\underline{x},t)}{\rho(\underline{x},t)}\,\mathrm{d}m = \int_{S} \frac{b(\underline{x},t)}{\rho(\underline{x},t)}\,\mathrm{d}m \;^5$$ [5.57]

and $\dfrac{\mathrm{d}}{\mathrm{d}t}I(S,t)$ is simply equal to $\dfrac{\mathrm{d}}{\mathrm{d}t}I(S,t) = \dfrac{\mathrm{d}}{\mathrm{d}t}i(S,t) = \int_{S}\dfrac{\mathrm{d}}{\mathrm{d}t}\left(\dfrac{b(\underline{x},t)}{\rho(\underline{x},t)}\right)\mathrm{d}m$

$$\frac{\mathrm{d}}{\mathrm{d}t}i(S,t) = \int_{S}\frac{1}{\rho(\underline{x},t)}\frac{\mathrm{d}}{\mathrm{d}t}(b(\underline{x},t))\,\mathrm{d}m + \int_{S} b(\underline{x},t)\frac{\mathrm{d}}{\mathrm{d}t}\left(\frac{1}{\rho(\underline{x},t)}\right)\mathrm{d}m.$$ [5.58]

Taking [5.50] into account, we obtain

$$\frac{\mathrm{d}}{\mathrm{d}t}i(S,t) = \int_{\Omega_t}\frac{\mathrm{d}}{\mathrm{d}t}b(\underline{x},t)\,\mathrm{d}\Omega_t + \int_{\Omega_t} b(\underline{x},t)\,\mathrm{div}\,\underline{U}(\underline{x},t)\,\mathrm{d}\Omega_t.$$ [5.59]

---

5 A notation introduced in Chapter 3 (section 3.4).

Unfortunately, this expression does not fit into the general pattern [5.42]. It needs being transformed using the explicit form [5.45] of $\dfrac{d}{dt}b(\underline{x},t)$ and taking advantage of the identity

$$\underline{\operatorname{grad}\, b}(\underline{x},t).\underline{U}(\underline{x},t)+b(\underline{x},t)\operatorname{div}\underline{U}(\underline{x},t)=\operatorname{div}\left(b(\underline{x},t)\underline{U}(\underline{x},t)\right). \qquad [5.60]$$

We obtain

$$\frac{d}{dt}i(S,t)=\int_{\Omega_t}\frac{\partial}{\partial t}b(\underline{x},t)\,d\Omega_t+\int_{\Omega_t}\operatorname{div}(b(\underline{x},t)\underline{U}(\underline{x},t))\,d\Omega_t \qquad [5.61]$$

where the first integral corresponds to the *"frozen" system* and the second one is the *convective term*.

Applying the divergence theorem to the second integral, this equation itself can be transformed into

$$\frac{d}{dt}i(S,t)=\int_{\Omega_t}\frac{\partial}{\partial t}b(\underline{x},t)\,d\Omega_t+\int_{\partial\Omega_t}b(\underline{x},t)\underline{U}(\underline{x},t).\underline{n}(\underline{x})\,da \qquad [5.62]$$

where comes out as the result of the sum of two terms shown in Figure 5.4, that $\dfrac{d}{dt}i(S,t)$ is the result of the sum of two terms:

– the first one represents the *change with time* of the integrand for *geometrical points* located in both Ω_t and Ω_{t+dt};

– the other is the *algebraic gain* to the integral due to the *geometrical points* that belong to Ω_t and not to Ω_{t+dt} and the *geometrical points* that do not belong to Ω_t and belong to Ω_{t+dt}.

The equations established here above are written in the case when $b(\underline{x},t)$ is a scalar. They remain valid whatever the rank of tensor density: [5.59] is unchanged while [5.61] and [5.62] become respectively

$$\frac{d}{dt}I(S,t)=\frac{d}{dt}i(S,t)=\int_{\Omega_t}\frac{\partial}{\partial t}b(\underline{x},t)\,d\Omega_t+\int_{\Omega_t}\operatorname{div}(b(\underline{x},t)\otimes\underline{U}(\underline{x},t))\,d\Omega_t \qquad [5.63]$$

and

$$\frac{d}{dt}I(S,t) = \frac{d}{dt}i(S,t) = \int_{\Omega_t} \frac{\partial}{\partial t}b(\underline{x},t)\,d\Omega_t + \int_{\partial\Omega_t}\left(b(\underline{x},t)\otimes\underline{U}(\underline{x},t)\right).\underline{n}(\underline{x})\,da. \qquad [5.64]$$

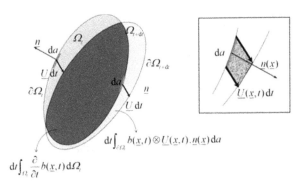

Figure 5.4. *Convective derivative of a volume integral*

5.2.6. *Euler's theorem*

With the expressions established in the preceding section for the acceleration field [5.54] and the convective derivative of a volume integral [5.63] and [5.64], we can now give a more explicit formulation to the momentum theorem stated in Chapter 3 (section 3.4.2):

$$\begin{cases} \text{in a Galilean frame,} \\ \forall S' \subseteq S, \\ \left[\mathcal{F}'_e\right] = \frac{d}{dt}\left[\mathcal{MU}'\right] \end{cases} \qquad [5.65]$$

where

$$\left[\mathcal{MU}'\right] = \left[0, \int_{S'}\underline{U}(\underline{x},t)\,dm, \int_{S'}\overline{OM}\wedge\underline{U}(\underline{x},t)\,dm\right] \qquad [5.66]$$

and

$$\frac{d}{dt}\left[\mathcal{MU}'\right] = \left[0, \int_{S'}\frac{d}{dt}\underline{U}(\underline{x},t)\,dm, \int_{S'}\overline{OM}\wedge\frac{d\underline{U}}{dt}(\underline{x},t)\,dm\right] = \left[\mathcal{Ma}'\right]. \qquad [5.67]$$

In this last equation, we observe that

$$\int_{S'} \frac{d}{dt} \underline{U}(\underline{x},t)\, dm = \frac{d}{dt}\int_{S'} \underline{U}(\underline{x},t)\, dm = \frac{d}{dt}\int_{\Omega'_t} \rho(\underline{x},t)\underline{U}(\underline{x},t)\, d\Omega_t \qquad [5.68]$$

then, through [5.63][6],

$$\int_{S'} \frac{d\underline{U}}{dt}\, dm = \int_{\Omega'_t}\left(\frac{\partial(\rho\underline{U})}{\partial t} + \operatorname{div}(\rho\underline{U}\otimes\underline{U})\right)d\Omega_t \qquad [5.69]$$

and finally with [5.64]

$$\int_{S'} \frac{d\underline{U}}{dt}\, dm = \int_{\Omega'_t}\frac{\partial(\rho\underline{U})}{\partial t}\, d\Omega_t + \int_{\partial\Omega'_t}(\rho\underline{U}\otimes\underline{U}).\underline{n}\, da. \qquad [5.70]$$

Regarding the second integral in [5.67], we recall that (Chapter 3, [3.24])

$$\int_{S'} \underline{OM}\wedge\frac{d\underline{U}}{dt}\, dm = \frac{d}{dt}\int_{S'} \underline{OM}\wedge\underline{U}\, dm = \frac{d}{dt}\int_{\Omega'_t} \underline{OM}\wedge\rho\underline{U}\, d\Omega_t, \qquad [5.71]$$

which can be expanded according to [5.63]

$$\int_{S'} \underline{OM}\wedge\frac{d\underline{U}}{dt}\, dm = \int_{\Omega'_t} \underline{OM}\wedge\frac{\partial(\rho\underline{U})}{\partial t}\, d\Omega_t + \int_{\Omega'_t} \operatorname{div}((\underline{OM}\wedge\rho\underline{U})\otimes\underline{U})\, d\Omega_t \qquad [5.72]$$

and, through [5.64],

$$\int_{S'} \underline{OM}\wedge\frac{d\underline{U}}{dt}\, dm = \int_{\Omega'_t} \underline{OM}\wedge\frac{\partial(\rho\underline{U})}{\partial t}\, d\Omega_t + \int_{\partial\Omega'_t} (\underline{OM}\wedge\rho\underline{U})\otimes\underline{U}.\underline{n}\, da \qquad [5.73]$$

also equal to

$$\int_{S'} \underline{OM}\wedge\frac{d\underline{U}}{dt}\, dm = \int_{\Omega'_t} \underline{OM}\wedge\frac{\partial(\rho\underline{U})}{\partial t}\, d\Omega_t + \int_{\partial\Omega'_t} \underline{OM}\wedge(\rho\underline{U}\otimes\underline{U}).\underline{n}\, da. \qquad [5.74]$$

Putting together [5.70] and [5.74], we obtain the famous result known as *Euler's theorem* that reads as follows:

In a Galilean frame, the wrench of external forces $\left[\mathcal{F}'_e\right]$ *applied to any subsystem S' equals the sum*

---

6 In this section, the arguments $\underline{x}$ and t are dropped for simplicity sake.

– of the wrench of forces $\dfrac{\partial(\rho \underline{U})}{\partial t}\,\mathrm{d}\Omega_t$ distributed throughout the volume Ω'_t;

– and the wrench of forces $(\rho \underline{U} \otimes \underline{U}).\underline{n}\,\mathrm{d}a$ distributed over the boundary $\partial\Omega'_t$.

Note that $(\rho \underline{U} \otimes \underline{U}).\underline{n}\,\mathrm{d}a$ represents the flux of momentum through the surface element $\mathrm{d}a$ and is always directed outwards.

5.2.7. *Kinetic energy theorem*

The kinetic energy theorem was established in Chapter 3 on the basis of the principle of virtual (rates of) work after noting that the rate of work by quantities of acceleration in the actual motion is just the convective derivative of the kinetic energy in that motion of the system or subsystem:

$$\begin{cases} K'(\underline{U}) = \dfrac{1}{2}\displaystyle\int_{S'} \underline{U}^2(\underline{x},t)\,\mathrm{d}m \\[2mm] \mathcal{A}'(\underline{U}) = \displaystyle\int_{S'} \dfrac{\mathrm{d}\underline{U}(\underline{x},t)}{\mathrm{d}t}.\underline{U}(\underline{x},t)\,\mathrm{d}m = \dfrac{\mathrm{d}}{\mathrm{d}t}K'(\underline{U}) \end{cases} \qquad [5.75]$$

then

$$\begin{cases} \text{in a Galilean frame,} \\ \forall S' \subseteq S \\ \underline{U} \text{ real motion of } S', \\ \mathcal{P}'_e(\underline{U}) + \mathcal{P}'_i(\underline{U}) = \dfrac{\mathrm{d}}{\mathrm{d}t}K'(\underline{U}). \end{cases} \qquad [5.76]$$

We can now give the explicit form of $K'(\underline{U})$ and $\mathcal{A}'(\underline{U})$ within the framework of the three-dimensional continuum:

$$\begin{cases} K'(\underline{U}) = \dfrac{1}{2}\displaystyle\int_{S'} \underline{U}^2(\underline{x},t)\,\mathrm{d}m = \dfrac{1}{2}\displaystyle\int_{\Omega'_t} \rho(\underline{x},t)\underline{U}^2(\underline{x},t)\,\mathrm{d}\Omega_t \\[2mm] \mathcal{A}'(\underline{U}) = \displaystyle\int_{S'} \dfrac{\mathrm{d}\underline{U}(\underline{x},t)}{\mathrm{d}t}.\underline{U}(\underline{x},t)\,\mathrm{d}m = \displaystyle\int_{\Omega'_t} \dfrac{\mathrm{d}\underline{U}(\underline{x},t)}{\mathrm{d}t}.\underline{U}(\underline{x},t)\rho(\underline{x},t)\,\mathrm{d}\Omega_t. \end{cases} \qquad [5.77]$$

thence

$$\frac{d}{dt}K'(\underline{U}) = \frac{1}{2}\int_{\Omega'_t}(\frac{\partial}{\partial t}(\rho\underline{U}^2) + \text{div}\,(\rho\underline{U}^2\underline{U}))\,d\Omega_t = \mathcal{A}'(\underline{U}) = \int_{\Omega'_t}\rho\underline{a}.\underline{U}\,d\Omega_t. \quad [5.78]$$

5.3. Piecewise continuity and continuous differentiability

5.3.1. *Convective derivative of a volume integral*

In the preceding section, both the velocity field $\underline{U}(\underline{x},t)$ and the quantity $b(\underline{x},t)$ were assumed to be continuous and continuously differentiable. As a matter of fact, the case of piecewise continuous and continuously differentiable fields need being investigated for practical applications, as already explained in Chapter 4. We will focus on the computation of the convective derivative of a volume integral and assume that either $\underline{U}(\underline{x},t)$ or $b(\underline{x},t)$, or both of them are piecewise continuous and continuously differentiable.

Σ_t will be used as a generic notation for any *geometrical surface* where a jump of these fields occurs when crossing Σ_t at point $\underline{x}$ following its normal $\underline{n}(\underline{x},t)$. The jump or discontinuity in b or $\underline{U}$ at that point is defined as the difference between the "downstream" value b_2 or $\underline{U}_2$ minus the upstream value b_1 or $\underline{U}_1$ and denoted by $[\![\ ]\!]$, the "flow" being defined by $\underline{n}(\underline{x},t)$:

$$[\![b(\underline{x},t)]\!] = b_2(\underline{x},t) - b_1(\underline{x},t), \quad [\![\underline{U}(\underline{x},t)]\!] = \underline{U}_2(\underline{x},t) - \underline{U}_1(\underline{x},t). \quad [5.79]$$

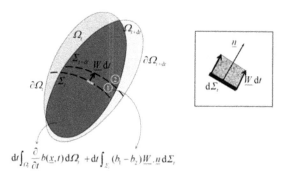

Figure 5.5. *Convective derivative of a volume integral in the discontinuous case*

The propagation velocity $\underline{W}(\underline{x},t)$ of the *geometrical* surface Σ_t is defined in the classical way: $W(\underline{x},t)\,dt$ is the distance between Σ_{t+dt} and Σ_t along $\underline{n}(\underline{x},t)$ with $W(\underline{x},t) \geq 0$ and $\underline{W}(\underline{x},t) = W(\underline{x},t)\underline{n}(\underline{x})$.

As in section 5.2.5, we refer to Figure 5.5 and follow a similar balance method. The change in Figure 5.5 with respect to Figure 5.4 consists in making the jump surface appear in both its locations at times t and $(t+dt)$. It points out that the balance equation which we wrote in [5.62] must be completed due to the necessity of carefully taking into account those *geometrical* points lying both in Ω_t and Ω_{t+dt} which were initially lying downstream the jump surface Σ_t and lie upstream Σ_{t+dt} at time $(t+dt)$ because the jump surface passed through them during the time lapse dt. These geometrical points correspond to the region that was swept out by the jump surface during dt. For them, the variation in the integrated quantity is the algebraic gain $b_1(\underline{x},t)-b_2(\underline{x},t)=-[\![b(\underline{x},t)]\!]$. It follows that the contribution to be added in the balance equation is equal to $-\int_{\Sigma_t}[\![b(\underline{x},t)]\!]\,\underline{W}(\underline{x},t).\underline{n}(\underline{x},t)d\Sigma_t$ as shown in Figure 5.5. Thence:

$$\frac{d}{dt}I(S,t)=\frac{d}{dt}i(S,t)=\int_{\Omega_t}\frac{\partial}{\partial t}b(\underline{x},t)d\Omega_t-\int_{\Sigma_t}[\![b(\underline{x},t)]\!]\,\underline{W}(\underline{x},t).\underline{n}(\underline{x},t)d\Sigma_t$$
$$+\int_{\partial\Omega_t}b(\underline{x},t)\underline{U}(\underline{x},t).\underline{n}(\underline{x})da. \qquad [5.80]$$

Applying the divergence theorem for piecewise continuous and continuously differentiable fields (Appendix 2, equation A2.24) to the last integral, this equation can be transformed into

$$\frac{d}{dt}I(S,t)=\int_{\Omega_t}\frac{\partial b}{\partial t}d\Omega_t-\int_{\Sigma_t}[\![b]\!]\,\underline{W}.\underline{n}d\Sigma_t$$
$$+\int_{\Sigma_t}[\![b\underline{U}]\!].\underline{n}d\Sigma_t+\int_{\Omega_t}\mathrm{div}(b\underline{U})d\Omega_t \qquad [5.81]$$

that is

$$\frac{d}{dt}I(S,t)=\frac{d}{dt}i(S,t)$$
$$=\int_{\Omega_t}(\frac{\partial b}{\partial t}+\mathrm{div}(b\underline{U}))d\Omega_t+\int_{\Sigma_t}[\![b(\underline{U}-\underline{W})]\!].\underline{n}d\Sigma_t. \qquad [5.82]$$

5.3.2. Equation of continuity

As a consequence of this result, we can now establish the Eulerian expression for the equation of continuity in the case of a piecewise continuous and continuously differentiable velocity field $\underline{U}(\underline{x},t)$, which physically corresponds to *shock waves*.

Let the considered integral in [5.82] be the mass of an arbitrary subsystem $S' \subseteq S$ with $\rho(\underline{x},t)$ the mass per unit volume:

$$\mathcal{M}(S',t) = \int_{\Omega_t'} \rho(\underline{x},t)\,d\Omega_t \; . \tag{5.83}$$

From [5.82], we derive

$$\frac{d}{dt}\mathcal{M}(S',t) = \int_{\Omega_t'} (\frac{\partial \rho}{\partial t} + \operatorname{div}(\rho\underline{U}))\,d\Omega_t + \int_{\Sigma_t} [\![\rho(\underline{U}-\underline{W})]\!].\underline{n}\,d\Sigma_t \; , \tag{5.84}$$

which must be equal to zero whatever the subsystem $S' \subseteq S$. We thus obtain the two equations that express the conservation of mass:

$$\begin{cases} \dfrac{\partial \rho}{\partial t} + \operatorname{div}(\rho\underline{U}) = 0 \;\; \text{in } \Omega_t \\[2mm] [\![\rho(\underline{U}-\underline{W})]\!].\underline{n} = 0 \;\; \text{on } \Sigma_t \, . \end{cases} \tag{5.85}$$

From the physical viewpoint, the first equation in [5.85] expresses the conservation of mass for the material element $dm = \rho(\underline{x},t)\,d\Omega_t$ conveyed by the motion as explained in section 5.2.3.

The physical meaning of the second equation comes from its explicit form

$$\rho_1(\underline{U}_1 - \underline{W}).\underline{n} = \rho_2(\underline{U}_2 - \underline{W}).\underline{n} \tag{5.86}$$

where we observe that $\underline{U}_1(\underline{x},t) - \underline{W}(\underline{x},t)$ and $\underline{U}_2(\underline{x},t) - \underline{W}(\underline{x},t)$ represent the velocity of the matter at point $\underline{x}$ and time t with respect to the jump surface in region 1 (upstream) and in region 2 (downstream) respectively (Figure 5.5). Thence, [5.86] expresses the conservation of mass for the matter that is swept by $d\Sigma_t$.

Since $\rho(\underline{x},t)$ cannot be zero, a consequence of [5.86] is that $(\underline{U}_1 - \underline{W}).\underline{n} = 0$ implies $(\underline{U}_2 - \underline{W}).\underline{n} = 0$ and $\underline{U}_1.\underline{n} = \underline{U}_2.\underline{n}$. Noting that $\underline{n}((\underline{W} - \underline{U}_1).\underline{n})$ and $\underline{n}((\underline{W} - \underline{U}_2).\underline{n})$ are the propagation velocities of the jump surface with respect to the matter in region 1 and 2 respectively, it proves that a velocity jump surface, i.e. a *shock wave*, cannot be stationary with respect to the matter. Another consequence of [5.86] is that a shock wave comes always with a jump in the mass per unit volume.

5.3.3. *Momentum theorem*

Under the assumption that $\rho(\underline{x},t)$ and $\underline{U}(\underline{x},t)$ were continuous and continuously differentiable, the momentum theorem [5.65] was established (Chapter 3, section 3.4.2) and given an explicit form as Euler's theorem (section 3.2.6). In the case when they are just piecewise continuous and continuously differentiable, the concept of quantity of acceleration that is at the root of these theorems remains valid where $\rho(\underline{x},t)$ and $\underline{U}(\underline{x},t)$ are continuous and continuously differentiable and make it possible to derive the acceleration field through [5.54], but obviously calls for a complementary definition to take the jumps into account.

On the contrary, the concept of momentum is by no way affected and the wrench of momentum is still written as in [5.66]. Thus, it is possible to take the momentum theorem [5.65] as the fundamental statement valid for piecewise continuous and continuously differentiable fields.

The explicit form of $\dfrac{d}{dt}[\mathcal{M}\mathcal{U}']$ taking the discontinuities into account proceeds from [5.82] and yields

$$\frac{d}{dt}[\mathcal{M}\mathcal{U}'] = \left[O, \frac{d}{dt}\int_{S'}\underline{U}(\underline{x},t)\,dm, \frac{d}{dt}\int_{S'}\underline{OM}\wedge\underline{U}(\underline{x},t)\,dm \right] \qquad [5.87]$$

where

$$\frac{d}{dt}\int_{S'}\underline{U}\,dm = \int_{\Omega_t'}\left(\frac{\partial(\rho\underline{U})}{\partial t} + \operatorname{div}(\rho\underline{U}\otimes\underline{U})\right)d\Omega_t$$
$$+ \int_{\Sigma_t}[\![\rho\underline{U}\otimes(\underline{U}-\underline{W})]\!].\underline{n}\,d\Sigma_t \qquad [5.88]$$

and

$$\frac{d}{dt}\int_{S'}\underline{OM}\wedge\underline{U}\,dm = \int_{\Omega_t'}\underline{OM}\wedge\frac{\partial(\rho\underline{U})}{\partial t}\,d\Omega_t + \int_{\Omega_t'}\operatorname{div}(\underline{OM}\wedge(\rho\underline{U}\otimes\underline{U}))\,d\Omega_t$$
$$+ \int_{\Sigma_t}\underline{OM}\wedge[\![\rho\underline{U}\otimes(\underline{U}-\underline{W})]\!].\underline{n}\,d\Sigma_t. \qquad [5.89]$$

Comparing [5.88] and [5.89] with [5.69] and [5.72] shows that the jumps in $\rho(\underline{x},t)$ and $\underline{U}(\underline{x},t)$ generate a *surface density of quantities of acceleration* on Σ_t equal to $[\![\rho\underline{U}\otimes(\underline{U}-\underline{W})]\!].\underline{n}$ that is, taking the continuity equation [5.86] into account,

$$[\![\rho \underline{U} \otimes (\underline{U} - \underline{W})]\!].\underline{n} = [\![\underline{U}]\!]\rho(\underline{U} - \underline{W}).\underline{n}. \tag{5.90}$$

5.3.4. *Euler's theorem*

Euler's theorem may now be obtained from the momentum theorem [5.65] through the divergence theorem applied to $\int_{\Omega_t'} \mathrm{div}(\rho \underline{U} \otimes \underline{U}) \mathrm{d}\Omega_t$ in [5.88] and [5.89]. It comes

$$\frac{\mathrm{d}}{\mathrm{d}t} \int_{S'} \underline{U}\, \mathrm{d}m = \int_{\Omega_t'} \frac{\partial(\rho \underline{U})}{\partial t} \mathrm{d}\Omega_t + \int_{\partial \Omega_t'} (\rho \underline{U} \otimes \underline{U}).\mathrm{d}a - \int_{\Sigma_t} [\![\rho \underline{U} \otimes \underline{U}]\!].\underline{n}\,\mathrm{d}\Sigma_t$$
$$+ \int_{\Sigma_t} [\![\rho \underline{U} \otimes (\underline{U} - \underline{W})]\!].\underline{n}\,\mathrm{d}\Sigma_t \tag{5.91}$$

and finally

$$\frac{\mathrm{d}}{\mathrm{d}t} \int_{S'} \underline{U}\, \mathrm{d}m = \int_{\Omega_t'} \frac{\partial(\rho \underline{U})}{\partial t} \mathrm{d}\Omega_t + \int_{\partial \Omega_t'} (\rho \underline{U} \otimes \underline{U}).\mathrm{d}a$$
$$- \int_{\Sigma_t} [\![\rho \underline{U})]\!] \otimes \underline{W}.\underline{n}\,\mathrm{d}\Sigma_t. \tag{5.92}$$

In the same way

$$\frac{\mathrm{d}}{\mathrm{d}t} \int_{S'} \overline{OM} \wedge \underline{U}\, \mathrm{d}m = \int_{\Omega_t'} \overline{OM} \wedge \frac{\partial(\rho \underline{U})}{\partial t} \mathrm{d}\Omega_t + \int_{\partial \Omega_t'} \overline{OM} \wedge (\rho \underline{U} \otimes \underline{U}).\mathrm{d}a$$
$$- \int_{\Sigma_t} \overline{OM} \wedge [\![\rho \underline{U} \otimes \underline{U}]\!].\underline{n}\,\mathrm{d}\Sigma_t \tag{5.93}$$
$$+ \int_{\Sigma_t} \overline{OM} \wedge [\![\rho \underline{U} \otimes (\underline{U} - \underline{W})]\!].\underline{n}\,\mathrm{d}\Sigma_t$$

thence

$$\frac{\mathrm{d}}{\mathrm{d}t} \int_{S'} \overline{OM} \wedge \underline{U}\, \mathrm{d}m = \int_{\Omega_t'} \overline{OM} \wedge \frac{\partial(\rho \underline{U})}{\partial t} \mathrm{d}\Omega_t + \int_{\partial \Omega_t'} \overline{OM} \wedge (\rho \underline{U} \otimes \underline{U}).\mathrm{d}a$$
$$- \int_{\Sigma_t} \overline{OM} \wedge [\![\rho \underline{U}]\!] \otimes \underline{W}.\underline{n}\,\mathrm{d}\Sigma_t. \tag{5.94}$$

The completed wording of Euler's theorem comes out:

In a Galilean frame, the wrench of external forces $\left[\mathcal{F}_e' \right]$ *applied to any subsystem S' equals the sum*

– of the wrench of forces $\dfrac{\partial(\rho \underline{U})}{\partial t} d\Omega_t$ *distributed throughout the volume* Ω_t' ;

– the wrench of forces $(\rho \underline{U} \otimes \underline{U}).\underline{n} da$ *distributed over the boundary* $\partial \Omega_t'$;

– and the wrench of forces $-[\![\rho \underline{U}]\!] W \, d\Sigma_t$ *distributed over the jump surface* Σ_t.

5.3.5. Kinetic energy theorem

Piecewise continuity and continuous differentiability of $\rho(\underline{x},t)$ and $\underline{U}(\underline{x},t)$ do not affect the definition of $K'(\underline{U})$ in the statement of the kinetic energy theorem while the definition of $\mathcal{A}'(\underline{U})$ obviously need being clarified and/or completed. In the same way, as for the momentum theorem, the kinetic energy theorem will be taken as a fundamental statement from which, incidentally, the completed expression of $\mathcal{A}'(\underline{U})$ will be derived.

The convective derivative of $K'(\underline{U})$ is obtained through [5.82]

$$\frac{d}{dt} K'(\underline{U}) = \frac{1}{2} \int_{\Omega_t'} (\frac{\partial}{\partial t}(\rho \underline{U}^2) + \mathrm{div}\,(\rho \underline{U}^2 \underline{U})) \, d\Omega_t$$
$$+ \frac{1}{2} \int_{\Sigma_t} [\![\rho \underline{U}^2 (\underline{U} - \underline{W})]\!].\underline{n} \, d\Sigma_t,$$

[5.95]

or, taking the continuity equation [5.86] into account,

$$\frac{d}{dt} K'(\underline{U}) = \frac{1}{2} \int_{\Omega_t'} (\frac{\partial}{\partial t}(\rho \underline{U}^2) + \mathrm{div}\,(\rho \underline{U}^2 \underline{U})) \, d\Omega_t$$
$$+ \int_{\Sigma_t} \frac{[\![\underline{U}^2]\!]}{2} \rho (\underline{U} - \underline{W}).\underline{n} \, d\Sigma_t.$$

[5.96]

Comparing [5.96] with [5.78], we recognize the first integral as the "regular part" of $\mathcal{A}'(\underline{U})$ and the second one comes out as the contribution of the surface density of quantities of acceleration [5.90] generated on Σ_t by the jumps in $\rho(\underline{x},t)$ and $\underline{U}(\underline{x},t)$ (section 5.3.3)

$$\frac{d}{dt}K'(\underline{U}) = \int_{\Omega'_t}\rho\underline{a}.\underline{U}\,d\Omega_t + \int_{\Sigma_t}\frac{[\![\underline{U}^2]\!]}{2}\rho(\underline{U}-\underline{W}).\underline{n}\,d\Sigma_t = \mathcal{A}'(\underline{U}), \qquad [5.97]$$

which gives the expression of the rate of work by quantities of acceleration in the real motion of the system or subsystem.

5.4. Comments

As stated at the beginning of this chapter, it comes out that the Eulerian description is the natural point of view when dealing with kinematics. Lagrangian kinematics may also be carried out, making it easier to follow material elements but unfortunately leading to heavy and often cumbersome calculations. Nevertheless, in the case of not feeling at ease with the Eulerian rationale as is proposed here, the Lagrangian approach will always provide a firm basis to rely on, as may be found, for instance, in [SAL 01].

The chapter does not aim at exhaustiveness: only those results that are necessary to the implementation of the virtual work method for the modeling of the three-dimensional continuum were presented together with some closely connected statements that were immediately available. Once again, the reader may refer to classical textbooks for more extensive presentations.

5.5. Explicit formulas in standard coordinate systems

5.5.1. *Orthonormal Cartesian coordinates*

$$\underline{U} = U_x\underline{e}_x + U_y\underline{e}_y + U_z\underline{e}_z$$

$$\underline{\underline{\mathrm{grad}\,U}} = \frac{\partial U_i}{\partial x_j}\underline{e}_i \otimes \underline{e}_j$$

$$\underline{\underline{d}} = d_{ij}\underline{e}_i \otimes \underline{e}_j \quad d_{ij} = \frac{1}{2}(\frac{\partial U_i}{\partial x_j}+\frac{\partial U_j}{\partial x_i})$$

$$\mathrm{div}\,\underline{U} = \frac{\partial U_i}{\partial x_i} = \frac{\partial U_x}{\partial x_x}+\frac{\partial U_y}{\partial x_y}+\frac{\partial U_z}{\partial x_z}$$

$$\frac{db}{dt} = \frac{\partial b}{\partial t}+\frac{\partial b}{\partial x_i}U_i \quad \text{(whatever the rank of the tensor quantity } b).$$

5.5.2. *Cylindrical coordinates*

$$\underline{U} = U_r \underline{e}_r + U_\theta \underline{e}_\theta + U_z \underline{e}_z$$

$$\underline{\underline{\text{grad}\, U}} = \frac{\partial U_r}{\partial r} \underline{e}_r \otimes \underline{e}_r + \frac{1}{r}(\frac{\partial U_r}{\partial \theta} - U_\theta)\underline{e}_r \otimes \underline{e}_\theta + \frac{\partial U_r}{\partial z} \underline{e}_r \otimes \underline{e}_z$$

$$+ \frac{\partial U_\theta}{\partial r} \underline{e}_\theta \otimes \underline{e}_r + \frac{1}{r}(\frac{\partial U_\theta}{\partial \theta} + U_r)\underline{e}_\theta \otimes \underline{e}_\theta + \frac{\partial U_\theta}{\partial z} \underline{e}_\theta \otimes \underline{e}_z$$

$$+ \frac{\partial U_z}{\partial r} \underline{e}_z \otimes \underline{e}_r + \frac{1}{r}\frac{\partial U_z}{\partial \theta} \underline{e}_z \otimes \underline{e}_\theta + \frac{\partial U_z}{\partial z} \underline{e}_z \otimes \underline{e}_z.$$

$$\text{div}\,\underline{U} = \frac{\partial U_r}{\partial r} + \frac{U_r}{r} + \frac{1}{r}\frac{\partial U_\theta}{\partial \theta} + \frac{\partial U_z}{\partial z}$$

$$\underline{\underline{d}} = d_{ij}\,\underline{e}_i \otimes \underline{e}_j$$

$$d_{rr} = \frac{\partial U_r}{\partial r} \quad d_{\theta\theta} = \frac{1}{r}\frac{\partial U_\theta}{\partial \theta} + \frac{U_r}{r} \quad d_{zz} = \frac{\partial U_z}{\partial z}$$

$$d_{r\theta} = \frac{1}{2}(\frac{\partial U_\theta}{\partial r} - \frac{U_\theta}{r} + \frac{1}{r}\frac{\partial U_r}{\partial \theta})$$

$$d_{\theta z} = \frac{1}{2}(\frac{1}{r}\frac{\partial U_z}{\partial \theta} + \frac{\partial U_\theta}{\partial z}) \quad d_{zr} = \frac{1}{2}(\frac{\partial U_r}{\partial z} + \frac{\partial U_z}{\partial r})$$

$$\frac{db}{dt} = \frac{\partial b}{\partial t} + \frac{\partial b}{\partial r}U_r + \frac{\partial b}{\partial \theta}\frac{U_\theta}{r} + \frac{\partial b}{\partial z}U_z \quad (b \text{ scalar})$$

$$\frac{d\underline{b}}{dt} = (\frac{db_r}{dt} - \frac{b_\theta U_\theta}{r})\underline{e}_r + (\frac{db_\theta}{dt} + \frac{b_r U_\theta}{r})\underline{e}_\theta + \frac{db_z}{dt}\underline{e}_z \quad (\underline{b} = b_r\underline{e}_r + b_\theta\underline{e}_\theta + b_z\underline{e}_z).$$

5.5.3. *Spherical coordinates*

$$\underline{U} = U_r \underline{e}_r + U_\theta \underline{e}_\theta + U_\varphi \underline{e}_\varphi$$

$$\underline{\underline{\text{grad}\, U}} = \frac{\partial U_r}{\partial r} \underline{e}_r \otimes \underline{e}_r + \frac{1}{r}(\frac{\partial U_r}{\partial \theta} - U_\theta)\underline{e}_r \otimes \underline{e}_\theta + \frac{1}{r}(\frac{1}{\sin\theta}\frac{\partial U_r}{\partial \varphi} - U_\varphi)\underline{e}_r \otimes \underline{e}_\varphi$$

$$+ \frac{\partial U_\theta}{\partial r} \underline{e}_\theta \otimes \underline{e}_r + \frac{1}{r}(\frac{\partial U_\theta}{\partial \theta} + U_r)\underline{e}_\theta \otimes \underline{e}_\theta + \frac{1}{r}(\frac{1}{\sin\theta}\frac{\partial U_\theta}{\partial \varphi} - U_\varphi \cot\theta)\underline{e}_\theta \otimes \underline{e}_\varphi$$

$$+ \frac{\partial U_\varphi}{\partial r} \underline{e}_\varphi \otimes \underline{e}_r + \frac{1}{r}\frac{\partial U_\varphi}{\partial \theta} \underline{e}_\varphi \otimes \underline{e}_\theta + \frac{1}{r}(\frac{1}{\sin\theta}\frac{\partial U_\varphi}{\partial \varphi} + U_\theta \cot\theta + U_r)\underline{e}_\varphi \otimes \underline{e}_\varphi.$$

$$\operatorname{div} \underline{U} = \frac{\partial U_r}{\partial r} + \frac{2U_r}{r} + \frac{1}{r}\frac{\partial U_\theta}{\partial \theta} + \frac{U_\theta}{r}\cot\theta + \frac{1}{r\sin\theta}\frac{\partial U_\varphi}{\partial \varphi}$$

$$\underline{\underline{d}} = d_{ij}\,\underline{e}_i \otimes \underline{e}_j$$

$$d_{rr} = \frac{\partial U_r}{\partial r} \quad d_{\theta\theta} = \frac{1}{r}\frac{\partial U_\theta}{\partial \theta} + \frac{U_r}{r} \quad d_{\varphi\varphi} = \frac{1}{r\sin\theta}\frac{\partial U_\varphi}{\partial \varphi} + \frac{U_\theta}{r}\cot\theta + \frac{U_r}{r}$$

$$d_{r\theta} = \frac{1}{2}\left(\frac{1}{r}\frac{\partial U_r}{\partial \theta} + \frac{\partial U_\theta}{\partial r} - \frac{U_\theta}{r}\right)$$

$$d_{\theta\varphi} = \frac{1}{2}\left(\frac{1}{r}\frac{\partial U_\varphi}{\partial \theta} + \frac{1}{r\sin\theta}\frac{\partial U_\theta}{\partial \varphi} - \frac{\cot\theta}{r}U_\varphi\right)$$

$$d_{\varphi r} = \frac{1}{2}\left(\frac{1}{r\sin\theta}\frac{\partial U_r}{\partial \varphi} + \frac{\partial U_\varphi}{\partial r} - \frac{U_\varphi}{r}\right)$$

$$\frac{db}{dt} = \frac{\partial b}{\partial t} + \frac{\partial b}{\partial r}U_r + \frac{\partial b}{\partial \theta}\frac{U_\theta}{r} + \frac{\partial b}{\partial \varphi}\frac{U_\varphi}{r\sin\theta} \quad (b \text{ scalar})$$

$$\underline{b} = b_r\underline{e}_r + b_\theta\underline{e}_\theta + b_z\underline{e}_z \text{ vector:}$$

$$\frac{d\underline{b}}{dt} = \left(\frac{db_r}{dt} - \frac{b_\theta U_\theta + b_\varphi U_\varphi}{r}\right)\underline{e}_r + \left(\frac{db_\theta}{dt} + \frac{b_r U_\theta}{r} - \cot\theta\frac{b_\varphi U_\varphi}{r}\right)\underline{e}_\theta$$

$$+ \left(\frac{db_\varphi}{dt} + \cot\theta\frac{b_\theta U_\varphi}{r} + \frac{b_r U_\varphi}{r}\right)\underline{e}_\varphi.$$

5.6. Practicing

5.6.1. *Rigid body motion*

As in Chapter 4 (section 4.5.4), we consider the motion defined, in orthonormal Cartesian coordinates, by $\underline{x} = \underline{\phi}(\underline{X},t) = \underline{a}(t) + \underline{\underline{\alpha}}(t).\underline{X}$ with $\underline{a}(t)$ continuous and continuously differentiable such that $\underline{a}(0) = 0$, and $\underline{\underline{\alpha}}(t)$ continuous and continuously differentiable such that $\underline{\underline{\alpha}}(0) = \underline{\underline{1}}$ and $^t\underline{\underline{\alpha}}(t).\underline{\underline{\alpha}}(t) = \underline{\underline{1}}$. Give the Eulerian representation of this motion and show that it can be written in the form

$$\underline{U}(\underline{x},t) = \underline{U}_0(t) + \underline{\underline{\Omega}}(t).\underline{x} = \underline{U}_0(t) + \underline{\Omega}(t) \wedge \underline{OM}$$

where the antisymmetric tensor $\underline{\underline{\Omega}}(t)$ is derived from $\underline{\underline{\alpha}}(t)$ and $\underline{\underline{\dot{\alpha}}}(t)$. Calculate the strain rate tensor and spin tensor.

Solution

– Differentiating $\underline{x} = \underline{\phi}(\underline{X},t) = \underline{a}(t) + \underline{\underline{\alpha}}(t).\underline{X}$ with respect to time, we obtain the velocity of the particle $\underline{X}$ at time t: $\underline{U}(\underline{X},t) = \underline{\dot{a}}(t) + \underline{\underline{\dot{\alpha}}}(t).\underline{X}$. Within the Eulerian framework, the velocity is expressed as a function of time and the present position of the particle; thence, taking ${}^t\underline{\underline{\alpha}}(t).\underline{\underline{\alpha}}(t) = \underline{\underline{1}}$ into account,

$$\underline{U}(\underline{x},t) = \underline{\dot{a}}(t) - \underline{\underline{\dot{\alpha}}}(t).\underline{\underline{\alpha}}^{-1}(t).\underline{a}(t) + \underline{\underline{\dot{\alpha}}}(t).\underline{\underline{\alpha}}^{-1}(t).\underline{x} = \underline{\dot{a}}(t) - \underline{\underline{\dot{\alpha}}}(t).{}^t\underline{\underline{\alpha}}(t) + \underline{\underline{\dot{\alpha}}}(t).{}^t\underline{\underline{\alpha}}(t).\underline{x}$$

which   can   be   written   as $\underline{U}(\underline{x},t) = \underline{U}_0(t) + \underline{\underline{\Omega}}(t).\underline{x}$ with $\underline{U}_0(t) = \underline{\dot{a}}(t) - \underline{\underline{\dot{\alpha}}}(t).{}^t\underline{\underline{\alpha}}(t).\underline{a}(t)$ and $\underline{\underline{\Omega}}(t) = \underline{\underline{\dot{\alpha}}}(t).{}^t\underline{\underline{\alpha}}(t)$.

– The tensor $\underline{\underline{\Omega}}(t)$ comes out as antisymmetric as a consequence of the differentiation of ${}^t\underline{\underline{\alpha}}(t).\underline{\underline{\alpha}}(t) = \underline{\underline{1}}$ that yields ${}^t\underline{\underline{\dot{\alpha}}}(t).\underline{\underline{\alpha}}(t) = -{}^t\underline{\underline{\alpha}}(t).\underline{\underline{\dot{\alpha}}}(t)$. Denoting by $\underline{\Omega}(t)$, the vector associated with $\underline{\underline{\Omega}}(t)$ through the external product, the Eulerian description of the motion becomes: $\underline{U}(\underline{x},t) = \underline{U}_0(t) + \underline{\underline{\Omega}}(t).\underline{x} = \underline{U}_0(t) + \underline{\Omega}(t) \wedge \underline{OM}$.

– The velocity gradient is $\operatorname{grad}\underline{U}(\underline{x},t) = \underline{\underline{\Omega}}(t)$. Thence, the strain rate tensor is identically zero $\underline{\underline{d}}(\underline{x},t) = 0$ and the spin tensor is constant $\underline{\underline{\Omega}}(\underline{x},t) = \underline{\underline{\Omega}}(t)$.

5.6.2. *Simple shear*

In orthonormal Cartesian coordinates, we consider the motion defined with the Eulerian representation: $\underline{U}(\underline{x},t) = \alpha x_2 \underline{e}_1 = \alpha (\underline{e}_1 \otimes \underline{e}_2).\underline{x}$ (Chapter 4, section 4.5.2). Calculate the rate of volume dilatation and the gradient of the velocity field, the spin tensor, the strain rate tensor with its principal axes and corresponding extension rates.

Solution

The rate of volume dilatation is obviously identically zero. The gradient of the velocity field is written as $\operatorname{grad}\underline{U}(\underline{x},t) = \alpha \underline{e}_1 \otimes \underline{e}_2$. Thence:

$$\underline{\underline{\Omega}}(\underline{x},t) = \frac{1}{2}\alpha(\underline{e}_1 \otimes \underline{e}_2 - \underline{e}_2 \otimes \underline{e}_1) \text{ and } \underline{\underline{d}}(\underline{x},t) = \frac{1}{2}\alpha(\underline{e}_1 \otimes \underline{e}_2 + \underline{e}_2 \otimes \underline{e}_1).$$

The principal axes are just the bisectors of $(\underline{e}_1,\underline{e}_2)$ and $\underline{e}_3$, with the principal extension rates $\frac{1}{2}\alpha$, $-\frac{1}{2}\alpha$ and 0.

5.6.3. "Lagrangian" Double shear

As in Chapter 4 (section 4.5.3), we consider the motion defined, in orthonormal Cartesian coordinates, by

$$\underline{x} = \underline{\phi}(\underline{X},t) \text{ with } x_1 = X_1 + \alpha t\, X_2, x_2 = X_2 + \alpha t\, X_1, x_3 = X_3.$$

Calculate the velocity field and give the Eulerian representation of this motion. Calculate the rate of volume dilatation and the gradient of the velocity field, the spin tensor, the strain rate tensor with its principal axes and corresponding extension rates.

Solution

– Differentiating $x_1 = X_1 + \alpha t\, X_2, x_2 = X_2 + \alpha t\, X_1, x_3 = X_3$ with respect to time, we obtain the velocity $\underline{U}(\underline{X},t)$ of the particle $\underline{X}$ at time t with components: $U_1(\underline{X},t) = \alpha X_2$, $U_2(\underline{X},t) = \alpha X_1$, $U_3(\underline{X},t) = 0$. Then, expressing $\underline{X}$ as a function of $(\underline{x},t)$ through $X_1 = \dfrac{x_1 - \alpha t\, x_2}{1-\alpha^2 t^2}, X_2 = \dfrac{x_2 - \alpha t\, x_1}{1-\alpha^2 t^2}, X_3 = x_3$, we obtain the velocity field $\underline{U}(\underline{x},t)$ in the Eulerian description:

$$\underline{U}(\underline{x},t) = \alpha\frac{x_2 - \alpha t\, x_1}{1-\alpha^2 t^2}\underline{e}_1 + \alpha\frac{x_1 - \alpha t\, x_2}{1-\alpha^2 t^2}\underline{e}_2.$$

– The rate of volume dilatation is $\text{div}\,\underline{U}(\underline{x},t) = \dfrac{-2\alpha^2 t}{1-\alpha^2 t^2}$ and the gradient of the velocity field $\text{grad}\,\underline{U}(\underline{x},t) = \dfrac{\alpha}{1-\alpha^2 t^2}(-\alpha t\,\underline{e}_1 \otimes \underline{e}_1 + \underline{e}_1 \otimes \underline{e}_2 + \underline{e}_2 \otimes \underline{e}_1 - \alpha t\,\underline{e}_2 \otimes \underline{e}_2)$ is symmetric: the spin tensor is identically zero, $\underline{\underline{\Omega}}(\underline{x},t) = 0$ and $\text{grad}\,\underline{U}(\underline{x},t) = \underline{\underline{d}}(\underline{x},t)$. The principal axes of $\underline{\underline{d}}(\underline{x},t)$ are defined by the orthonormal basis $(\underline{u}_1, \underline{u}_2, \underline{u}_3)$:

$$\underline{u}_1 = \frac{\sqrt{2}}{2}(\underline{e}_1 + \underline{e}_2), \, \underline{u}_2 = \frac{\sqrt{2}}{2}(\underline{e}_1 - \underline{e}_2), \, \underline{u}_3 = \underline{e}_3$$

and $\underline{\underline{d}}(\underline{x},t)$ is written as $\underline{\underline{d}}(\underline{x},t) = \frac{\alpha}{1+\alpha t}\underline{u}_1 \otimes \underline{u}_1 - \frac{\alpha}{1-\alpha t}\underline{u}_2 \otimes \underline{u}_2.$

5.6.4. *"Eulerian" Double shear*

In orthonormal Cartesian coordinates, we consider the motion defined by $\underline{U} = \underline{U}(\underline{x},t)$ with $U_1 = \alpha x_2, U_2 = \alpha x_1, U_3 = 0$. Calculate the rate of volume dilatation and the gradient of the velocity field. Determine the spin tensor and the strain rate tensor with its principal axes and corresponding extension rates.

Solution

The rate of volume dilatation is identically zero: $\text{div}\,\underline{U}(\underline{x},t) = 0$. The motion is isochoric. The gradient of the velocity field is just: $\underline{\underline{\text{grad}}}\,\underline{U}(\underline{x},t) = \alpha(\underline{e}_1 \otimes \underline{e}_2 + \underline{e}_2 \otimes \underline{e}_1)$. Thence, the spin tensor is identically zero, $\underline{\underline{\Omega}}(\underline{x},t) = 0$, and the strain rate tensor is written as: $\underline{\underline{d}}(\underline{x},t) = \alpha(\underline{e}_1 \otimes \underline{e}_2 + \underline{e}_2 \otimes \underline{e}_1)$. Its principal axes are defined by the orthogonal unit vectors $(\underline{u}_1, \underline{u}_2, \underline{u}_3)$:

$$\underline{u}_1 = \frac{\sqrt{2}}{2}(\underline{e}_1 + \underline{e}_2), \, \underline{u}_2 = \frac{\sqrt{2}}{2}(\underline{e}_1 - \underline{e}_2), \, \underline{u}_3 = \underline{e}_3$$

and $\underline{\underline{d}}(\underline{x},t)$ is written as $\underline{\underline{d}}(\underline{x},t) = \alpha(\underline{u}_1 \otimes \underline{u}_1 - \underline{u}_2 \otimes \underline{u}_2)$.

Comment

Comparison with section 5.6.3, we observe that in making $t \rightarrow 0$ the expressions obtained for the "Lagrangian" Double shear reduce to those obtained for the "Eulerian" Double shear, (which is no surprise!) and the volume dilatation rate tends to zero.

5.6.5. *Irrotational and isochoric motions*

We give the general form of the velocity fields $\underline{U} = \underline{U}(\underline{x},t)$ for which the spin vector is identically zero, thus defining irrotational motions. We give the additional condition for such motions to be isochoric (no volume change).

Solution

– From the Curl theorem (Appendix 2, section A2.6), $\underline{U} = \underline{U}(\underline{x},t)$ is a conservative vector field (meaning that its line integral is path independent). Then, from the converse of the gradient theorem (Appendix 2, section A2.7.2), $\underline{U} = \underline{U}(\underline{x},t)$ must be the gradient of an arbitrary twice continuously differentiable scalar function $\varphi(\underline{x},t)$ known as the *velocity potential*:

$$\underline{U}(\underline{x},t) = \operatorname{grad} \varphi(\underline{x},t).$$

– The additional condition for such motions to be isochoric is $\operatorname{div}\underline{U}(\underline{x},t) = 0$ which implies $\Delta\varphi(\underline{x},t) = 0$: the velocity potential must be a *harmonic* function of the space variables.

5.6.6. *Point vortex*

In cylindrical coordinates, we consider the motion defined for $r \neq 0$ by

$$\underline{U}(\underline{x},t) = \underline{U}(r,\theta,z,t) = \frac{\Gamma(t)}{2\pi r}\underline{e}_\theta,$$

where $\Gamma(t)$ is a real function of time. Determine the pathlines. Calculate the gradient of the velocity field, the spin tensor, the strain rate tensor with principal axes and corresponding extension rates, and the rate of volume dilatation. Show that the velocity field is the gradient of a velocity potential that is an harmonic function of the space variables.

Solution

– The pathlines are circular lines in planes orthogonal to Oz that are centered on Oz. The motion can be understood as the rotation of concentric cylindrical layers about Oz, with angular speeds increasing when approaching this axis.

– The gradient of the velocity field is written as

$$\underline{\underline{\operatorname{grad}}}\,U(r,\theta,z,t) = -\frac{1}{r}U_\theta\,\underline{e}_r \otimes \underline{e}_\theta + \frac{\partial U_\theta}{\partial r}\underline{e}_\theta \otimes \underline{e}_r = -\frac{\Gamma(t)}{2\pi r^2}(\underline{e}_r \otimes \underline{e}_\theta + \underline{e}_\theta \otimes \underline{e}_r)$$

and happens to be symmetric. It follows that the spin tensor is zero: $\underline{\underline{\Omega}}(r,\theta,z,t) = 0$. The motion is irrotational. The strain rate tensor

$$\underline{\underline{d}}(r,\theta,z) = -\frac{\Gamma(t)}{2\pi r^2}(\underline{e}_r \otimes \underline{e}_\theta + \underline{e}_\theta \otimes \underline{e}_r)$$

with principal axes along the bisectors of $(\underline{e}_r, \underline{e}_\theta)$ and $\underline{e}_z$, with principal extension rates $-\frac{\Gamma(t)}{2\pi r^2}, \frac{\Gamma(t)}{2\pi r^2}, 0$ (double shear as in section 5.6.4).

The rate of volume dilatation is zero: $\mathrm{tr}\,\underline{\underline{d}}(r,\theta,z,t) = 0$.

– The motion is both isochoric and irrotational (section 5.6.5) and we can check that the velocity field is the gradient of the velocity potential $\varphi(\underline{x},t) = \frac{\Gamma(t)}{2\pi}\theta$ that is actually a harmonic function of the space variables: $\frac{\Gamma(t)}{2\pi}\Delta\theta = 0$.

Comment

Although the particles move along circular lines about the axis Oz, the spin vector is identically zero in the system, which means that mean angular velocity of a small spherical material region attached to the particle (section 5.1.8) is zero; thence, the mean rigid body motion of that infinitesimal element is a translation similar to the motion of a cabin in a Ferris wheel.

5.6.7. *Fluid sink (or source)*

In cylindrical coordinates, we consider the motion defined for $r \neq 0$ by

$$\underline{U}(\underline{x},t) = \underline{U}(r,\theta,z,t) = \frac{\beta(t)}{r}\underline{e}_r.$$

Determine the pathlines. Calculate the gradient of the velocity field, the spin tensor, the strain rate tensor with principal axes and corresponding extension rates, and the rate of volume dilatation. Show that the velocity field is the gradient of a velocity potential that is an harmonic function of the space variables.

Solution

– The pathlines are straight lines along the radii to the axis Oz in planes orthogonal to Oz. The particles move towards the axis or from the axis depending on the sign of the scalar function $\beta(t)$: $\beta(t) > 0$ corresponds to a source and $\beta(t) < 0$ to a sink. The gradient of the velocity field is simply

$$\underline{\underline{\text{grad}}}\, U(r,\theta,z,t) = -\frac{\beta(t)}{r^2}(\underline{e}_r \otimes \underline{e}_r - \underline{e}_\theta \otimes \underline{e}_\theta)$$ that is symmetric. Thence, the motion

is irrotational, $\underline{\underline{\Omega}}(r,\theta,z,t) = 0$, the strain rate tensor has principal axes along $\underline{e}_r$, $\underline{e}_\theta$

and $\underline{e}_z$ with principal extension rates $-\dfrac{\beta(t)}{r^2}, \dfrac{\beta(t)}{r^2}, 0$: the rate of volume dilatation

is zero.

– The velocity field is the gradient of the velocity potential $\varphi(\underline{x},t) = \beta(t)\ln r$ that is actually a harmonic function of the space variables: $\beta(t)\Delta \ln r = 0$.

Comment

Making $\beta(t) = \beta < 0$ constant and $\Gamma(t) = \Gamma$ constant in section 5.6.6, and

taking the sum of the corresponding velocity potentials $\varphi(\underline{x},t) = \beta \ln r + \dfrac{\Gamma}{2\pi}\theta$ as a

velocity potential yields the plane mathematical modeling of the "sink vortex" studied experimentally in [KLI 63] where the pathlines are log spirals about Oz.

5.6.8. *Geometrical compatibility of a thermal strain rate field*

Consider a system S with volume Ω and boundary $\partial\Omega$ that is subject in the present configuration to a temperature change rate field $\tau(\underline{x},t)$. The constituent material of the system is supposed to be isotropic and for the infinitesimal element at any point $\underline{x}$ of the system its response to a temperature change rate $\tau(\underline{x},t)$ is a "thermal" strain rate tensor $\underline{\underline{d}}_\tau(\underline{x},t)$ that depends linearly on $\tau(\underline{x},t)$ through the isotropic tensor $\alpha(\underline{x})\underline{\underline{1}}$; for a homogeneous constituent material, throughout the system, this tensor is constant: $\alpha(\underline{x})\underline{\underline{1}} = \alpha\underline{\underline{1}}$.

What condition should be satisfied by $\tau(\underline{x},t)$ to generate a geometrically compatible "thermal" strain rate field $\underline{\underline{d}}_\tau(\underline{x},t)$?

Assuming no boundary conditions are imposed on the velocity field $\underline{U}(\underline{x},t)$, determine the velocity field generated by such a temperature change rate field.

Solution

– With the simplified notations introduced in section 5.1.10, the compatibility conditions [5.40] can be written explicitly here in the form

$$(\alpha\tau)_{,22} + (\alpha\tau)_{,33} = 0, \ (\alpha\tau)_{,11} + (\alpha\tau)_{,22} = 0, \ (\alpha\tau)_{,11} + (\alpha\tau)_{,33} = 0,$$

$$(\alpha\tau)_{,21} = 0, \ (\alpha\tau)_{,23} = 0, \ (\alpha\tau)_{,31} = 0$$

and it follows that the field $\alpha(\underline{x},t)\tau(\underline{x},t)$ must be an affine function of the Cartesian coordinates.

– Assuming that this condition is satisfied, using orthonormal Cartesian coordinates, we write $\alpha(\underline{x},t)\dot{\tau}(\underline{x},t)$ in the form $\alpha(\underline{x},t)\tau(\underline{x},t) = a_1(t)x_1 + a_2(t)x_2 + a_3(t)x_3 + b.$ Integrating $\underline{\underline{d}}_{=\tau}(\underline{x},t) = \left(a_1(t)x_1 + a_2(t)x_2 + a_3(t)x_3\right)\underline{\underline{1}} + b\underline{\underline{1}}$ following the method given in section 5.1.10, we determine the velocity field $\underline{U}_{\tau}(\underline{x},t)$ with components

$$\begin{cases} U_1 = \dfrac{a_1}{2}(x_1^2 - x_2^2 - x_3^2) + a_2 x_1 x_2 + a_3 x_1 x_3 + b x_1 \\[2mm] U_2 = \dfrac{a_1}{2}(x_2^2 - x_3^2 - x_1^2) + a_3 x_2 x_3 + a_1 x_2 x_1 + b x_2 \\[2mm] U_3 = \dfrac{a_1}{2}(x_3^2 - x_1^2 - x_2^2) + a_1 x_3 x_1 + a_2 x_3 x_2 + b x_3 \end{cases}$$

up to the addition of an arbitrary rigid body motion.

Comments

We first note that the result is valid independently of Ω being simply connected or not.

Also, if the constituent material is homogeneous, the geometrical compatibility condition falls upon $\tau(\underline{x},t)$ alone; however, we see that in the case of heterogeneity, a linear or affine temperature change rate field $\tau(\underline{x},t)$ may generate a geometrically incompatible "thermal" strain rate field.

As already stated in section 5.1.10, a consequence of the geometrical incompatibility of the "thermal" strain rate field will be that internal forces will come into action and contribute to a continuous velocity field in the system, provided they can be sustained by the constituent material. If they cannot be sustained, the solution will be a piecewise continuous velocity field with localized fracture in the system.

Classical Force Modeling for the Three-dimensional Continuum

6.1. Virtual motions

The next step in the virtuous circle for the mechanical modeling of the three-dimensional continuum consists in building up the force model on the geometrical description which was presented in the preceding chapters. Setting the principle of virtual (rates of) work as the fundamental statement, the force modeling process will be carried out through the virtual work method as presented in Chapter 3 (Figure 6.1).

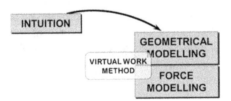

Figure 6.1. *Force modeling*

It calls first for the definition of the vector spaces of virtual motions for the considered system S and its subsystems S' following the rules set in Chapter 3 (section 3.2.2).

The system and subsystems are defined at a given instant of time and their geometrical modeling is described by a volume Ω (respectively Ω')[1] and the vector field $\underline{x}$ to characterize the positions of the particles. Thence, the virtual

---

1 From now on, the index t is unambiguously dropped out for simplicity sake.

motions of S or S' are defined by vector fields denoted by $\hat{\underline{U}}(\underline{x})$ which represent the virtual rates of variation of $\underline{x}$. They generate a vector space with infinite dimension, which obviously includes rigid body motions as required by the method.

The vector space of virtual motions must also *encompass the actual (or real) motions*, a condition that leads us to examine the mathematical regularity conditions to be imposed on the vector fields $\hat{\underline{U}}$.

Regarding the actual motions, we concluded in Chapter 4 (section 4.4.1) that continuity and continuous differentiability of the actual velocity fields, which had been initially assumed as the natural transcription of the experimental intuitive concept of material continuity, had to be softened and weakened into *piecewise* conditions in order to enlarge the domain of relevance of the geometrical model. Consistently with the requirement recalled here above, only piecewise continuity and continuous differentiability will finally be imposed on virtual velocity fields $\hat{\underline{U}}$ (section 6.4.3). However, implementing the virtual work method on the vector spaces of virtual motions generated by continuous and continuously differentiable virtual velocity fields in a first run proves sufficient for completing the force modeling process and avoids "mathematical technicalities" that might overshadow the line of thought. Thence, until section 6.4.3, virtual motions will be continuous and continuously differentiable.

In Chapter 3 (section 3.2.2), we adopted the generic notation $\hat{\mathbf{U}}$ for the virtual motions in order to acknowledge their possible definition by the rates of variation of geometrical or non-geometrical parameters. In the present case, we will identify the notation for the virtual motion with the virtual velocity field that defines it and denote it by $\hat{\underline{U}}$.

6.2. Virtual rates of work

6.2.1. *Virtual rate of work by quantities of acceleration*

As anticipated in Chapter 3 (section 3.2.3), writing down the continuous linear forms $\mathcal{A}(\hat{\underline{U}})$ and $\mathcal{A}'(\hat{\underline{U}})$ which express the virtual rate of work by quantities of acceleration for S (respectively S') follows easily from the geometrical modeling itself. Assuming that *the actual velocity field is continuous and continuously differentiable*, the acceleration field is obtained from Chapter 5 in the form:

$$\underline{a}(\underline{x},t) = \frac{d}{dt}U(\underline{x},t) = \frac{\partial}{\partial t}U(\underline{x},t) + \left(\operatorname{grad} U(\underline{x},t)\right).\underline{U}(\underline{x},t).$$

[6.1]

The quantity of acceleration of the element $dm = \rho(\underline{x},t)\,d\Omega$ is just

$$\underline{a}(\underline{x},t)\,dm = \rho(\underline{x},t)\,\underline{a}(\underline{x},t)\,d\Omega$$

[6.2]

and consequently the linear forms $\mathcal{A}(\hat{\underline{U}})$ and $\mathcal{A}'(\hat{\underline{U}})$ are the integrals on Ω (respectively Ω') of the volume density $\rho(\underline{x},t)\,\underline{a}(\underline{x},t).\hat{\underline{U}}(\underline{x})\,d\Omega$:

$$\mathcal{A}(\hat{\underline{U}}) = \int_{\Omega} \rho(\underline{x},t)\,\underline{a}(\underline{x},t).\hat{\underline{U}}(\underline{x})\,d\Omega$$

[6.3]

$$\mathcal{A}'(\hat{\underline{U}}) = \int_{\Omega'} \rho(\underline{x},t)\,\underline{a}(\underline{x},t).\hat{\underline{U}}(\underline{x})\,d\Omega.$$

[6.4]

We noted in Chapter 5 (section 5.3.3) that, in the case of a piecewise continuous and continuously differentiable actual velocity field, a surface density of quantity of acceleration should be taken into account in addition to [6.2]: this circumstance will be considered later on (section 6.4.2).

6.2.2. *Virtual rate of work by external forces for the system*

Writing down the continuous linear form $\mathcal{P}_e(\hat{\underline{U}})$ which expresses the virtual rate of work by external forces *for the system*, we bear in mind that the classical three-dimensional continuum model aims essentially at dealing with a system made of "diluted" particles instead of point masses. Regarding external forces exerted on the elements $dm = \rho(\underline{x},t)\,d\Omega$ by the *world external to S*, the intuitive concept is to assume that they retain the same vector modeling in the form of a *mass density* $\underline{F}(\underline{x},t)$. This leads to a contribution to $\mathcal{P}_e(\hat{\underline{U}})$ given by the volume integral:

$$\int_{\Omega} \rho(\underline{x},t)\,\underline{F}(\underline{x},t).\hat{\underline{U}}(\underline{x})\,d\Omega.$$

[6.5]

These forces are called *body forces*.

Obviously, they are not the only contribution which should be introduced. External forces may also be exerted on the boundary of the system itself, as shown by everyday experience. These *surface forces* may be contact forces or forces at a distance and will be modeled through a *surface density* $\underline{T}_\Omega(\underline{x},t)$. The force acting on the surface element da of $\partial\Omega$ at point $\underline{x}$ with outward normal $\underline{n}(\underline{x})$ is $\underline{df}_\Omega = \underline{T}_\Omega(\underline{x},t)\,da$ and the contribution to the virtual rate of work by external forces takes the form of the surface integral

$$\int_{\partial\Omega} \underline{T}_\Omega(\underline{x},t).\underline{\hat{U}}(\underline{x})\,da. \qquad [6.6]$$

As a result of this intuitive approach, the virtual rate of work by external forces for the system is written (Figure 6.2):

$$\mathcal{P}_e(\underline{\hat{U}}) = \int_\Omega \rho(\underline{x},t)\underline{F}(\underline{x},t).\underline{\hat{U}}(\underline{x})\,d\Omega + \int_{\partial\Omega} \underline{T}_\Omega(\underline{x},t).\underline{\hat{U}}(\underline{x})\,da. \qquad [6.7]$$

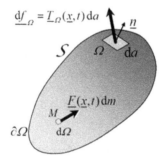

Figure 6.2. *External forces for the system*

An additional contribution to $\mathcal{P}_e(\underline{\hat{U}})$ in the form of surface integrals over surfaces Σ_F internal to Ω might also be written in order to take external force surface densities $\underline{F}_\Sigma(\underline{x},t)$ into account

$$\int_{\Sigma_F} \underline{F}_\Sigma(\underline{x},t).\underline{\hat{U}}(\underline{x})\,d\Sigma, \qquad [6.8]$$

a circumstance which will be explicitly considered in section 6.4.2.

6.2.3. *Virtual rate of work by external forces for subsystems*

Let $S' \subset S$ be a subsystem of S *stricto sensu* (Figure 6.3). The external forces acting on any of its constituent particles are the forces exerted by the world external to S and the forces exerted by the *constituent particles of S external to S'*. We assume that the sum of these forces results in a mass density of body forces $\underline{F}_{\Omega'}(\underline{x},t)$ where the index Ω' insists on the fact that, for the same particle at point $\underline{x}$, the mass density of external body forces depends on the considered subsystem:

$$\underline{F}_{\Omega'}(\underline{x},t)\,dm = \rho(\underline{x},t)\,\underline{F}_{\Omega'}(\underline{x},t)\,d\Omega. \qquad [6.9]$$

External forces are also acting on the boundary surface of the subsystem due to the world external to S'. They are assumed to take the form of a surface density $\underline{T}_{\Omega'}(\underline{x},t)$ on $\partial\Omega'$ where the index Ω' underscores the fact that for the same surface element da with normal $\underline{n}(\underline{x})$ the force

$$\underline{df}_{\Omega'} = \underline{T}_{\Omega'}(\underline{x},t)\,da \qquad [6.10]$$

depends on the considered subsystem. Consistently with [6.7], we may observe that these forces are only exerted by the *constituent particles of S external to S'*. In the particular case, when external force surface densities are exerted on surfaces Σ_F internal to Ω' (section 6.2.2), they contribute to $\mathcal{P}'_e(\hat{\underline{U}})$ in the same form as [6.8].

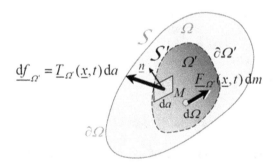

Figure 6.3. *External forces for a subsystem*

Thence, the virtual rate of work by external forces for a subsystem is written as

$$\mathcal{P}'_e(\hat{\underline{U}}) = \int_{\Omega'}\rho(\underline{x},t)\,\underline{F}_{\Omega'}(\underline{x},t).\hat{\underline{U}}(\underline{x})\,d\Omega + \int_{\partial\Omega'}\underline{T}_{\Omega'}(\underline{x},t).\hat{\underline{U}}(\underline{x})\,da. \qquad [6.11]$$

6.2.4. *Virtual rate of work by internal forces*

The classical force modeling for the three-dimensional continuum states as a fundamental assumption that *the virtual rate of work by internal forces for the system or any subsystem can be written as the integral of a volume density* $p_i(\hat{U})$ *independent of the considered subsystem, which is a linear form in the local values of the virtual velocity field and its first gradient:*

$$\forall S' \subseteq S, \ \mathcal{P}_i'(\hat{U}) = \int_{\Omega'} p_i(\hat{U}) \,\mathrm{d}\Omega \tag{6.12}$$

with

$$p_i(\hat{U}) = -\underline{A}(\underline{x},t).\hat{\underline{U}}(\underline{x}) - \underline{\underline{t}}(\underline{x},t) : \operatorname{grad}\hat{U}(\underline{x}), \tag{6.13}$$

where the cofactors $\underline{A}$ and $\underline{\underline{t}}$ are respectively a vector field and a second rank tensor field which model internal forces. The notation " : " stands for the *doubly contracted product* of the two second rank tensors $\operatorname{grad}\hat{U}(\underline{x})$ and $\underline{\underline{t}}(\underline{x},t)$ as defined in Appendix 1 (section A1.7.5).

As a first comment on this fundamental assumption, we may recall that in Chapter 2 (section 2.5.2), when dealing with a system made of discrete particles, we indeed found that the virtual rate of work by internal forces for the system or any subsystem was the sum of discrete densities but these densities were *dependent* on the considered subsystem.

6.3. Implementation of the principle of virtual work

6.3.1. *Specifying the virtual rate of work by internal forces*

Implementing the principle of virtual work, we first refer to the dual statement of the law of mutual actions (Chapter 3, section 3.2.4)

$$\begin{cases} \forall S' \subseteq S \\ \forall \hat{\underline{U}} \text{ r.b.v.m. of } S', \ \mathcal{P}_i'(\hat{U}) = 0 \end{cases} \tag{6.14}$$

to be satisfied by [6.12] and [6.13].

Any rigid body virtual motion (r.b.v.m.) of S' is generated by an *arbitrary vector* $\hat{\underline{U}}_0$ and an *arbitrary antisymmetric tensor* $\hat{\underline{\underline{\omega}}}_0$ and written (Chapter 5, section 5.1.9)

$$\hat{\underline{U}}(\underline{x}) = \hat{\underline{U}}_0 + \hat{\underline{\underline{\omega}}}_0 . \underline{OM}, \qquad [6.15]$$

from which we derive

$$\underline{\underline{\text{grad}}}\, \hat{\underline{U}}(\underline{x}) = \hat{\underline{\underline{\omega}}}_0 . \qquad [6.16]$$

Thence, [6.14] takes the form

$$\begin{cases} \forall S' \subseteq S \\ \forall \hat{\underline{U}}_0 \in \mathbb{R}^3, \ \forall \hat{\underline{\underline{\omega}}}_0 \in \mathbb{R}^3, \\ P_i'(\hat{\underline{U}}) = \int_{\Omega'} (-\underline{A}(\underline{x},t).\hat{\underline{U}}_0 - \underline{\underline{t}}(\underline{x},t):\hat{\underline{\underline{\omega}}}_0)\, \mathrm{d}\Omega = 0. \end{cases} \qquad [6.17]$$

Making $\hat{\underline{\underline{\omega}}}_0 = 0$ first, we see that [6.17] reduces to

$$\begin{cases} \forall \Omega' \subseteq \Omega \\ \forall \hat{\underline{U}}_0 \in \mathbb{R}^3, \\ \hat{\underline{U}}_0 . \int_{\Omega'} -\underline{A}(\underline{x},t)\, \mathrm{d}\Omega = 0, \end{cases} \qquad [6.18]$$

which proves that the field $\underline{A}(\underline{x},t)$ is *identically zero* over Ω. Then, splitting the tensor field $\underline{\underline{t}}(\underline{x},t)$ into its symmetric and antisymmetric parts, $\underline{\underline{\sigma}}(\underline{x},t)$ and $\underline{\underline{\alpha}}(\underline{x},t)$ respectively (Appendix 1, section A1.3.5), we note that [6.17] becomes

$$\begin{cases} \forall \Omega' \subseteq \Omega \\ \forall \hat{\underline{\underline{\omega}}}_0 \in \mathbb{R}^3, \\ \hat{\underline{\underline{\omega}}}_0 : \int_{\Omega'} -\underline{\underline{\alpha}}(\underline{x},t)\, \mathrm{d}\Omega = 0 \end{cases} \qquad [6.19]$$

from which we derive that the antisymmetric tensor field is *identically zero* over Ω.

It follows that the most general form, [6.12] and [6.13], proposed for $\mathcal{P}_i'(\hat{U})$ according to the initial fundamental assumption is specified by the principle of virtual work and written as

$$\forall S' \subseteq S, \ \mathcal{P}_i'(\hat{U}) = \int_{\Omega'} -\underline{\underline{\sigma}}(\underline{x},t):\operatorname{grad}\hat{U}(\underline{x})\,d\Omega \qquad [6.20]$$

where $\underline{\underline{\sigma}}(\underline{x},t)$ is an *arbitrary symmetric tensor field*.

The second rank symmetric tensor field $\underline{\underline{\sigma}}(\underline{x},t)$ is called the *Cauchy stress tensor field*.

It is worth noting that, defining the *virtual strain rate* $\underline{\underline{\hat{d}}}(\underline{x})$ as the symmetric part of $\operatorname{grad}\hat{U}(\underline{x})$, in the same way as $\underline{\underline{d}}(\underline{x},t)$

$$\underline{\underline{\hat{d}}}(\underline{x}) = \frac{1}{2}(\operatorname{grad}\hat{U}(\underline{x}) + {}^t\operatorname{grad}\hat{U}(\underline{x})), \qquad [6.21]$$

the virtual rate of work by internal forces may equivalently be written as

$$\forall S' \subseteq S, \ \mathcal{P}_i'(\hat{U}) = \int_{\Omega'} -\underline{\underline{\sigma}}(\underline{x},t):\underline{\underline{\hat{d}}}(\underline{x})\,d\Omega. \qquad [6.22]$$

6.3.2. *Equations of motion for the system*

Having specified the expression of the virtual rate of work by internal forces, we now apply the second statement of the principle of virtual work to the system, which takes the form

$$\left\{ \begin{array}{l} \text{in a Galilean frame,} \\ \text{for } S, \\ \forall \underline{\hat{U}} \text{ v.m. }, \ \mathcal{P}_e(\hat{U}) + \mathcal{P}_i(\hat{U}) = \mathcal{A}(\hat{U}) \end{array} \right. \qquad [6.23]$$

that is

$$\left\{ \begin{aligned} &\text{in a Galilean frame,} \\ &\forall \hat{\underline{U}} \text{ v.m.,} \\ &\int_{\Omega} \rho(\underline{x},t)\,\underline{F}(\underline{x},t).\hat{\underline{U}}(\underline{x})\,d\Omega + \int_{\Omega} \hat{\underline{U}}(\underline{x}).\text{div}^{\,t}\underline{\underline{\sigma}}(\underline{x},t)\,d\Omega \\ &+ \int_{\partial\Omega} \underline{T}_{\Omega}(\underline{x},t)\cdot\hat{\underline{U}}(\underline{x})\,da - \int_{\partial\Omega} (\underline{\underline{\sigma}}(\underline{x},t)\cdot\hat{\underline{U}}(x))\cdot\underline{n}(\underline{x})\,da \\ &= \int_{\Omega} \rho(\underline{x},t)\,\underline{a}(\underline{x},t).\hat{\underline{U}}(\underline{x})\,d\Omega. \end{aligned} \right. \tag{6.24}$$

In order to carry out the dualization procedure on [6.24] and derive the equations of motion, the expression of $\mathcal{P}_i(\hat{\underline{U}})$ must be transformed into integrals whose integrands come out as scalar products of $\hat{\underline{U}}(\underline{x})$ with a vectorial cofactor. Assuming the second rank symmetric tensor field $\underline{\underline{\sigma}}(\underline{x},t)$ to be continuous and continuously differentiable, we can transform the integrand in $\mathcal{P}_i(\hat{\underline{U}})$ through the identity

$$\text{div}(\underline{\underline{\sigma}}.\hat{\underline{U}}) = \underline{\underline{\sigma}}:\text{grad}\,\hat{\underline{U}} + \hat{\underline{U}}.\text{div}\,^t\underline{\underline{\sigma}}\,^2 \tag{6.25}$$

and obtain

$$\int_{\Omega} -\underline{\underline{\sigma}}(\underline{x},t):\text{grad}\,\hat{\underline{U}}(\underline{x})\,d\Omega = \int_{\Omega}(-\text{div}(\underline{\underline{\sigma}}.\hat{\underline{U}})+\hat{\underline{U}}.\text{div}\,^t\underline{\underline{\sigma}})\,d\Omega \tag{6.26}$$

so that, applying the *divergence theorem*, [6.24] becomes

$$\left\{ \begin{aligned} &\text{in a Galilean frame,} \\ &\forall \hat{\underline{U}} \text{ v.m.,} \\ &\int_{\Omega} \rho(\underline{x},t)\,\underline{F}(\underline{x},t).\hat{\underline{U}}(\underline{x})\,d\Omega + \int_{\Omega} \hat{\underline{U}}(\underline{x}).\text{div}\,^t\underline{\underline{\sigma}}(\underline{x},t)\,d\Omega \\ &+ \int_{\partial\Omega} \underline{T}_{\Omega}(\underline{x},t).\hat{\underline{U}}(\underline{x})\,da - \int_{\partial\Omega} (\underline{\underline{\sigma}}(\underline{x},t).\hat{\underline{U}}(\underline{x})).\underline{n}(\underline{x})\,da \\ &= \int_{\Omega} \rho(\underline{x},t)\,\underline{a}(\underline{x},t).\hat{\underline{U}}(\underline{x})\,d\Omega. \end{aligned} \right. \tag{6.27}$$

---

2 We have $\text{grad}(\underline{\underline{\sigma}}.\hat{\underline{U}}) = \underline{\underline{\sigma}}.\text{grad}\,\hat{\underline{U}} + \hat{\underline{U}}.\text{grad}\,^t\underline{\underline{\sigma}}$, thence [6.19]; (see Appendix 2, section A2.9).

We note that this equation contains three volume integrals over Ω and two surface integrals on $\partial\Omega$. Taking advantage of the arbitrariness of $\hat{\underline{U}}$, we can apply [6.27] with virtual velocity fields that are continuous and continuously differentiable and *equal to zero* on $\partial\Omega$ in such a way that the surface integrals vanish:

$$\begin{cases} \text{in a Galilean frame,} \\ \forall \hat{\underline{U}} \text{ v.m.,} \\ \displaystyle\int_\Omega \rho(\underline{x},t)\underline{F}(\underline{x},t).\hat{\underline{U}}(\underline{x})\,d\Omega + \int_\Omega \hat{\underline{U}}(\underline{x}).\text{div}\,{}^t\underline{\underline{\sigma}}(\underline{x},t)\,d\Omega \\ \qquad\qquad = \displaystyle\int_\Omega \rho(\underline{x},t)\,\underline{a}(\underline{x},t).\hat{\underline{U}}(\underline{x})\,d\Omega. \end{cases} \qquad [6.28]$$

Then, from the *dualization* procedure and taking the symmetry of $\underline{\underline{\sigma}}(\underline{x},t)$ into account, we derive *the field equation of motion*:

$$\begin{cases} \text{in a Galilean frame, } \forall \underline{x} \in \Omega, \\ \text{div}\,\underline{\underline{\sigma}}(\underline{x},t) + \rho(\underline{x},t)(\underline{F}(\underline{x},t)-\underline{a}(\underline{x},t)) = 0. \end{cases} \qquad [6.29]$$

With this result, [6.27] now reduces to

$$\begin{cases} \forall \hat{\underline{U}} \text{ v.m.,} \\ \displaystyle\int_{\partial\Omega} \underline{T}_\Omega(\underline{x},t).\hat{\underline{U}}(\underline{x})\,da - \int_{\partial\Omega} (\underline{\underline{\sigma}}(\underline{x},t).\hat{\underline{U}}(\underline{x})).\underline{n}(\underline{x})\,da = 0. \end{cases} \qquad [6.30]$$

which yields the *boundary equation* for the symmetric tensor field $\underline{\underline{\sigma}}(\underline{x},t)$:

$$\forall \underline{x} \in \partial\Omega,\ \underline{\underline{\sigma}}(\underline{x},t).\underline{n}(\underline{x}) = \underline{T}_\Omega(\underline{x},t). \qquad [6.31]$$

6.3.3. *Equations of motion for a subsystem*

The second statement of the principle of virtual work for a subsystem is written in a form quite similar to [6.24] and can be transformed in the same way into

in a Galilean frame,

$\forall \hat{\underline{U}}$ v.m. of S',

$$\int_{\Omega'} \rho(\underline{x},t)\, \underline{F}_{\Omega'}(\underline{x},t).\hat{\underline{U}}(\underline{x})\,d\Omega + \int_{\Omega'} \hat{\underline{U}}(\underline{x}).\text{div}\,{}^t\underline{\underline{\sigma}}(\underline{x},t)\,d\Omega \qquad [6.32]$$

$$+ \int_{\partial\Omega'} \underline{T}_{\Omega'}(\underline{x},t).\hat{\underline{U}}(\underline{x})\,da - \int_{\partial\Omega'} (\underline{\underline{\sigma}}(\underline{x},t).\hat{\underline{U}}(\underline{x})).\underline{n}(\underline{x})\,da$$

$$= \int_{\Omega'} \rho(\underline{x},t)\,\underline{a}(\underline{x},t).\hat{\underline{U}}(\underline{x})\,d\Omega.$$

With the same arguments as in the preceding section, we obtain the *field equation of motion for* S'

in a Galilean frame, $\forall \underline{x} \in \Omega'$,

$$\text{div}\,\underline{\underline{\sigma}}(\underline{x},t) + \rho(\underline{x},t)(\underline{F}_{\Omega'}(\underline{x},t) - \underline{a}(\underline{x},t)) = 0. \qquad [6.33]$$

and the equation on the boundary surface $\partial\Omega'$

$$\forall \underline{x} \in \partial\Omega',\ \underline{\underline{\sigma}}(\underline{x},t).\underline{n}(\underline{x}) = \underline{T}_{\Omega'}(\underline{x},t). \qquad [6.34]$$

6.3.4. *The model*

Although they look quite similar to [6.29] and [6.31] respectively, equations [6.33] and [6.34] bear completely different meanings since they are referring to a *subsystem*, the result of a thought experiment.

– Putting together [6.29] and [6.33], we note that these equations are compatible if and only if

$$\forall \Omega' \subset \Omega,\ \forall \underline{x} \in \Omega',\ \underline{F}_{\Omega'}(\underline{x},t) = \underline{F}(\underline{x},t). \qquad [6.35]$$

In plain words, it means that for a given particle at point $\underline{x}$, the mass density of external body forces *does not depend* on the subsystem this particle is considered to belong to. From a physical point of a view: the *constituent particles of S external to a subsystem S' do not exert any action on the constituent particles of S'* (see section 6.2.3).

The classical force model for the three-dimensional continuum assumes that the particles of the system do not exert any action at a distance on each other that can be modeled as a mass density.[3]

– Comparing [6.34] with [6.31], we understand the completely different role played by these two equations. While [6.31], where $T_\Omega(\underline{x},t)$ holds the status of a data, stands as the boundary condition imposed on the stress field $\underline{\underline{\sigma}}(\underline{x},t)$ that satisfies the field equation [6.29], equation [6.34] *determines the surface density* $T_{\Omega'}(\underline{x},t)$ *of external forces acting on* $\partial\Omega'$. As already pointed out in (section 6.2.3), these forces are only exerted by the constituent particles of S external to S'.

– Moreover, it comes out from [6.34] that $T_{\Omega'}(\underline{x},t)$ actually depends on $\partial\Omega'$, as anticipated, but only through the normal $\underline{n}(\underline{x})$. The physical meaning of this result is that the surface density of external body forces acting on a given subsystem is the result of *local contact actions* between the particles. More precisely, considering the surface element da at point $\underline{x}$ with $\underline{n}(\underline{x})$ the outward normal to $\partial\Omega'$, the force $\underline{df}_{\Omega'}$ acting on this element is

$$\underline{df}_{\Omega'} = \underline{\underline{\sigma}}(\underline{x},t).\underline{n}(\underline{x})\,da = \underline{df}(\underline{x},\underline{n},t) = \underline{T}(\underline{x},\underline{n},t)\,da, \qquad [6.36]$$

the result of *"contact actions"* exerted on the particles infinitely close to da inside S' by the particles infinitely close to da outside S' (Figure 6.4).

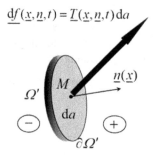

Figure 6.4. *Contact forces at the boundary of a subsystem*

---

3 In [SAL 01] this assumption is made at the very beginning, when writing the virtual rate of work by external forces for a subsystem.

6.3.5. *Consistency*

The field equation of motion [6.29] is a system of three partial derivative equations (explicit formulas given in section 6.9) for the six scalar components of the symmetric tensor field $\underline{\underline{\sigma}}(\underline{x},t)$ with respect to the space variables, while [6.31] is a system of three boundary conditions for these components. The condition for these equations to be mathematically compatible is that the data $\underline{F}(\underline{x},t)$ and $\underline{a}(\underline{x},t)$ in Ω and $\underline{T}_\Omega(\underline{x},t)$ on $\partial\Omega$ must be compatible with the fundamental law of dynamics (Chapter 3, sections 3.2.5 and 3.3.3) as a necessary condition:

$$\begin{cases} \text{in a Galilean frame,} \\ \text{for } S, \\ [\mathcal{F}_e] = [\mathcal{M}a]. \end{cases} \qquad [6.37]$$

Note that for any subsystem $S' \subset S$, the equation

$$\begin{cases} \text{in a Galilean frame,} \\ \text{for } S', \\ [\mathcal{F}'_e] = [\mathcal{M}a'] \end{cases} \qquad [6.38]$$

is then valid as a consequence of [6.29] and [6.31].

It is also recalled that the necessary condition [6.35] expresses the consistency of the physical assumptions regarding external forces on the particles whether they are considered as part of the system or a subsystem.

6.4. Piecewise continuous and continuously differentiable fields

6.4.1. *The issues*

The model in the preceding section was built upon the continuity and continuous differentiability assumptions for the three fields involved: chronologically the actual velocity field $\underline{U}(\underline{x},t)$, the virtual velocity fields that generate the vector space of virtual motions $\underline{\hat{U}}(\underline{x})$ and the symmetric tensor field (the Cauchy stress field) $\underline{\underline{\sigma}}(\underline{x},t)$.

In this section 6.4.2, we will first keep on assuming the virtual velocity field to be continuous and continuously differentiable and consider the case when the actual velocity field and the Cauchy stress field are piecewise continuous and continuously differentiable.

Piecewise continuous and continuously differentiable virtual velocity fields will be considered afterwards (section 6.4.3).

6.4.2. Piecewise continuous and continuously differentiable $\underline{U}(\underline{x},t)$ and $\underline{\underline{\sigma}}(\underline{x},t)$

The actual velocity field $\underline{U}(\underline{x},t)$ is now assumed to be *piecewise continuous and continuously differentiable* with Σ_t denoting the corresponding jump surfaces and the same notations as in Chapter 5 (section 5.3.1). We recall that the continuity equation written on the jump surface

$$[\![\rho(\underline{x},t)(\underline{U}(\underline{x},t) - \underline{W}(\underline{x},t))]\!] . \underline{n}(\underline{x}) = 0 \quad \text{on } \Sigma_t \qquad [6.39]$$

implies that a *shock wave* (actual velocity jump) always comes with a jump in the mass per unit volume $[\![\rho(\underline{x},t)]\!]$ when crossing Σ_t.

The Cauchy stress field is now also assumed to be piecewise continuous and continuously differentiable with Σ_σ denoting its jump surfaces, but this new context has obviously no influence on the expression of the virtual rate of work by internal forces for the system and subsystems, derived from the fundamental assumption in section 6.2.4 and the first statement of the principle of virtual work. We still write

$$\forall S' \subseteq S, \ \mathcal{P}'_i(\hat{U}) = \int_{\Omega'} -\underline{\underline{\sigma}}(\underline{x},t) : \operatorname{grad} \hat{U}(\underline{x}) \, d\Omega \qquad [6.40]$$

where $\underline{\underline{\sigma}}(\underline{x},t)$ is the Cauchy stress field (symmetric tensor field).

Resuming with the implementation of the virtual work method, we note that the rate of work by external forces for the system and subsystems is not affected by the discontinuities in $\rho(\underline{x},t)$:

$$\mathcal{P}_e(\hat{\underline{U}}) = \int_{\Omega} \rho(\underline{x},t)\underline{F}(\underline{x},t).\hat{\underline{U}}(\underline{x})\,\mathrm{d}\Omega + \int_{\partial\Omega}\underline{T}_{\Omega}(\underline{x},t).\hat{\underline{U}}(\underline{x})\,\mathrm{d}a$$

$$\mathcal{P}'_e(\hat{\underline{U}}) = \int_{\Omega'} \rho(\underline{x},t)\underline{F}_{\Omega'}(\underline{x},t).\hat{\underline{U}}(\underline{x})\,\mathrm{d}\Omega + \int_{\partial\Omega'}\underline{T}_{\Omega'}(\underline{x},t).\hat{\underline{U}}(\underline{x})\,\mathrm{d}a.$$

[6.41]

In contrast, the expressions of the *virtual rate of work by quantities of acceleration must be reconsidered* due to the discontinuities in the actual velocity field. Referring to Chapter 5 (section 5.3.3), we recall that they generate the surface density of quantities of acceleration on Σ_t

$$[[\rho\underline{U}\otimes(\underline{U}-\underline{W})]].\underline{n} = [[\underline{U}]]\rho(\underline{U}-\underline{W}).\underline{n}.$$

[6.42]

It follows that the expressions of the virtual rate of work by quantities of acceleration shall be completed and become

$$\mathcal{A}(\hat{\underline{U}}) = \int_{\Omega}\rho(\underline{x},t)\underline{a}(\underline{x},t).\hat{\underline{U}}(\underline{x})\,\mathrm{d}\Omega + \int_{\Sigma_t}\hat{\underline{U}}.[[\underline{U}]]\rho(\underline{U}-\underline{W}).\underline{n}\,\mathrm{d}\Sigma$$

$$\mathcal{A}'(\hat{\underline{U}}) = \int_{\Omega'}\rho(\underline{x},t)\underline{a}(\underline{x},t).\hat{\underline{U}}(\underline{x})\,\mathrm{d}\Omega + \int_{\Sigma_t\cap\Omega'}\hat{\underline{U}}.[[\underline{U}]]\rho(\underline{U}-\underline{W}).\underline{n}\,\mathrm{d}\Sigma$$

[6.43]

where $\Sigma_t\cap\Omega'$ stands for the part of the jump surface Σ_t which lies inside Ω'.

With these modified expressions, the principle of virtual work is written for the system

$$\left\{\begin{array}{l} \text{in a Galilean frame,} \\[4pt] \forall\hat{\underline{U}} \text{ v.m.,} \\[4pt] \displaystyle\int_{\Omega}\rho(\underline{x},t)\underline{F}(\underline{x},t).\hat{\underline{U}}(\underline{x})\,\mathrm{d}\Omega + \int_{\partial\Omega}\underline{T}_{\Omega}(\underline{x},t).\hat{\underline{U}}(\underline{x})\,\mathrm{d}a \\[10pt] \displaystyle -\int_{\Omega}\underline{\underline{\sigma}}(\underline{x},t):\mathrm{grad}\,\hat{\underline{U}}(\underline{x})\,\mathrm{d}\Omega \\[10pt] \displaystyle =\int_{\Omega}\rho(\underline{x},t)\underline{a}(\underline{x},t).\hat{\underline{U}}(\underline{x})\,\mathrm{d}\Omega \\[10pt] \displaystyle +\int_{\Sigma_t}\hat{\underline{U}}(\underline{x}).[[\underline{U}(\underline{x},t)]]\rho(\underline{x},t)(\underline{U}(\underline{x},t)-\underline{W}(\underline{x},t)).\underline{n}(\underline{x})\,\mathrm{d}\Sigma \end{array}\right.$$

[6.44]

(analogous equation for any subsystem) and transformed, using the identity [6.25], into

$$
\left\{
\begin{aligned}
&\text{in a Galilean frame,} \\
&\forall \underline{\hat{U}} \text{ v.m.,} \\
&\int_{\Omega} \rho(\underline{x},t)\,\underline{F}(\underline{x},t).\underline{\hat{U}}(\underline{x})\,\mathrm{d}\Omega + \int_{\Omega} \underline{\hat{U}}(\underline{x}).\mathrm{div}\,{}^{t}\underline{\underline{\sigma}}(\underline{x},t)\,\mathrm{d}\Omega \\
&- \int_{\Omega} \mathrm{div}\,(\underline{\underline{\sigma}}(\underline{x},t).\underline{\hat{U}}(\underline{x}))\,\mathrm{d}\Omega + \int_{\partial\Omega} \underline{T}_{\Omega}(\underline{x},t).\underline{\hat{U}}(\underline{x})\,\mathrm{d}a \\
&= \int_{\Omega} \rho(\underline{x},t)\,\underline{a}(\underline{x},t).\underline{\hat{U}}(\underline{x})\,\mathrm{d}\Omega \\
&+ \int_{\Sigma_{t}} \underline{\hat{U}}(\underline{x}).[\![\underline{U}(\underline{x},t)]\!]\,\rho(\underline{x},t)(\underline{U}(\underline{x},t)-\underline{W}(\underline{x},t)).\underline{n}(\underline{x})\,\mathrm{d}\Sigma.
\end{aligned}
\right.
\qquad [6.45]
$$

With the *divergence theorem*, taking the discontinuity of the stress field into account (Appendix 2, section A2.5.2), we have the final expression conveniently expressed for the dualization procedure:

$$
\left\{
\begin{aligned}
&\text{in a Galilean frame,} \\
&\forall \underline{\hat{U}} \text{ v.m.,} \\
&\int_{\Omega} \rho(\underline{x},t)\,\underline{F}(\underline{x},t).\underline{\hat{U}}(\underline{x})\,\mathrm{d}\Omega + \int_{\Omega} \underline{\hat{U}}(\underline{x}).\mathrm{div}\,{}^{t}\underline{\underline{\sigma}}(\underline{x},t)\,\mathrm{d}\Omega \\
&+ \int_{\partial\Omega} \underline{T}_{\Omega}(\underline{x},t).\underline{\hat{U}}(\underline{x})\,\mathrm{d}a - \int_{\partial\Omega} (\underline{\underline{\sigma}}(\underline{x},t).\underline{\hat{U}}(\underline{x})).\underline{n}(\underline{x})\,\mathrm{d}a \\
&+ \int_{\Sigma_{\sigma}} ([\![\underline{\underline{\sigma}}(\underline{x},t)]\!].\underline{\hat{U}}(\underline{x})).\underline{n}(\underline{x})\,\mathrm{d}\Sigma \\
&= \int_{\Omega} \rho(\underline{x},t)\,\underline{a}(\underline{x},t).\underline{\hat{U}}(\underline{x})\,\mathrm{d}\Omega \\
&+ \int_{\Sigma_{t}} \underline{\hat{U}}(\underline{x}).[\![\underline{U}(\underline{x},t)]\!]\,\rho(\underline{x},t)(\underline{U}(\underline{x},t)-\underline{W}(\underline{x},t)).\underline{n}(\underline{x})\,\mathrm{d}\Sigma.
\end{aligned}
\right.
\qquad [6.46]
$$

Thence, we can derive the field equations of motion considering the integrals over Ω, $\partial\Omega$, Σ_{t} and Σ_{σ} separately and taking the symmetry of $\underline{\underline{\sigma}}(\underline{x},t)$ into account.

Equations [6.29] in Ω and [6.31] on $\partial\Omega$ are unchanged and completed on the jump surfaces Σ_{t} and Σ_{σ}, i.e. on $\Sigma_{t} \cup \Sigma_{\sigma}$, by

$$\begin{cases} \text{in a Galilean frame, } \forall \underline{x} \in \Sigma_t \cup \Sigma_\sigma, \\ [\![\underline{\underline{\sigma}}(\underline{x},t)]\!].\underline{n}(\underline{x}) = \rho(\underline{x},t)[\![\underline{U}(\underline{x},t)]\!](\underline{U}(\underline{x},t) - \underline{W}(\underline{x},t)).\underline{n}(\underline{x}). \end{cases} \qquad [6.47]$$

In plain words, this means that

– when crossing a jump surface for the actual velocity field (*shock wave*), the vector $\underline{\underline{\sigma}}(\underline{x},t).\underline{n}(\underline{x})$, which is called the *stress vector on the surface*, is *discontinuous* with a jump given by [6.47];

– if the Cauchy stress field is discontinuous when crossing a surface Σ_σ, which is *not a shock wave*, the stress vector $\underline{\underline{\sigma}}(\underline{x},t).\underline{n}(\underline{x})$ on that surface is *continuous*.

The same reasoning can be implemented to take external force surface densities as introduced in section 6.2.2 into account and yield:

$$\forall \underline{x} \in \Sigma_F, [\![\underline{\underline{\sigma}}(\underline{x},t)]\!].\underline{n}(\underline{x}) + \underline{F}_\Sigma(\underline{x},t) = 0. \qquad [6.48]$$

6.4.3. *Piecewise continuous and continuously differentiable virtual velocity fields*

After the completion of the force modeling, for the three-dimensional continuum using continuous and continuously differentiable virtual velocity fields as mathematical test functions, we examine what changes should be made in the expressions of the continuous linear forms $\mathcal{P}_e(\hat{\underline{U}})$, $\mathcal{P}'_e(\hat{\underline{U}})$, $\mathcal{P}_i(\hat{\underline{U}})$, $\mathcal{P}'_i(\hat{\underline{U}})$, $\mathcal{A}(\hat{\underline{U}})$ and $\mathcal{A}'(\hat{\underline{U}})$ in order to express the principle of virtual work with this model on the vector space of *piecewise continuous and continuously differentiable virtual velocity fields*.

It is clear that neither $\mathcal{P}_e(\hat{\underline{U}})$ and $\mathcal{P}'_e(\hat{\underline{U}})$ nor $\mathcal{A}(\hat{\underline{U}})$ and $\mathcal{A}'(\hat{\underline{U}})$ are affected by possible jumps of the field $\hat{\underline{U}}$. Only $\mathcal{P}_i(\hat{\underline{U}})$ and $\mathcal{P}'_i(\hat{\underline{U}})$ need to be modified.

Let $\hat{\underline{U}}$ be a piecewise continuous and continuously differentiable virtual velocity field with a jump surface $\Sigma_{\hat{U}}$ transversally oriented by the unit vector $\underline{n}(\underline{x})$ and the virtual velocity jump $[\![\hat{\underline{U}}(\underline{x})]\!] = \hat{\underline{U}}_2(\underline{x}) - \hat{\underline{U}}_1(\underline{x})$ at point $\underline{x}$.

Consider a subsystem S' whose boundary $\partial\Omega'$ caps $\Sigma_{\hat{U}}$ on both sides ① and ②. Using $\Sigma_{\hat{U}}$ as a boundary, we can divide S' into two complementary subsystems, namely:

– S_1', on the ① side of $\Sigma_{\hat{U}}$, with boundary $\partial\Omega_1' \cup \Sigma_{\hat{U}}$ and outward normal $\underline{n}(\underline{x})$ at point $\underline{x}$ of $\Sigma_{\hat{U}}$.

– S_2', on the ② side, with boundary $\partial\Omega_2' \cup \Sigma_{\hat{U}}$ and outward normal $-\underline{n}(\underline{x})$ at point $\underline{x}$ of $\Sigma_{\hat{U}}$ (Figure 6.5).

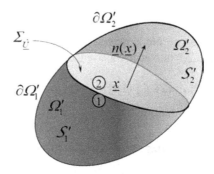

Figure 6.5. *Discontinuity of the virtual velocity field*

The virtual velocity field $\hat{\underline{U}}$ is continuous and continuously differentiable in Ω_1' and Ω_2' so that we can write the principle of virtual work in each subsystem in the same form as in section 6.3.1.

The external forces to be taken into account for S_1' are the body forces $\rho(\underline{x},t)\underline{F}(\underline{x},t)d\Omega$ in Ω_1', the surface forces $\underline{T}_{\Omega_1'}(\underline{x},t)da$ on $\partial\Omega_1'$ and the surface forces on $\Sigma_{\hat{U}}$ with outward normal $\underline{n}(\underline{x})$. These latter forces are just the contact forces exerted by S_2' on S_1' that are equal to $\underline{\underline{\sigma}}_2(\underline{x},t).\underline{n}(\underline{x})d\Sigma$. Thence, the principle of virtual work for S_1' is written as

in a Galilean frame,

$\forall \hat{\underline{U}}$ v.m. of S_1',

$$\int_{\Omega_1'} \rho(\underline{x},t)\,\underline{F}(\underline{x},t).\hat{\underline{U}}(\underline{x})\,d\Omega + \int_{\partial\Omega_1'} \underline{T}_{\Omega_1'}(\underline{x},t).\hat{\underline{U}}(\underline{x})\,da \qquad [6.49]$$

$$+ \int_{\Sigma_{\hat{U}}} \hat{\underline{U}}_1(\underline{x}).\underline{\underline{\sigma}}_2(\underline{x},t).\underline{n}(\underline{x})\,d\Sigma - \int_{\Omega_1'} \underline{\underline{\sigma}}(\underline{x},t):\operatorname{grad}\hat{\underline{U}}(\underline{x})\,d\Omega$$

$$= \int_{\Omega_1'} \rho(\underline{x},t)\,\underline{a}(\underline{x},t).\hat{\underline{U}}(\underline{x})\,d\Omega.$$

The same reasoning applies to S_2' but for the fact that the external forces on $\Sigma_{\hat{U}}$ are equal to $-\underline{\underline{\sigma}}_1(\underline{x},t).\underline{n}(\underline{x})d\Sigma$. Thence, for the principle of virtual work:

in a Galilean frame,

$\forall \hat{\underline{U}}$ v.m. of S_2',

$$\int_{\Omega_2'} \rho(\underline{x},t)\,\underline{F}(\underline{x},t).\hat{\underline{U}}(\underline{x})\,d\Omega + \int_{\partial\Omega_2'} \underline{T}_{\Omega_2'}(\underline{x},t).\hat{\underline{U}}(\underline{x})\,da \qquad [6.50]$$

$$- \int_{\Sigma_{\hat{U}}} \hat{\underline{U}}_2(\underline{x}).\underline{\underline{\sigma}}_1(\underline{x},t).\underline{n}(\underline{x})\,d\Sigma - \int_{\Omega_2'} \underline{\underline{\sigma}}(\underline{x},t):\operatorname{grad}\hat{\underline{U}}(\underline{x})\,d\Omega$$

$$= \int_{\Omega_2'} \rho(\underline{x},t)\,\underline{a}(\underline{x},t).\hat{\underline{U}}(\underline{x})\,d\Omega.$$

Summing up [6.49] and [6.50], the integrals on Ω_1' and Ω_2', $\partial\Omega_1'$ and $\partial\Omega_2'$ result in integrals on the whole system that are precisely $\mathcal{P}_e'(\hat{\underline{U}})$, change the expression into: $-\int_{\Omega'} \underline{\underline{\sigma}}(\underline{x},t):\operatorname{grad}\hat{\underline{U}}(\underline{x})\,d\Omega$ and $\mathcal{A}'(\hat{\underline{U}})$ while the two integrals on $\Sigma_{\hat{U}}$ result in

$$\int_{\Sigma_{\hat{U}}} (\hat{\underline{U}}_1(\underline{x}).\underline{\underline{\sigma}}_2(\underline{x},t) - \hat{\underline{U}}_2(\underline{x}).\underline{\underline{\sigma}}_1(\underline{x},t)).\underline{n}(\underline{x})\,d\Sigma \qquad [6.51]$$

If $\Sigma_{\hat{U}}$ is *not a shock wave*, we know from [6.47] that *the stress vector* $\underline{\underline{\sigma}}(\underline{x},t).\underline{n}(\underline{x})$ is continuous, thence

$$\int_{\Sigma_{\hat{U}}} (\hat{U}_1(\underline{x}).\underline{\underline{\sigma}}_2(\underline{x},t) - \hat{U}_2(\underline{x}).\underline{\underline{\sigma}}_1(\underline{x},t)).\underline{n}(\underline{x})\,d\Sigma$$

$$= -\int_{\Sigma_{\hat{U}}} [\![\hat{U}(\underline{x})]\!].\underline{\underline{\sigma}}(\underline{x},t).\underline{n}(\underline{x})\,d\Sigma. \qquad [6.52]$$

As a result, summing up [6.49] and [6.50] finally yields

$$\begin{cases} \text{in a Galilean frame,} \\ \forall \hat{\underline{U}} \text{ v.m. of } S', \\ \int_{\Omega'} \rho(\underline{x},t)\,\underline{F}(\underline{x},t).\hat{\underline{U}}(\underline{x})\,d\Omega + \int_{\partial\Omega'} \underline{T}_{\Omega'}(\underline{x},t).\hat{\underline{U}}(\underline{x})\,da \\ -\int_{\Omega'} \underline{\underline{\sigma}}(\underline{x},t):\text{grad}\,\hat{\underline{U}}(\underline{x})\,d\Omega - \int_{\Sigma_{\hat{U}}} [\![\hat{\underline{U}}(\underline{x})]\!].\underline{\underline{\sigma}}(\underline{x},t).\underline{n}(\underline{x})\,d\Sigma \\ = \int_{\Omega} \rho(\underline{x},t)\,\underline{a}(\underline{x},t).\hat{\underline{U}}(\underline{x})\,d\Omega \end{cases} \qquad [6.53]$$

where the first line of the equation is just $\mathcal{P}'_e(\hat{\underline{U}})$ and the last one $\mathcal{A}'(\hat{\underline{U}})$.

This equation provides the expression of the principle of virtual work for the subsystem when $\hat{\underline{U}}$ is piecewise continuous and continuously differentiable, where the complete expression of $\mathcal{P}'_i(\hat{\underline{U}})$ is

$$\mathcal{P}'_i(\hat{\underline{U}}) = \int_{\Omega'} -\underline{\underline{\sigma}}(\underline{x},t):\text{grad}\,\hat{\underline{U}}(\underline{x})\,d\Omega - \int_{\Sigma_{\hat{U}}} [\![\hat{\underline{U}}(\underline{x})]\!].\underline{\underline{\sigma}}(\underline{x},t).\underline{n}(\underline{x})\,d\Sigma. \qquad [6.54]$$

Thus, virtual velocity jumps contribute to the virtual rate of work by internal forces through the surface density $-[\![\hat{\underline{U}}(\underline{x})]\!].\underline{\underline{\sigma}}(\underline{x},t).\underline{n}(\underline{x})$ on the velocity jump surface $\Sigma_{\hat{U}}$ (provided it does not coincide with a shock wave). This density may also be written as

$$-[\![\hat{\underline{U}}(\underline{x})]\!].\underline{\underline{\sigma}}(\underline{x},t).\underline{n}(\underline{x}) = -\underline{\underline{\sigma}}(\underline{x},t):[\![\hat{\underline{U}}(\underline{x})]\!] \otimes \underline{n}(\underline{x}), \qquad [6.55]$$

a result that can be interpreted within the mathematical framework of the *Distribution theory* where the gradient field in the case of a piecewise continuous and continuously differentiable vector field $\hat{\underline{U}}$ must be understood as the regular gradient field $\text{grad}\,\hat{\underline{U}}(\underline{x})$ to be integrated over Ω' as a volume density and the tensor field $[[\hat{\underline{U}}(\underline{x})]] \otimes \underline{n}(\underline{x})$ over the jump surface to be integrated as a surface density (see Appendix 2, section A2.5.2).

6.5. The stress vector approach

6.5.1. *The stress vector*

The term *stress vector* was incidentally introduced in section 6.4.2 to name the density of contact actions on a surface element da with outward normal $\underline{n}$ at the boundary $\partial\Omega'$ of a subsystem S'. Referring to [6.36] and Figure 6.4, the stress vector is

$$\underline{T}(\underline{x},\underline{n},t) = \underline{\underline{\sigma}}(\underline{x},t).\underline{n}. \tag{6.56}$$

Thus, at a given point $\underline{x}$ inside Ω', the Cauchy stress tensor $\underline{\underline{\sigma}}(\underline{x},t)$ defines the *linear mapping* that determines the stress vector acting on the surface element da with outward normal $\underline{n}$.

Such a surface element will be called the *facet with normal $\underline{n}$* at point $\underline{x}$.

Through [6.56], the stress vector can also be defined at the boundary $\partial\Omega$ of Ω and the boundary condition [6.31] may be written as a condition on the stress vector acting on the boundary facet:

$$\forall \underline{x} \in \partial\Omega, \ \underline{T}(\underline{x},t,\underline{n}) = \underline{T}_\Omega(\underline{x},t). \tag{6.57}$$

6.5.2. *The stress vector as the historical fundamental concept*

6.5.2.1. *Physical approach of the stress vector*

The concept of stress tensor was introduced by Cauchy in the 1820s (although the term *tensor* would be introduced later on by Hamilton) from the intuitive idea of the stress vector as reported in the introduction of Love's *Treatise on the Mathematical Theory of Elasticity*[4]:

---

4 [CAU 29, LOV 44].

"By the autumn of 1822 Cauchy had discovered most of the elements of the pure theory of elasticity. He had introduced the notion of stress at a point determined by the tractions per unit of area across all facets through the point..."

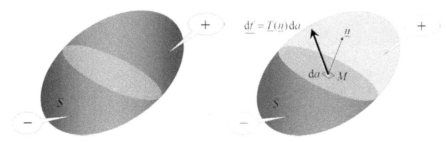

Figure 6.6. *Internal contact actions modeled by the stress vector*

It seems worth outlining here the way forces can be modeled on the basis of the stress vector following the spirit of Cauchy's original discovery. We take Figure 6.6 as a starting point. Considering a system S, we imagine, in a thought experiment, that it is divided into two subsystems denoted by (+) and (−), and that the (+) subsystem vanishes and is substituted by its actions on the (−) one. The fundamental intuition is that these actions are just *contact actions*. Moreover, it is assumed that these contact actions can be reduced, at each point of the contact surface, to an *infinitesimal concentrated force* $\underline{df}$ *proportional* to the area da of the infinitesimal surface element $\underline{df} = \underline{T}\,da$. As a rule, $\underline{T}$, the stress vector acting on da at point M, depends on the orientation of the surface element characterized by its outward normal. Thus, $\underline{df}$ will be written as $\underline{df} = \underline{T}(\underline{n})\,da$.

Obviously, the roles played by the two subsystems could be interchanged and, as a result of the law of mutual actions, we may write:

$$\underline{T}(\underline{n})\,da = -\underline{T}(-\underline{n})\,da. \tag{6.58}$$

The cornerstone of the stress vector approach, which is sometimes considered as "more physical" than the virtual work method, is then to establish the force model for the continuum by implementing the *fundamental law of dynamics* (Chapter 3, sections 3.3.3 and 3.3.6) on *infinitesimal subsystems*. Such a subsystem is submitted to the external forces exerted by the world external to the system and the contact forces exerted on its boundary.

6.5.2.2. *The stress vector depends linearly on $\underline{n}$ through the stress tensor*

Considering three mutually orthogonal directions at point M, we define the components of the stress vectors acting on the surface elements with outward normal $\underline{e}_1$, $\underline{e}_2$, $\underline{e}_3$ as follows:

$$\begin{cases} \underline{T}(\underline{e}_1) = \sigma_{11}\underline{e}_1 + \sigma_{21}\underline{e}_2 + \sigma_{31}\underline{e}_3 \\ \underline{T}(\underline{e}_2) = \sigma_{12}\underline{e}_1 + \sigma_{22}\underline{e}_2 + \sigma_{32}\underline{e}_3 \\ \underline{T}(\underline{e}_3) = \sigma_{13}\underline{e}_1 + \sigma_{23}\underline{e}_2 + \sigma_{33}\underline{e}_3. \end{cases}$$ [6.59]

$$\underline{T}(\underline{e}_j) = \sigma_{ij}\underline{e}_i.$$ [6.60]

The first result to be established is that $\underline{T}(\underline{n})$ depends linearly on $\underline{n}$. For this purpose, the infinitely small subsystem to be considered is a tetrahedron built at point M on the mutually orthogonal directions chosen here above, with lengths dx_1, dx_2, dx_3 along $\underline{e}_1$, $\underline{e}_2$, $\underline{e}_3$ (Figure 6.7).

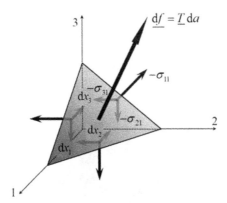

Figure 6.7. *Small tetrahedron argument*

Recalling the fundamental law of dynamics,

$$\begin{cases} \text{in a Galilean frame,} \\ \text{for } S', \\ [\mathcal{F}'_e] = [\mathcal{M}a'] \end{cases}$$ [6.61]

we write it down for the resultant of external forces and quantity of acceleration to the first significant order in dx_i, which turns out to be the second order. To this order, the body force and quantity of acceleration of the subsystem can be neglected (third order). Thence, the overall resultant of surface forces on the boundary of the tetrahedron (second order) must be equal to zero:

$$\underline{T}(\underline{n}) - \underline{T}(\underline{e}_j)\underline{e}_j \cdot \underline{n} = 0. \tag{6.62}$$

This result proves that $\underline{T}(\underline{n})$ depends linearly on $\underline{n}$ and that $\underline{\underline{\sigma}} = \sigma_{ij}\underline{e}_i \otimes \underline{e}_j$ is actually a tensor:

$$\underline{T}(\underline{n}) = \sigma_{ij}\underline{e}_i \otimes \underline{e}_j \cdot \underline{n} = \underline{\underline{\sigma}}.\underline{n}. \tag{6.63}$$

6.5.2.3. *The stress tensor is symmetric*

The next steps will be to prove the symmetry of the tensor $\underline{\underline{\sigma}} = \sigma_{ij}\underline{e}_i \otimes \underline{e}_j \cdot \underline{n}$ and to establish the equation of motion for the tensor field $\underline{\underline{\sigma}} = \sigma_{ij}\underline{e}_i \otimes \underline{e}_j$. The convenient infinitely small subsystem for the implementation of the fundamental law of dynamics is a parallelepiped with lengths dx_1, dx_2, dx_3 along $\underline{e}_1$, $\underline{e}_2$, $\underline{e}_3$.

We first examine the implication of [6.61] as regards the resultant moment of external forces and quantity of acceleration (Figure 6.8).

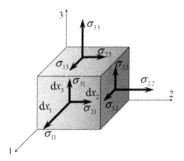

Figure 6.8. *Small parallelepiped argument for the symmetry proof*

Considering the moments about the axes, the first significant order is the third order in dx_i. To this order the moments of the body force and quantity of acceleration shall be neglected as they are of the fourth order; thence, only the moment of surface forces on the boundary of the element shall be considered. As an example, for the moment about $\underline{e}_1$, the only contributions come from the moments of $\sigma_{32}\,\underline{e}_3\,dx_1\,dx_3$ with the lever arm $\underline{e}_2\,dx_2$ and $\sigma_{23}\,\underline{e}_2\,dx_1\,dx_2$ with the lever arm $\underline{e}_3\,dx_3$ and we obtain the symmetry relation

$$(\sigma_{23}-\sigma_{32})\,dx_1\,dx_2\,dx_3 = 0. \qquad [6.64]$$

6.5.2.4. Field equation of motion

The final step is to derive the field equation of motion. Writing down the implication of [6.61] regarding the overall resultant of forces and quantity of acceleration, we find that, as a consequence of [6.58], the first significant order is the third order in dx_i. To this order we take into account the body force, quantity of acceleration and difference between surface forces exerted on parallel faces with opposite outward normals. As an example, along $\underline{e}_1$ (Figure 6.9), we obtain

$$(\frac{\partial\sigma_{11}}{\partial x_1}+\frac{\partial\sigma_{12}}{\partial x_2}+\frac{\partial\sigma_{13}}{\partial x_3})\,dx_1\,dx_2\,dx_3 + \rho\,F_1\,dx_1\,dx_2\,dx_3 = \rho\,a_1\,dx_1\,dx_2\,dx_3 \qquad [6.65]$$

that yields the component along $\underline{e}_1$ of the equation of motion [6.29].

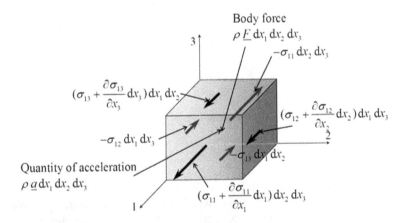

Figure 6.9. *Small parallelepiped argument for the equation of motion*

6.5.2.5. Continuity of the stress vector

In establishing the field equation of motion here above, the Cauchy stress field is implicitly assumed continuous and continuously differentiable. However, the stress vector approach which we have been implementing may also account for piecewise continuous and continuously differentiable stress fields.

Let Σ_σ be a jump surface for the Cauchy stress field transversally oriented as in Figure 6.10 (Notations are the same as in section 6.4.2.). Splitting this surface into two parallel surfaces infinitely close to each other, we consider the infinitely flat small parallelepiped volume bounded by these two surfaces. In the absence of surface density of body force or quantity of acceleration (shock wave) within this volume, equilibrium implies, to the order zero in λ, that the forces exerted on both faces must balance

$$\underline{df}_2 - \underline{df}_1 = (\underline{\underline{\sigma}}_2 - \underline{\underline{\sigma}}_1).\underline{n}\,da = [\![\underline{\underline{\sigma}}]\!].\underline{n}\,da = 0 \qquad [6.66]$$

which is the expression of the continuity of the stress vector in the absence of surface density of body force or shock wave as in [6.47].

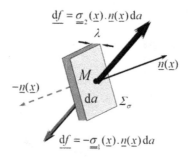

Figure 6.10. *Discontinuity of the stress field*

Comments

Besides its great historical importance, the stress vector approach presents a high pedagogical interest. This intuitive and heuristic reasoning is a quite valuable tool when trying to make the concept of stress tensor field understandable at a first reading. Its presentation is almost unescapable in any textbook. Also, the same guiding principle, that is to implement the fundamental law of dynamics on infinitesimal subsystems after having introduced a contact model for the internal forces, can be adopted when building up force models for other continuum geometrical descriptions such as the one-dimensional continuum (beams, etc.) or the two-dimensional continuum (plates, etc.)[5].

---

5 e.g. [SAL 13].

When such an approach is adopted, the principle of virtual work is not a principle any longer but a theorem that comes out, we may say, rather unexpectedly since the kinematics and the force model are developed independently from each other. For this reason, the virtual work method may intellectually appear more satisfactory. Also, it plays the role of a checking tool that confirms the consistency of a heuristic model. This is especially true when trying to build up more sophisticated force models for the three-dimensional continuum, such as micropolar media, where producing a consistent model on the only basis of intuition may prove difficult.

6.6. Local analysis

6.6.1. Components of the stress vector and stress tensor

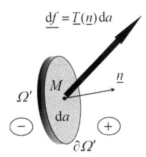

Figure 6.11. *Stress vector acting on a facet*

Let da be an infinitesimal surface element transversally oriented by its outward normal $\underline{n}$, thus defining a *facet*, and $\underline{T}(\underline{n})$ the stress vector acting on that facet[6]. The fundamental equation relating $\underline{T}(\underline{n})$ with the stress tensor (Figure 6.11)

$$\underline{df} = \underline{T}(\underline{n})\,da = \underline{\underline{\sigma}}:\underline{n}\,da \qquad [6.67]$$

is written explicitly in an orthonormal basis as

$$T_i\,\underline{e}_i = \sigma_{ij}(\underline{e}_i \otimes \underline{e}_j):n_k\underline{e}_k = \underline{e}_i\,\sigma_{ij}\,n_j. \qquad [6.68]$$

---

6 As only properties of the various functions at a given point and given instant of time will be considered, the notation will be simplified by not mentioning the space and time variables.

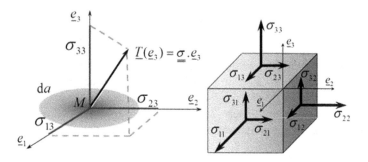

Figure 6.12. *Components of the stress tensor in an orthonormal basis*

In plain words, this equation states the practical interpretation of the components of the stress tensor: *the component* σ_{ij} *of the stress tensor is the component in the* $\underline{e}_i$ *direction of the stress vector on the facet with normal* $\underline{e}_j$ (Figure 6.12).

As far as physical units are concerned, it follows from [6.67] that the components of $\underline{\underline{\sigma}}$ have the dimensions of a force per unit area, for which the units are usually the pascal Pa and its multiples kPa, MPa and GPa.

6.6.2. *Normal stress, shear or tangential stress*

It is easily understood that the components of the stress vector on a given facet that are important from a physical point of view are the components in an orthonormal basis related to the facet itself. Such a basis will be denoted by $(\underline{t}_1, \underline{t}_2, \underline{n})$ where $\underline{t}_1$ and $\underline{t}_2$ are two orthogonal unit vectors in the plane of the facet (Figure 6.13). The stress vector is thus written as

$$\underline{T}(\underline{n}) = \sigma\,\underline{n} + \sigma_{t_1}\underline{t}_1 + \sigma_{t_2}\underline{t}_2 = \sigma\,\underline{n} + \underline{\tau}. \qquad [6.69]$$

In this equation, σ is the *normal stress*. Putting [6.67] and [6.69] together, we have $\underline{df} = (\sigma\,\underline{n} + \underline{\tau})\,da$ that shows that the normal stress is positive when the normal component of $\underline{df}$ is a tension. A positive normal stress is said to be *tensile*. A negative normal stress is *compressive*.

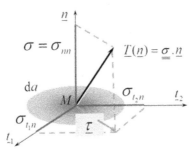

Figure 6.13. *Normal stress and tangential components*

The components $\underline{\tau}_1 = \sigma_{t_1} \underline{t}_1$ and $\underline{\tau}_2 = \sigma_{t_2} \underline{t}_2$ are the *tangential components* of the stress vector. They define the *shear stress* $\underline{\tau} = \sigma_{t_1} \underline{t}_1 + \sigma_{t_2} \underline{t}_2$ also called *tangential stress*. The shear stress is often only characterized by its magnitude τ :

$$\tau = |\underline{\tau}| = (\sigma_{t_1}^2 + \sigma_{t_2}^2)^{\frac{1}{2}}. \tag{6.70}$$

Referring to [6.67] or Figure 6.12, we see that the normal and tangential components of the stress vector are easily related to the components of the stress tensor in the orthonormal basis $(\underline{t}_1, \underline{t}_2, \underline{n})$:

$$\begin{cases} \sigma = \underline{n}.\underline{\underline{\sigma}}.\underline{n} = \sigma_{nn} \\ \tau_1 = \underline{t}_1.\underline{\underline{\sigma}}.\underline{n} = \sigma_{t_1 n} \\ \tau_2 = \underline{t}_2.\underline{\underline{\sigma}}.\underline{n} = \sigma_{t_2 n} \end{cases} \tag{6.71}$$

thence

$$\begin{cases} \sigma = \sigma_{nn} \\ \tau = ((\underline{\underline{\sigma}}.\underline{n})^2 - (\underline{n}.\underline{\underline{\sigma}}.\underline{n})^2)^{\frac{1}{2}}. \end{cases} \tag{6.72}$$

6.6.3. *Principal axes of the stress tensor. Principal stresses*

Being a second rank symmetric tensor, the stress tensor $\underline{\underline{\sigma}}$ has three real principal values and admits three orthogonal principal axes (Appendix 1, section A1.4.3). Let $\underline{u}_1, \underline{u}_2, \underline{u}_3$ be a triad of unit vectors along the principal axes and

$\sigma_1, \sigma_2, \sigma_3$ the corresponding principal values, also called *principal stresses*. The stress tensor $\underline{\underline{\sigma}}$ can then be written in the "diagonal" form

$$\underline{\underline{\sigma}} = \sigma_1 \underline{u}_1 \otimes \underline{u}_1 + \sigma_2 \underline{u}_2 \otimes \underline{u}_2 + \sigma_3 \underline{u}_3 \otimes \underline{u}_3 \qquad [6.73]$$

and, for a facet with normal $\underline{u}_i$, the stress vector is just

$$\underline{T}(\underline{u}_i) = \sigma_i \underline{u}_i, \qquad [6.74]$$

which is purely normal to the facet (Figure 6.14). This is the characterizing property of the principal axes of $\underline{\underline{\sigma}}$: *on a facet oriented perpendicularly to a principal axis of $\underline{\underline{\sigma}}$, the stress vector is purely normal (no shear stress on that facet) and its magnitude is equal to the corresponding principal stress.*

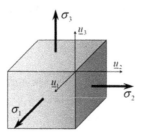

Figure 6.14. *Stress vectors on facets perpendicular to the principal axes*

Equation [6.73] shows that the stress tensor is completely defined once its principal axes, defined by three orientation parameters, and the three corresponding principal stresses are known. Here, we may again quote Love's introduction to his *Treatise on the Mathematical Theory of Elasticity*:

> *"For this purpose he [Cauchy] had generalized the notion of hydrostatic pressure, and he had shown that the stress is expressible by means of six component stresses, and also by means of three purely normal tractions across a certain triad of planes which cut each other at right angles – the principal planes of stress".*

6.6.4. *Isotropic Cauchy stress tensor*

As pointed out by Love in that paragraph, the Cauchy stress tensor concept comes out as a generalization of the notion of hydrostatic pressure. More precisely, if the three principal values of $\underline{\underline{\sigma}}$ are equal to each other, the stress tensor is isotropic

$$\sigma_1 = \sigma_2 = \sigma_3 = -p \;\Rightarrow\; \underline{\underline{\sigma}} = -p\underline{\underline{1}} \tag{6.75}$$

and the stress vector on any facet in Figure 6.12 is always normal to the facet:

$$\forall \underline{n}, \quad \underline{T}(\underline{n}) = -p\,\underline{n}, \tag{6.76}$$

which corresponds exactly to the physical notion of *hydrostatic pressure* (Figure 6.15).

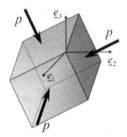

Figure 6.15. *Stress vectors in the case of an isotropic stress tensor*

6.7. The hydrostatic pressure force modeling

When the Cauchy tensor field is assumed to be isotropic all over the system

$$\forall \underline{x} \in \Omega, \quad \underline{\underline{\sigma}}(\underline{x},t) = -p(\underline{x},t)\underline{\underline{1}} \tag{6.77}$$

the force modeling reduces to the hydrostatic pressure model where the internal forces depend only on the scalar field p. Referring to section 6.2.4, the model is built up on the same vector space of virtual motions, with the same assumptions regarding the virtual rates of work, but the volume density of virtual rate of work by internal forces is given the restrictive expression

$$p_i(\hat{\underline{U}}) = p(\underline{x},t)\,\mathrm{div}\,\hat{\underline{U}}(\underline{x}), \tag{6.78}$$

which depends on the *scalar field* $p(\underline{x},t)$.

The equations of motion are obtained in the same way as before:

$$\begin{cases} \text{in a Galilean frame, } \forall \underline{x} \in \Omega, \\ -\operatorname{grad} \underline{p(\underline{x},t)} + \rho(\underline{x},t)\big(\underline{F}(\underline{x},t) - \underline{a}(\underline{x},t)\big) = 0. \end{cases}$$ [6.79]

$$\forall \underline{x} \in \partial\Omega, \ -p(\underline{x},t)\underline{n}(\underline{x}) = \underline{T}_\Omega(\underline{x},t)$$ [6.80]

$$\forall \underline{x} \in \partial\Omega', \ -p(\underline{x},t)\underline{n}(\underline{x}) = \underline{T}_{\Omega'}(\underline{x},t).$$ [6.81]

This model is convenient for inviscid fluids. This gives us a good opportunity to underscore the importance of the condition imposed on virtual motions that they should not comply with any geometrical restriction. It is quite frequently assumed that the actual behavior of some fluids, for which the model is relevant, is isochoric: $\operatorname{div}\underline{U}(\underline{x},t) = 0$. If this condition was imposed on the virtual motions, the virtual work method could not be implemented!

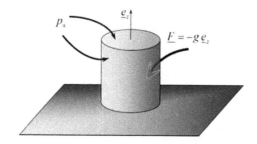

Figure 6.16. *Equilibrium of a solid under gravity*

As an example to illustrate the shortcomings of the hydrostatic pressure model and the need for a more elaborated one, let us consider the equilibrium of a cylindrically-shaped solid standing on its base on a horizontal plane surface and subject only to the effects of its weight, atmospheric pressure and the reaction force from the surface supposed to be perfectly smooth (Figure 6.16). From the field equations of motion [6.79] and the boundary condition [6.80] written on the upper surface of the solid, the pressure field in the whole solid is fully determined

$$p(\underline{x},t) = p_{\mathrm{a}} - \rho g z,$$ [6.82]

which does not comply with the boundary condition on the cylindrical surface!

As a matter of fact, this result could have been anticipated by investigating the *consistency* of the model. The partial derivative system [6.79] for the scalar field $p(\underline{x},t)$ requires, as a consequence of the *Curl theorem*, that the data $\underline{F}(\underline{x},t)$ and $\underline{a}(\underline{x},t)$ satisfy the necessary condition

$$\underline{\text{curl}}\left(\rho(\underline{F}-\underline{a})\right)=0. \tag{6.83}$$

Moreover, the boundary data must be compatible with [6.80] which implies that they are of the form

$$\underline{T}_{\Omega}(\underline{x},t)=T_{\Omega}(\underline{x},t)\,\underline{n}(\underline{x}); \tag{6.84}$$

they must also be compatible with the field equation, which implies that, for any points P and P' on $\partial\Omega$ where $T_{\Omega}(\underline{x},t)$ is specified, we have

$$T_{\Omega}(P')-T_{\Omega}(P)=\int_{\widehat{PP'}}\rho(\underline{F}-\underline{a}).\underline{dx} \tag{6.85}$$

where $\widehat{PP'}$ is any arc joining P and P' in Ω or on $\partial\Omega$. (In Figure 6.16, this latter condition is obviously not satisfied on the cylindrical boundary.)

This consistency analysis shows how the internal forces appear to be somehow "cramped" by the hydrostatic pressure model that depends on a scalar field only. To provide the necessary "freedom", the Cauchy stress field model must jump from a *one scalar parameter* field to a *six independent scalar parameter* model: a symmetric tensor field!

6.8. Validation and implementation

6.8.1. *Relevance of the model*

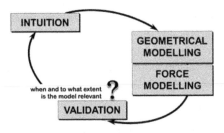

Figure 6.17. *The complete "virtuous circle"*

Whatever the method chosen for building it up, the classical force modeling for the three-dimensional continuum relies on the physical intuition/assumption that internal forces are the result of contact actions between particles that are infinitely close to each other, and can be modeled by a concentrated force per unit area.

Thence, in the last step of the virtuous circle (Figure 6.17), the question to be answered comes out as "in what circumstances, for what applications and to what extent can this physical assumption be considered valid?"

Almost two centuries after Cauchy introduced the pivotal concept of *stress vector* and what was to be named afterwards the *stress tensor*, we may say that the domain of validity of that model is amazingly wide, provided *problems* are considered at the relevant *scale*. In this sentence, the two words "problems" and "scale" are most important. They both refer to the application under concern, from the available data to the expected results and insist that no general answer to the relevance question should be expected. In that respect, not surprisingly, we retrieve the same conditions as in Chapter 4, section 4.4.2 about the relevance of the three-dimensional continuum geometrical modeling and we will not repeat the list already given of such relevant problems at the right scale. In addition, because it sounds quite enlightening, let us mention that the same concept of stress vector is used in everyday problems of fluid mechanics (viscous or inviscid fluid), metal or polymer forming, civil engineering and geotectonics where stresses are considered at the fault level in earthquake analysis.

6.8.2. Implementation

As a final result of the modeling process, the field of internal forces for the three-dimensional continuum is defined by *six* scalar fields, the components of the Cauchy stress field, which are governed, at each instant of time, by *three* linear partial differential equations and boundary conditions written in the actual configuration $\Omega, \partial\Omega$. In order that this mathematical problem be compatible, the data in the form of body forces and quantities of acceleration in Ω, surface forces on $\partial\Omega$, must comply with compatibility conditions recalled in section 6.3.5. Then, for a given set of data, solving the problem does not yield a unique solution. Actually, solutions to the problem generate an infinite dimensional affine space whose associated vector space is generated by the self-equilibrating stress fields (section 6.10.5).

This conclusion may first appear somewhat frustrating from a physical viewpoint but is actually quite satisfactory. Equations are missing to make the complete determination of the internal force field possible; they describe the *specific mechanical response of the constituent material* of the system under concern. They are called the *constitutive equations*, obtained from experimental results and formulated within the framework of theories such as *Thermoelasticity, Viscoelasticity, Elastoplasticity*, etc.

A particular case, historically important [SAL 13], is the theory of yield design where the mechanical response of the constituent material is limited to defining the internal stress states that can be sustained at any point within Ω. Through a direct implementation of the virtual work approach, it is possible to derive valuable information about the bearing capacity of the whole studied system, in the form of upper and lower bounds, without determining the internal force field completely. Simple examples are given in sections 6.10.6–6.10.10.

6.9. Explicit formulas for the equation of motion in standard coordinate systems

Although intrinsic notations for tensors and differential operators prove very convenient in simplifying formulas and equations that can be better understood at a glance, practical applications do often require the use of the explicit expressions that are given here under in standard coordinate systems.

6.9.1. Orthonormal Cartesian coordinates

$$
\begin{cases}
\dfrac{\partial \sigma_{xx}}{\partial x} + \dfrac{\partial \sigma_{xy}}{\partial y} + \dfrac{\partial \sigma_{xz}}{\partial z} + \rho(F_x - a_x) = 0. \\[2mm]
\dfrac{\partial \sigma_{yx}}{\partial x} + \dfrac{\partial \sigma_{yy}}{\partial y} + \dfrac{\partial \sigma_{yz}}{\partial z} + \rho(F_y - a_y) = 0 \\[2mm]
\dfrac{\partial \sigma_{zx}}{\partial x} + \dfrac{\partial \sigma_{zy}}{\partial y} + \dfrac{\partial \sigma_{zz}}{\partial z} + \rho(F_z - a_z) = 0
\end{cases}
$$

6.9.2. Cylindrical coordinates

$$
\begin{cases}
\dfrac{\partial \sigma_{rr}}{\partial r} + \dfrac{1}{r}\dfrac{\partial \sigma_{r\theta}}{\partial \theta} + \dfrac{\partial \sigma_{rz}}{\partial z} + \dfrac{\sigma_{rr} - \sigma_{\theta\theta}}{r} + \rho(F_r - a_r) = 0 \\[2mm]
\dfrac{\partial \sigma_{\theta r}}{\partial r} + \dfrac{1}{r}\dfrac{\partial \sigma_{\theta\theta}}{\partial \theta} + \dfrac{\partial \sigma_{yz}}{\partial z} + 2\dfrac{\sigma_{r\theta}}{r} + \rho(F_\theta - a_\theta) = 0 \\[2mm]
\dfrac{\partial \sigma_{zr}}{\partial r} + \dfrac{1}{r}\dfrac{\partial \sigma_{z\theta}}{\partial \theta} + \dfrac{\partial \sigma_{zz}}{\partial z} + \dfrac{\sigma_{rz}}{r} + \rho(F_z - a_z) = 0.
\end{cases}
$$

6.9.3. *Spherical coordinates*

$$\begin{cases} \dfrac{\partial \sigma_{rr}}{\partial r} + \dfrac{1}{r}\dfrac{\partial \sigma_{r\theta}}{\partial \theta} + \dfrac{1}{r\sin\theta}\dfrac{\partial \sigma_{r\varphi}}{\partial \varphi} + \dfrac{1}{r}(2\sigma_{rr} - \sigma_{\theta\theta} - \sigma_{\varphi\varphi} + \sigma_{r\theta}\cot\theta) + \rho(F_r - a_r) = 0 \\[2mm] \dfrac{\partial \sigma_{\theta r}}{\partial r} + \dfrac{1}{r}\dfrac{\partial \sigma_{\theta\theta}}{\partial \theta} + \dfrac{1}{r\sin\theta}\dfrac{\partial \sigma_{\theta\varphi}}{\partial \varphi} + \dfrac{1}{r}((\sigma_{\theta\theta} - \sigma_{\varphi\varphi})\cot\theta + 3\sigma_{r\theta}) + \rho(F_\theta - a_\theta) = 0 \\[2mm] \dfrac{\partial \sigma_{\varphi r}}{\partial r} + \dfrac{1}{r}\dfrac{\partial \sigma_{\varphi\theta}}{\partial \theta} + \dfrac{1}{r\sin\theta}\dfrac{\partial \sigma_{\varphi\varphi}}{\partial \varphi} + \dfrac{1}{r}(3\sigma_{\varphi r} + 2\sigma_{\varphi\theta}\cot\theta) + \rho(F_\varphi - a_\varphi) = 0. \end{cases}$$

6.10. Practicing

6.10.1. *Spherical and deviatoric parts of the stress tensor*

– It is common practice to split a stress tensor into its "spherical" and "deviatoric" parts, $\sigma_m \underline{\underline{1}}$ and $\underline{\underline{s}}$, the trace of the latter being equal to zero: $\underline{\underline{\sigma}} = \sigma_m \underline{\underline{1}} + \underline{\underline{s}}$, $s_{ii} = 0$. Prove that this decomposition can be made in one and only one way. Determine the principal axes and principal values of $\underline{\underline{s}}$.

– The "deviatoric" part $\hat{\underline{\underline{d}}}'$ of the virtual strain rate $\hat{\underline{\underline{d}}}$ being defined through $\hat{\underline{\underline{d}}} = \dfrac{1}{3}(\mathrm{tr}\,\hat{\underline{\underline{d}}})\underline{\underline{1}} + \hat{\underline{\underline{d}}}'$, write down the volume density of virtual rate of work by internal forces for a continuous medium, separating the "deviatoric" and "spherical" contributions.

Solution

Taking the trace of $\underline{\underline{\sigma}} = \sigma_m \underline{\underline{1}} + \underline{\underline{s}}$, $s_{ii} = 0$, we determine σ_m in one and only one way: $\sigma_m = \dfrac{\sigma_{ii}}{3} = \dfrac{\sigma_1 + \sigma_2 + \sigma_3}{3}$. σ_m is called the *mean normal stress*. Thence, we derive the components of $\underline{\underline{s}}$, called the deviator of $\underline{\underline{\sigma}}$:

$$\begin{cases} i = j : s_{ij} = \sigma_{ij} - \sigma_m \\ i \neq j : s_{ij} = \sigma_{ij}. \end{cases}$$

In an orthonormal basis arranged to coincide with the principal axes of $\underline{\underline{\sigma}}$, the deviator is written $\underline{\underline{s}} = \sigma_1 \underline{u}_1 \otimes \underline{u}_1 + \sigma_2 \underline{u}_2 \otimes \underline{u}_2 + \sigma_3 \underline{u}_3 \otimes \underline{u}_3 - \sigma_m \underline{\underline{1}}$, which shows that it has the same principal axes as $\underline{\underline{\sigma}}$ and principal values equal to

$$s_1 = \frac{2\sigma_1 - \sigma_2 - \sigma_3}{3}, \quad s_2 = \frac{2\sigma_2 - \sigma_3 - \sigma_1}{3}, \quad s_3 = \frac{2\sigma_3 - \sigma_1 - \sigma_2}{3}.$$

– Introducing $\hat{\underline{\underline{d}}}'$, the volume density of the virtual rate of work by internal forces becomes $p_i(\hat{\underline{U}}) = -(\sigma_m \underline{\underline{1}} + \underline{\underline{s}}):(\frac{1}{3}(\mathrm{tr}\,\hat{\underline{\underline{d}}})\underline{\underline{1}} + \hat{\underline{\underline{d}}}') = -(\sigma_m \underline{\underline{1}} + \underline{\underline{s}}):(\frac{1}{3}(\mathrm{div}\,\hat{\underline{U}})\underline{\underline{1}} + \hat{\underline{\underline{d}}}')$.

Thence

$$p_i(\hat{\underline{U}}) = -\frac{1}{3}\sigma_m \,\mathrm{div}\,\hat{\underline{U}}\,\underline{\underline{1}}:\underline{\underline{1}} - \sigma_m \underline{\underline{1}}:\hat{\underline{\underline{d}}}' - \frac{1}{3}\underline{\underline{s}}:(\mathrm{div}\,\hat{\underline{U}})\underline{\underline{1}} - \underline{\underline{s}}:\hat{\underline{\underline{d}}}'$$

where $\sigma_m \underline{\underline{1}}:\hat{\underline{\underline{d}}}' = \sigma_m \,\mathrm{tr}\,\hat{\underline{\underline{d}}}' = 0$, $\underline{\underline{s}}:(\mathrm{div}\,\hat{\underline{U}})\underline{\underline{1}} = (\mathrm{tr}\,\underline{\underline{s}})(\mathrm{div}\,\hat{\underline{U}}) = 0$

and $\frac{1}{3}\sigma_m \,\mathrm{div}\,\hat{\underline{U}}\,\underline{\underline{1}}:\underline{\underline{1}} = \sigma_m \,\mathrm{div}\,\hat{\underline{U}}$.

Finally, $p_i(\hat{\underline{U}}) = -\sigma_m \,\mathrm{div}\,\hat{\underline{U}} - \underline{\underline{s}}:\hat{\underline{\underline{d}}}'$ where the "spherical" contribution is $-\sigma_m \,\mathrm{div}\,\hat{\underline{U}}$, similar to [6.78], and the deviatoric contribution $-\underline{\underline{s}}:\hat{\underline{\underline{d}}}'$.

6.10.2. *Extremal values of the normal stress*

Show that if the principal stresses of the Cauchy stress tensor $\underline{\underline{\sigma}}$ are numbered and ordered according to $\sigma_I \geq \sigma_{II} \geq \sigma_{III}$, the maximal and minimal values of the normal stress acting on an arbitrary facet are respectively σ_I and σ_{III}.

Solution

Let $(\underline{u}_I, \underline{u}_{II}, \underline{u}_{III})$ be an orthonormal basis along the principal axes of the stress tensor numbered and ordered according to $\sigma_I \geq \sigma_{II} \geq \sigma_{III}$. For an arbitrary facet with outward normal $\underline{n}$, the normal stress is written as $\sigma = \underline{n}.\underline{\underline{\sigma}}.\underline{n} = \sigma_I n_I^2 + \sigma_{II} n_{II}^2 + \sigma_{III} n_{III}^2$ where $\underline{n} = n_I \underline{u}_I + n_{II} \underline{u}_{II} + n_{III} \underline{u}_{III}$.

As $n_{\mathrm{I}}^2 + n_{\mathrm{II}}^2 + n_{\mathrm{III}}^2 = 1$, we conclude straightforwardly:

$$\mathrm{Max}\{\sigma = \underline{n}.\underline{\underline{\sigma}}.\underline{n}, \forall \underline{n}\} = \sigma_{\mathrm{I}}, \text{ which is reached for } n_{\mathrm{I}} = 1, \ n_{\mathrm{II}} = n_{\mathrm{III}} = 0.$$

$$\mathrm{Min}\{\sigma = \underline{n}.\underline{\underline{\sigma}}.\underline{n}.\forall \underline{n}\} = \sigma_{\mathrm{III}}, \text{ which is reached for } n_{\mathrm{I}} = n_{\mathrm{II}} = 0, \ n_{\mathrm{III}} = 1.$$

6.10.3. *Stress vector acting on the "octahedral" facet*

Let $(\underline{u}_1, \underline{u}_2, \underline{u}_3)$ be a triad of unit vectors along the principal axes of the stress tensor and consider a facet with outward unit normal $\underline{n}$ along the symmetry axis of the triad: $\underline{n} = \dfrac{\sqrt{3}}{3}(\underline{u}_1 + \underline{u}_2 + \underline{u}_3)$. Determine the normal stress and the magnitude of the shear stress on this facet which is currently called the "octahedral" facet.

Solution

The stress vector acting on the facet is

$$\underline{T}(\underline{n}) = (\sigma_1 \underline{u}_1 \otimes \underline{u}_1 + \sigma_2 \underline{u}_2 \otimes \underline{u}_2 + \sigma_3 \underline{u}_3 \otimes \underline{u}_3).\dfrac{\sqrt{3}}{3}(\underline{u}_1 + \underline{u}_2 + \underline{u}_3).$$

Thence, the normal stress is

$$\sigma = \underline{T}(\underline{n}).\underline{n} = \sigma_1 n_1^2 + \sigma_2 n_2^2 + \sigma_3 n_3^2 = \dfrac{\sigma_1 + \sigma_2 + \sigma_3}{3} = \sigma_{\mathrm{m}},$$

which is the mean normal stress.

The magnitude of the shear stress is given by $|\underline{\tau}|^2 = \underline{T}(\underline{n}).\underline{T}(\underline{n}) - \sigma^2$ where $\sigma = \sigma_{\mathrm{m}}$ and $\underline{T}(\underline{n}).\underline{T}(\underline{n}) = (\sigma_1^2 + \sigma_2^2 + \sigma_3^2)/3$. Hence:

$$|\underline{\tau}|^2 = (\sigma_1^2 + \sigma_2^2 + \sigma_3^2)/3 - \sigma_{\mathrm{m}}^2 = ((\sigma_1 - \sigma_{\mathrm{m}})^2 + (\sigma_2 - \sigma_{\mathrm{m}})^2 + (\sigma_3 - \sigma_{\mathrm{m}})^2)/3$$

or, referring to section 6.10.1, $|\underline{\tau}|^2 = (s_1^2 + s_2^2 + s_3^2)/3$ with s_1, s_2, s_3 the principal values of the stress deviator ($s_1 + s_2 + s_3 = 0$) and $(s_1^2 + s_2^2 + s_3^2)$ its second invariant (Appendix 1, section A1.5.2). This magnitude is also called the *octahedral shear stress* and its most convenient expression for practical applications is often:

$$|\underline{\tau}|^2 = ((\sigma_1 - \sigma_2)^2 + (\sigma_2 - \sigma_3)^2 + (\sigma_3 - \sigma_1)^2)/3 .$$

6.10.4. *Mohr circles*

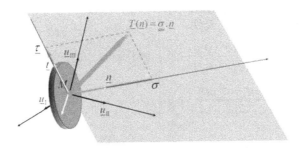

Figure 6.18. *Facet and Mohr's plane*

Consider at point M an arbitrary facet with outward normal $\underline{n}$ and the stress vector acting at point M on this facet: $\underline{T}(\underline{n}) = \sigma\underline{n} + \underline{\tau}$ (Figure 6.18). In the plane, defined by $\underline{n}$ and $\underline{T}(\underline{n})$, let $\underline{t}$ denote the unit vector orthogonal to $\underline{n}$. The plane $(\underline{n},\underline{T}(\underline{n}))$ (often called Mohr's plane) is transversally oriented in such a way that $(\underline{n},\underline{t}) = +\pi/2$. Prove that, as $\underline{n}$ varies, the extremity of the stress vector $\underline{T}(\underline{n}) = \sigma\underline{n} + \tau\underline{t}$, with τ a signed quantity, generates a region in the plane $(\underline{n},\underline{T}(\underline{n}))$ that is bounded by three circles whose diameters are defined from the values of the principal stresses.

Solution

Let $(\underline{u}_{\mathrm{I}},\underline{u}_{\mathrm{II}},\underline{u}_{\mathrm{III}})$ be an orthonormal basis along the principal axes of the stress tensor numbered and ordered according to $\sigma_{\mathrm{I}} \geq \sigma_{\mathrm{II}} \geq \sigma_{\mathrm{III}}$.

With $\underline{n} = n_{\mathrm{I}}\underline{u}_{\mathrm{I}} + n_{\mathrm{II}}\underline{u}_{\mathrm{II}} + n_{\mathrm{III}}\underline{u}_{\mathrm{III}}$, we have

$$n_{\mathrm{I}}^2 + n_{\mathrm{II}}^2 + n_{\mathrm{III}}^2 = 1$$
$$\sigma = \sigma_{\mathrm{I}} n_{\mathrm{I}}^2 + \sigma_{\mathrm{II}} n_{\mathrm{II}}^2 + \sigma_{\mathrm{III}} n_{\mathrm{III}}^2$$
$$\tau^2 + \sigma^2 = \sigma_{\mathrm{I}}^2 n_{\mathrm{I}}^2 + \sigma_{\mathrm{II}}^2 n_{\mathrm{II}}^2 + \sigma_{\mathrm{III}}^2 n_{\mathrm{III}}^2 .$$

These three equations must be compatible with the positivity conditions $n_{\mathrm{I}}^2 \geq 0$, $n_{\mathrm{II}}^2 \geq 0$ and $n_{\mathrm{III}}^2 \geq 0$, which can be written explicitly:

$$
\left.
\begin{aligned}
n_{\mathrm{I}}^2 &= \frac{\tau^2 + (\sigma - \sigma_{\mathrm{II}})(\sigma - \sigma_{\mathrm{III}})}{(\sigma_{\mathrm{I}} - \sigma_{\mathrm{II}})(\sigma_{\mathrm{I}} - \sigma_{\mathrm{III}})} \geq 0 \\
n_{\mathrm{II}}^2 &= \frac{\tau^2 + (\sigma - \sigma_{\mathrm{I}})(\sigma - \sigma_{\mathrm{III}})}{(\sigma_{\mathrm{II}} - \sigma_{\mathrm{I}})(\sigma_{\mathrm{II}} - \sigma_{\mathrm{III}})} \geq 0 \\
n_{\mathrm{III}}^2 &= \frac{\tau^2 + (\sigma - \sigma_{\mathrm{I}})(\sigma - \sigma_{\mathrm{II}})}{(\sigma_{\mathrm{III}} - \sigma_{\mathrm{I}})(\sigma_{\mathrm{III}} - \sigma_{\mathrm{II}})} \geq 0
\end{aligned}
\right\}
\text{ and imply }
\left\{
\begin{aligned}
&\tau^2 + (\sigma - \sigma_{\mathrm{II}})(\sigma - \sigma_{\mathrm{III}}) \geq 0 \\
&\tau^2 + (\sigma - \sigma_{\mathrm{I}})(\sigma - \sigma_{\mathrm{III}}) \leq 0 \\
&\tau^2 + (\sigma - \sigma_{\mathrm{I}})(\sigma - \sigma_{\mathrm{II}}) \geq 0.
\end{aligned}
\right.
$$

These inequalities show that, in the $(\underline{n}, \underline{T}(\underline{n}))$ plane, the end of the stress vector lies in the region delimited by three circles with diameters defined by σ_{I}, σ_{II}, σ_{III} on the $\underline{n}$ axis (Figure 6.19). These circles are usually named Mohr circles[7].

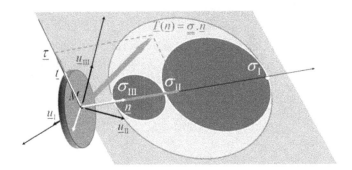

Figure 6.19. *Facet and Mohr circles*

Moreover, we see that each Mohr's circle is described by the end of the stress vector when the facet rotates about a principal axis of the stress tensor. The largest Mohr's circle is described when the facet rotates about the principal axis of the intermediate principal stress σ_{II}. Given its practical importance, for instance, for studying the yield conditions of metal or granular materials, the largest Mohr's circle is often called *the* Mohr circle. More precisely[8], when the facet rotates through a given angle about this axis, the end of the stress vector moves on the Mohr circle through twice that angle around the center in the opposite direction as shown in Figure 6.20.

---

7 C.O. Mohr (1835–1918).
8 See [SAL 01].

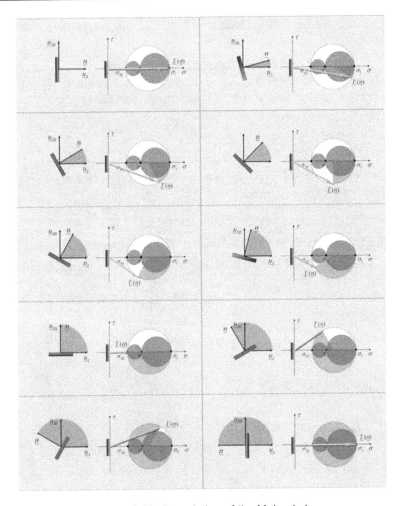

Figure 6.20. *Description of the Mohr circle*

6.10.5. *Self-equilibrating stress fields*

Consider a system S with volume Ω and boundary $\partial\Omega$ that is subject to zero body forces, $\underline{F}(\underline{x},t) = 0$ in Ω, zero surface forces $\underline{T}_\Omega(\underline{x},t) = 0$ on $\partial\Omega$ and where the acceleration field is zero $\underline{a}(\underline{x},t) = 0$. Let $\underline{\underline{\sigma}}(\underline{x},t) = 0$ be any stress field, piecewise continuous and continuously differentiable, satisfying the field and boundary equations of motion for this system with these data.

– Prove that the integral of such a *self-equilibrating* stress field over Ω is always zero.

– Prove also that, apart from the case of the zero stress field, for any self-equilibrating stress field we always find points and facets in Ω where the normal stress is strictly tensile $(\sigma > 0)$ and points and facets where the normal stress is strictly compressive $(\sigma < 0)$.

Solution

The proofs are based upon the use of the principle of virtual work with well-chosen virtual velocity fields.

– First, consider virtual velocity fields defined in Ω by $\hat{\underline{U}}(\underline{x}) = \hat{\underline{\underline{A}}}.\underline{x}$ where $\hat{\underline{\underline{A}}}$ is an arbitrary symmetric second rank tensor. The corresponding virtual strain rate field is constant in Ω: $\hat{\underline{\underline{d}}}(\underline{x}) = \hat{\underline{\underline{A}}}$. It follows that $\mathcal{P}_i(\hat{\underline{U}}) = -\hat{\underline{\underline{A}}} : \int_\Omega \underline{\underline{\sigma}}(\underline{x}) \, d\Omega$. The data imply that $\mathcal{P}_e(\hat{\underline{U}})$ and $\mathcal{A}(\hat{\underline{U}})$ are zero whatever $\hat{\underline{U}}$. Thence, from the principle of virtual work, we derive:

$$\forall \hat{\underline{\underline{A}}} \text{ symmetric, } \mathcal{P}_i(\hat{\underline{U}}) = -\hat{\underline{\underline{A}}} : \int_\Omega \underline{\underline{\sigma}}(\underline{x},t) \, d\Omega = 0.$$

Recalling that $\underline{\underline{\sigma}}(\underline{x},t)$ is symmetric, this implies

$$\int_\Omega \underline{\underline{\sigma}}(\underline{x},t) \, d\Omega = 0, \text{ i.e. } \forall i,j, \int_\Omega \sigma_{ij}(\underline{x},t) \, d\Omega = 0:$$

any component of the stress tensor cannot have a constant sign over Ω but for being identically zero.

– Specifying now $\hat{\underline{\underline{A}}} = \underline{\underline{1}}$, the virtual velocity field is $\hat{\underline{U}}(\underline{x}) = \underline{x}$ and $\hat{\underline{\underline{d}}}(\underline{x}) = \hat{\underline{\underline{A}}} = \underline{\underline{1}}$. From the principle of virtual work, we derive: $\mathcal{P}_i(\hat{\underline{U}}) = -\int_\Omega \text{tr} \, \underline{\underline{\sigma}}(\underline{x},t) \, d\Omega = 0$. With the principal stresses numbered and ordered according to $\sigma_I \geq \sigma_{II} \geq \sigma_{III}$, this may be written $\int_\Omega (\sigma_I(\underline{x},t) + \sigma_{II}(\underline{x},t) + \sigma_{III}(\underline{x},t)) \, d\Omega = 0$. Thence, putting aside the case of the zero stress field, it follows that neither $\sigma_I(\underline{x},t) < 0$ nor $\sigma_{III}(\underline{x},t) > 0$ can be satisfied $\forall \underline{x} \in \Omega$: this proves (see section 6.10.2) the existence of tensile normal stresses (respectively compressive) at certain points on certain facets in Ω.

Comments

Self-equilibrating stress fields are commonly encountered as the result of incompatible deformations generated by temperature changes (see Chapter 5, section 5.6.8), drying, metal forming processes and also in biological elements. Practically, this explains why some structures made of fragile material may part into pieces when submitted to sharp temperature changes, either positive or negative (also cracks due to drying observed in concrete or clay structures, wood, etc.).

6.10.6. *Tresca's strength condition*

The Tresca[9] strength condition for isotropic materials states that, numbering and ordering the principal stresses according to $\sigma_I \geq \sigma_{II} \geq \sigma_{III}$, the stress tensor at a given point $\underline{x}$ must comply with $\sigma_I - \sigma_{III} \leq \sigma_0$ where σ_0 is a positive physical constant determined experimentally which characterizes the resistance of the material.

– Prove that this condition is equivalent to stating that the magnitude τ of the shear stress on any facet at that point satisfies $\tau \leq \sigma_0/2$.

– Consider two stress tensors $\underline{\underline{\sigma}}^\alpha$ and $\underline{\underline{\sigma}}^\beta$ which both comply with the Tresca strength condition, prove that any stress tensor in the form of a convex combination of $\underline{\underline{\sigma}}^\alpha$ and $\underline{\underline{\sigma}}^\beta$, namely $\underline{\underline{\sigma}} = \lambda \underline{\underline{\sigma}}^\alpha + (1-\lambda)\underline{\underline{\sigma}}^\beta, 0 \leq \lambda \leq 1$, also satisfies that condition.

Solution

– The result can be obtained directly from section 6.10.4. The Tresca strength condition amounts to stating that the diameter of *the* Mohr circle must be inferior or equal to zero. Recalling that the equation of this circle is $\tau^2 + (\sigma - \sigma_I)(\sigma - \sigma_{III}) = 0$, we see that the maximum value of τ is $\mathrm{Max}\,\tau = (\sigma_I - \sigma_{III})/2$ (half the diameter of the circle) reached for $\sigma = (\sigma_I + \sigma_{III})/2$ as it appears in Figure 6.19, which proves the equivalence: $\sigma_I - \sigma_{III} \leq \sigma_0 \Leftrightarrow \mathrm{Max}\,\tau \leq \sigma_0/2$.

– With reference to section 6.10.2, the Tresca strength condition may equivalently be written as: $\mathrm{Max}\{\underline{n}.\underline{\underline{\sigma}}.\underline{n},\forall \underline{n}\} - \mathrm{Min}\{\underline{n}.\underline{\underline{\sigma}}.\underline{n}.\forall \underline{n}\} \leq \sigma_0$.

---

9 H. Tresca (1814–1885).

For a convex combination of $\underline{\underline{\sigma}}^{\alpha}$ and $\underline{\underline{\sigma}}^{\beta}$, we have

$$\underline{n}.\underline{\underline{\sigma}}.\underline{n} = \lambda\,\underline{n}.\underline{\underline{\sigma}}^{\alpha}.\underline{n} + (1-\lambda)\,\underline{n}.\underline{\underline{\sigma}}^{\beta}.\underline{n}$$

and, as both λ and $(1-\lambda)$ are non-negative:

$$\mathrm{Max}\left\{\underline{n}.\underline{\underline{\sigma}}.\underline{n}, \forall\underline{n}\right\} \leq \lambda\,\mathrm{Max}\left\{\underline{n}.\underline{\underline{\sigma}}^{\alpha}.\underline{n}, \forall\underline{n}\right\} + (1-\lambda)\,\mathrm{Max}\left\{\underline{n}.\underline{\underline{\sigma}}^{\beta}.\underline{n}, \forall\underline{n}\right\}.$$

In the same way:

$$\mathrm{Min}\left\{\underline{n}.\underline{\underline{\sigma}}.\underline{n}, \forall\underline{n}\right\} \geq \lambda\,\mathrm{Min}\left\{\underline{n}.\underline{\underline{\sigma}}^{\alpha}.\underline{n}, \forall\underline{n}\right\} + (1-\lambda)\,\mathrm{Min}\left\{\underline{n}.\underline{\underline{\sigma}}^{\beta}.\underline{n}, \forall\underline{n}\right\}.$$

Thence:

$$\begin{aligned}
\mathrm{Max}\left\{\underline{n}.\underline{\underline{\sigma}}.\underline{n}, \forall\underline{n}\right\} - \mathrm{Min}\left\{\underline{n}.\underline{\underline{\sigma}}.\underline{n}.\forall\underline{n}\right\} &\leq \lambda\,\mathrm{Max}\left\{\underline{n}.\underline{\underline{\sigma}}^{\alpha}.\underline{n}, \forall\underline{n}\right\} \\
&+ (1-\lambda)\,\mathrm{Max}\left\{\underline{n}.\underline{\underline{\sigma}}^{\beta}.\underline{n}, \forall\underline{n}\right\} \\
&- \lambda\,\mathrm{Min}\left\{\underline{n}.\underline{\underline{\sigma}}^{\alpha}.\underline{n}, \forall\underline{n}\right\} \\
&+ (1-\lambda)\,\mathrm{Min}\left\{\underline{n}.\underline{\underline{\sigma}}^{\beta}.\underline{n}, \forall\underline{n}\right\}
\end{aligned}$$

and finally $\mathrm{Max}\left\{\underline{n}.\underline{\underline{\sigma}}.\underline{n}, \forall\underline{n}\right\} - \mathrm{Min}\left\{\underline{n}.\underline{\underline{\sigma}}.\underline{n}.\forall\underline{n}\right\} \leq \sigma_0$.

This result establishes the *convexity* of the Tresca strength condition. It follows that the set of all Cauchy stress tensors which satisfy the Tresca strength condition is a convex of $\mathbb{R}^6$.

6.10.7. *Maximum resisting rate of work for Tresca's strength condition*

Consider all Cauchy stress tensors that satisfy the Tresca strength condition $\sigma_{\mathrm{I}} - \sigma_{\mathrm{III}} \leq \sigma_0$. From section 6.10.6, we know that they generate a convex set $G \subset \mathbb{R}^6$.

– Prove that G is not bounded in $\mathbb{R}^6$ following the direction of isotropic tensors. Let $\hat{\underline{\underline{d}}}$ be any virtual strain rate tensor, calculate the quantity $\mathrm{Sup}\left\{\underline{\underline{\sigma}}:\hat{\underline{\underline{d}}} \,\middle|\, \underline{\underline{\sigma}} \in G\right\}$ as a function of $\hat{\underline{\underline{d}}}$ and σ_0.

– Let $\hat{\underline{U}}$ be a piecewise continuous and continuously differentiable virtual velocity field with jump surfaces denoted by $\Sigma_{\hat{U}}$. As defined in Chapter 3, section 3.5.1, the virtual resisting rate of work by a stress field $\underline{\underline{\sigma}}$ in such a virtual velocity field is $-\mathcal{P}_i(\hat{\underline{U}})$ with the expression [6.54] $-\mathcal{P}_i(\hat{\underline{U}}) = \int_{\Omega} \underline{\underline{\sigma}}(\underline{x}) : \hat{\underline{\underline{d}}}(\underline{x}) \, d\Omega + \int_{\Sigma_{\hat{U}}} [\![\hat{\underline{U}}(\underline{x})]\!] \cdot \underline{\underline{\sigma}}(\underline{x}) \cdot \underline{n}(\underline{x}) \, d\Sigma$. Show that for any stress field $\underline{\underline{\sigma}}$ complying with the Tresca strength condition in Ω, the virtual resisting rate of work in $\hat{\underline{U}}$ admits an upper bound (finite or infinite) that is a function of $\hat{\underline{U}}$, σ_0 and the geometrical data.

Solution

– Splitting $\underline{\underline{\sigma}}$ into its spherical and deviatoric parts (section 6.10.1), $\underline{\underline{\sigma}} = \underline{\underline{s}} + \frac{1}{3}(\mathrm{tr}\,\underline{\underline{\sigma}})\underline{\underline{1}}$ shows that the Tresca strength condition only depends on the deviator and is written as $s_{\mathrm{I}} - s_{\mathrm{III}} \leq \sigma_0$. It follows that if $\underline{\underline{\sigma}} \in G$ all tensors with the same deviator as $\underline{\underline{\sigma}}$ belong to G, which is therefore not bounded following the direction of isotropic tensors $\lambda\underline{\underline{1}}$.

– Splitting $\hat{\underline{\underline{d}}}$ into its deviatoric and spherical parts, $\hat{\underline{\underline{d}}} = \frac{1}{3}(\mathrm{tr}\,\hat{\underline{\underline{d}}})\underline{\underline{1}} + \hat{\underline{\underline{d}}}'$, we know from section 6.10.1 that $\underline{\underline{\sigma}} : \hat{\underline{\underline{d}}} = \frac{1}{3}(\mathrm{tr}\,\underline{\underline{\sigma}})(\mathrm{tr}\,\hat{\underline{\underline{d}}}) + \underline{\underline{s}} : \hat{\underline{\underline{d}}}'$. It follows that

- if $\mathrm{tr}\,\hat{\underline{\underline{d}}} \neq 0$, since G is not bounded in the direction of isotropic tensors $\lambda\underline{\underline{1}}$, we have $\mathrm{Sup}\{\underline{\underline{\sigma}} : \hat{\underline{\underline{d}}} \,|\, \underline{\underline{\sigma}} \in G\} = +\infty$;

- if $\mathrm{tr}\,\hat{\underline{\underline{d}}} = 0$, i.e. $\hat{\underline{\underline{d}}} = \hat{\underline{\underline{d}}}'$, we write $\underline{\underline{\sigma}} : \hat{\underline{\underline{d}}}$ explicitly in the orthonormal basis of the principal axes of $\hat{\underline{\underline{d}}}$ ordered according to $\hat{d}_1 \geq \hat{d}_2 \geq \hat{d}_3$ where $\hat{d}_1$ and $\hat{d}_3$ have opposite signs: $\underline{\underline{\sigma}} : \hat{\underline{\underline{d}}} = \underline{\underline{s}} : \hat{\underline{\underline{d}}} = (s_{11} - s_{22})\hat{d}_1 + (s_{33} - s_{22})\hat{d}_3$. Then

if $\hat{d}_1 \geq -\hat{d}_3$, i.e. $|\hat{d}_1| \geq |\hat{d}_3|$, we have $\mathrm{Sup}\{\underline{\underline{\sigma}} : \hat{\underline{\underline{d}}} \,|\, \underline{\underline{\sigma}} \in G\} = \sigma_0\,\hat{d}_1 = \sigma_0\,|\hat{d}_1|$,

if $\hat{d}_1 \leq -\hat{d}_3$, i.e. $|\hat{d}_1| \leq |\hat{d}_3|$, we have $\mathrm{Sup}\{\underline{\underline{\sigma}} : \hat{\underline{\underline{d}}} \,|\, \underline{\underline{\sigma}} \in G\} = -\sigma_0\,\hat{d}_3 = \sigma_0\,|\hat{d}_3|$, which can be written in the general form $\text{Sup}\left\{\underline{\underline{\sigma}}:\underline{\underline{\hat{d}}}\;\middle|\;\underline{\underline{\sigma}}\in G\right\}=\sigma_0\,\text{Max}\left\{\hat{d}_i\;\middle|\;i=1,2,3\right\}.$

This expression can be transformed, taking $\text{tr}\,\underline{\underline{\hat{d}}}=0$ into account, into

$$\text{Sup}\left\{\underline{\underline{\sigma}}:\underline{\underline{\hat{d}}}\;\middle|\;\underline{\underline{\sigma}}\in G\right\}=\pi(\underline{\underline{\hat{d}}})=\frac{\sigma_0}{2}\left(\left|\hat{d}_1\right|+\left|\hat{d}_2\right|+\left|\hat{d}_3\right|\right)\;\text{if }\text{tr}\,\underline{\underline{\hat{d}}}=0$$

$$\text{Sup}\left\{\underline{\underline{\sigma}}:\underline{\underline{\hat{d}}}\;\middle|\;\underline{\underline{\sigma}}\in G\right\}=\pi(\underline{\underline{\hat{d}}})=+\infty\;\text{if }\text{tr}\,\underline{\underline{\hat{d}}}\neq0.$$

This is the expression of the *Maximum resisting rate of work* by the stress tensor $\underline{\underline{\sigma}}$ in the virtual strain rate $\underline{\underline{\hat{d}}}$ (see Chapter 3, section 3.5.1) under the Tresca strength condition.

– From the preceding results, it comes out that the first integral in $-\mathcal{P}_i(\hat{U})$ is bounded by $\int_{\Omega}\pi(\underline{\underline{\hat{d}}}(\underline{x}))\,\mathrm{d}\Omega$. Introducing the stress vector on $\Sigma_{\hat{U}}$ at point $\underline{x}$ with its normal and shear components, $\underline{T}(\underline{n},\underline{x})=\underline{\underline{\sigma}}(\underline{x}).\underline{n}=\sigma(\underline{x})\underline{n}+\underline{\tau}(\underline{x})$, the integrand in the second integral is written as $[\![\underline{\hat{U}}(\underline{x})]\!].\underline{\underline{\sigma}}(\underline{x}).\underline{n}(\underline{x})=\sigma(\underline{x})[\![\underline{\hat{U}}(\underline{x})]\!].\underline{n}+[\![\underline{\hat{U}}(\underline{x})]\!].\underline{\tau}(\underline{x})$. As established in section 6.10.6, the Tresca strength condition is equivalent to stating that $\left|\underline{\tau}(\underline{x})\right|\leq\sigma_0/2$, which does not impose any limitation on the normal stress $\sigma(\underline{x})$. It follows that:

if $[\![\underline{\hat{U}}(\underline{x})]\!].\underline{n}\neq0$, $\text{Sup}[\![\underline{\hat{U}}(\underline{x})]\!].\underline{\underline{\sigma}}(\underline{x}).\underline{n}(\underline{x})=\pi([\![\underline{\hat{U}}(\underline{x})]\!],\underline{n}(\underline{x}))=+\infty\,;$

if $[\![\underline{\hat{U}}(\underline{x})]\!].\underline{n}=0$, $\text{Sup}[\![\underline{\hat{U}}(\underline{x})]\!].\underline{\underline{\sigma}}(\underline{x}).\underline{n}(\underline{x})=\pi([\![\underline{\hat{U}}(\underline{x})]\!],\underline{n}(\underline{x}))=\frac{\sigma_0}{2}\left|[\![\underline{\hat{U}}(\underline{x})]\!]\right|.$

As a result, the maximum value of the resisting rate of work by a stress field $\underline{\underline{\sigma}}$ complying with the Tresca strength condition in a virtual velocity field is

$$\mathcal{P}_{\mathrm{mr}}(\hat{U})=\int_{\Omega}\pi(\underline{\underline{\hat{d}}}(\underline{x}))\,\mathrm{d}\Omega+\int_{\Sigma_{\hat{U}}}\pi([\![\underline{\hat{U}}(\underline{x})]\!],\underline{n}(\underline{x}))\,\mathrm{d}\Sigma.$$

This *maximum resisting rate of work* is finite if and only if the virtual velocity field $\underline{\hat{U}}$ is *isochoric*, which implies both $\forall\underline{x}\in\Omega,\text{tr}\,\underline{\underline{\hat{d}}}(\underline{x})=0$ and $\forall\underline{x}\in\Sigma_{\hat{U}},[\![\underline{\hat{U}}(\underline{x})]\!].\underline{n}=0$.

6.10.8. *Spherical shell*

Consider a spherical shell, with inner and outer radii r_1 and r_2 respectively, that is subject to a fixed uniform external pressure p_2 on $\partial\Omega_2$ and a variable uniform pressure p_1 on $\partial\Omega_1$ with $p_1 \geq p_2$. No body forces are considered. The solid is made of a homogeneous isotropic material for which the stress state at any point within the body must comply with the Tresca strength condition $\sigma_I - \sigma_{III} \leq \sigma_0$ (section 6.10.6). The analysis aims at finding out what limitation is imposed on the internal pressure p_1 by this condition in order that it can be sustained by the shell in equilibrium with p_2 and no body forces.

– Determine the general form of the radial virtual velocity fields, functions of the spherical coordinate r alone, $\hat{\underline{U}}(\underline{x}) = \hat{U}_r(r)\underline{e}_r$, that generate isochoric virtual motions.

– Through the principle of virtual work, show that if p_1 exceeds a certain value, a function of the geometrical data and the material constant σ_0, the equilibrium of the shell with $p_2 = 0$ on $\partial\Omega_2$ and $\underline{F}(\underline{x}) = 0$ in Ω cannot be maintained.

– When p_1 reaches this maximum value, determine a stress field in equilibrium with the data and complying with the Tresca strength condition.

Solution

– In spherical coordinates, the condition $\operatorname{div}\hat{\underline{U}}(\underline{x}) = 0$ for the radial velocity fields $\hat{\underline{U}}(\underline{x}) = \hat{U}_r(r)\underline{e}_r$ to be isochoric reduces to $\dfrac{\partial \hat{U}_r}{\partial r} + 2\dfrac{\hat{U}_r}{r} = 0$. Thence $\hat{U}_r(r) = \alpha/r^2$, with α an arbitrary constant, and

$$\hat{\underline{d}}(r) = \frac{\partial \hat{U}_r}{\partial r}\underline{e}_r \otimes \underline{e}_r + \frac{\hat{U}_r}{r}\underline{e}_\theta \otimes \underline{e}_\theta + \frac{\hat{U}_r}{r}\underline{e}_\varphi \otimes \underline{e}_\varphi$$
$$= -\frac{2\alpha}{r^3}\underline{e}_r \otimes \underline{e}_r + \frac{\alpha}{r^3}\underline{e}_\theta \otimes \underline{e}_\theta + \frac{\alpha}{r^3}\underline{e}_\varphi \otimes \underline{e}_\varphi.$$

– The virtual rate of work by external forces in this class of virtual motions is equal to $\mathcal{P}_e(\hat{\underline{U}}) = -4\pi\alpha(p_2 - p_1)$, while the virtual rate of work by internal forces is $\mathcal{P}_i(\hat{\underline{U}}) = -\alpha\displaystyle\int_\Omega \frac{-2\sigma_{rr} + \sigma_{\theta\theta} + \sigma_{\varphi\varphi}}{r^3}r^2\sin\theta\,dr\,d\theta\,d\varphi$. Applying the principle of virtual work $P_e(\hat{U}) + P_i(\hat{U}) = A(\hat{U})$ with such a virtual velocity field for any stress field in equilibrium with the data, we obtain, with $A(\hat{U}) = 0$,

$$(p_1 - p_2) = \frac{1}{4\pi} \int_{\Omega} \frac{-2\sigma_{rr} + \sigma_{\theta\theta} + \sigma_{\varphi\varphi}}{r} \sin\theta \, dr \, d\theta \, d\varphi.$$

Under Tresca's strength condition, the magnitude of the integrand cannot exceed $2\sigma_0 / r$, a value that is reached when $\sigma_{\theta\theta} = \sigma_{\varphi\varphi}$ and $\sigma_{\theta\theta} - \sigma_{rr} = \sigma_0$, and the other components are equal to zero. Performing the integration, it follows that $(p_1 - p_2) \leq 2\sigma_0 \ln\frac{r_2}{r_1}$. This proves that if $p_1 > p_2 + 2\sigma_0 \ln\frac{r_2}{r_1}$, the *equilibrium* of the shell with the given data *cannot be maintained* while complying with the Tresca strength condition.

– Looking for a possible stress field in equilibrium with the data and complying with the Tresca strength condition when $p_1 = p_2 + 2\sigma_0 \ln\frac{r_2}{r_1}$, we assume that $\sigma_{rr}, \sigma_{\theta\theta}, \sigma_{\varphi\varphi}$ are principal stresses with $\sigma_{\theta\theta} = \sigma_{\varphi\varphi}$ and $\sigma_{\theta\theta} - \sigma_{rr} = \sigma_0$. The equilibrium equations reduce to

$$\begin{cases} \dfrac{\partial \sigma_{rr}}{\partial r} + \dfrac{1}{r}(2\sigma_{rr} - \sigma_{\theta\theta} - \sigma_{\varphi\varphi}) = 0 \\[2mm] \dfrac{1}{r}\dfrac{\partial \sigma_{\theta\theta}}{\partial \theta} = 0 \\[2mm] \dfrac{1}{r\sin\theta}\dfrac{\partial \sigma_{\varphi\varphi}}{\partial \varphi} = 0 \end{cases}$$

and prove that the stress field depends on r only. With the boundary condition on $\partial\Omega_2$, we determine the stress field $\underline{\underline{\sigma}} = (-p_2 + 2\sigma_0 \ln\frac{r}{r_2})\underline{\underline{1}} + \sigma_0 (\underline{e}_\theta \otimes \underline{e}_\theta + \underline{e}_\varphi \otimes \underline{e}_\varphi)$ that satisfies the boundary condition on $\partial\Omega_1$ as anticipated.

Comments

This analysis proves that a uniform internal pressure $p_1 > p_2 + 2\sigma_0 \ln\frac{r_2}{r_1}$ is not compatible with the equilibrium of the shell under Tresca's strength condition.

Moreover, it has also been proven that for $p_1 = p_2 + 2\sigma_0 \ln \frac{r_2}{r_1}$ a stress field can be constructed that complies with the equilibrium equations and Tresca's strength condition. The result obviously holds for $p_2 \le p_1 < p_2 + 2\sigma_0 \ln \frac{r_2}{r_1}$. Nevertheless, it must be underscored that this is *not sufficient* to prove that any uniform internal pressure p_1 such that $p_2 < p_1 \le p_2 + 2\sigma_0 \ln \frac{r_2}{r_1}$ will be actually sustained! Such a conclusion requires the full knowledge of the history of the loading process and the constitutive equations defining the behavior of the constituent material (see [SAL 02, SAL 13]).

6.10.9. *Rotating circular ring*

Consider a circular cylindrical ring Ω with axis Oz, inner and outer radii a and b respectively and height h (Figure 6.21). In the Galilean frame of the observer, the ring rotates with the constant angular velocity $\underline{\omega} = \omega \underline{e}_z$ about Oz. No surface forces are exerted on the boundary nor any body forces exerted in the volume. The constituent material is homogeneous and isotropic with ρ the mass per unit volume and the stress state at any point in the body must comply with the Tresca strength condition: $\sigma_I - \sigma_{III} \le \sigma_0$ (section 6.10.6). The analysis aims at finding out the limitation on ω imposed by the strength condition.

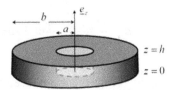

Figure 6.21. *Rotating circular ring*

– Using cylindrical coordinates with axis Oz, consider the virtual velocity fields defined by $\underline{\hat{U}}(r,\theta,z) = \frac{\hat{\alpha}}{r}\underline{e}_r$ where $\hat{\alpha}$ is an arbitrary constant. Calculate the corresponding virtual strain rate field and rate of volume dilatation.

– Implementing the principle of virtual work with this class of virtual velocity fields determines an upper bound for the angular velocities ω that can be sustained by the ring.

Solution

– The virtual strain rate tensor field corresponding to $\underline{\hat{U}}(r,\theta,z) = \dfrac{\hat{\alpha}}{r}\underline{e}_r$ is written

as $\underline{\underline{\hat{d}}}(r,\theta,z) = \dfrac{\hat{\alpha}}{r^2}(\underline{e}_\theta \otimes \underline{e}_\theta - \underline{e}_r \otimes \underline{e}_r)$, which shows that the virtual motion is

isochoric: $\mathrm{tr}\,\underline{\underline{\hat{d}}}(r,\theta,z) = \mathrm{div}\,\underline{\hat{U}}(r,\theta,z) = 0$ (section 6.10.7).

– Explicitly, with the acceleration field given by $\hat{a}(r,\theta,z) = -\omega^2 r\underline{e}_r$, we have:

$$\mathcal{A}(\hat{U}) = -\int_\Omega \rho\omega^2 \hat{a}\,r\mathrm{d}r\,\mathrm{d}\theta\,\mathrm{d}z = -\pi h\hat{\alpha}\rho\omega^2(b^2 - a^2).$$

In the absence of external forces, the principle of virtual work reduces to $\mathcal{P}_i(\hat{U}) = \mathcal{A}(\hat{U})$ for any stress field that satisfies the equations of motion. Thence for such a stress field, $\forall \hat{\alpha},\ \pi h\hat{\alpha}\rho\omega^2(b^2 - a^2) = -\mathcal{P}_i(\hat{U})$, the virtual resisting work by internal forces (Chapter 3, section 3.5.1).

With $\underline{\underline{\hat{d}}}(r,\theta,z) = \hat{\alpha}(\underline{e}_\theta \otimes \underline{e}_\theta - \underline{e}_r \otimes \underline{e}_r)/r^2$, we get

$$-\mathcal{P}_i(\hat{U}) = \hat{\alpha}\int_\Omega \frac{(\sigma_{\theta\theta} - \sigma_{rr})}{r}\,\mathrm{d}r\,\mathrm{d}\theta\,\mathrm{d}z .$$

Thence, from the virtual work equation:

$$\forall \hat{\alpha},\ \pi h\hat{\alpha}\rho\omega^2(b^2 - a^2) = \hat{\alpha}\int_\Omega \frac{(\sigma_{\theta\theta} - \sigma_{rr})}{r}\,\mathrm{d}r\,\mathrm{d}\theta\,\mathrm{d}z ,$$

where the Tresca strength condition caps the magnitude of the integrand by $\dfrac{\sigma_0}{r}$. It follows that, for this equation to be mathematically compatible with $\sigma_I(\underline{x}) - \sigma_{III}(\underline{x}) \leq \sigma_0, \forall \underline{x} \in \Omega$, the angular velocity should not exceed the maximum value ω_c given by $\omega_c^2 = \dfrac{2\sigma_0}{\rho(b^2 - a^2)}\ln\dfrac{b}{a}$.

Comments

This analysis has just proven that a constant angular velocity $\omega > \omega_c$ is not compatible with the Tresca strength condition. As in section 6.10.8, it does not provide a sufficient condition for sustainability and, moreover, in this case it can even be proven that the condition is actually not sufficient, whatever the complete knowledge of the history of the loading process and the constitutive equations defining the behavior of the constituent material. This result can be obtained through the implementation of the principle of virtual work with more sophisticated virtual velocity fields such as $\hat{\underline{U}}(r,z) = \hat{U}_r(r,z)\underline{e}_r + \hat{U}_z(r,z)\underline{e}_z$ with $\operatorname{div}\hat{\underline{U}}(r,z) = 0$ (in order to be isochoric as required by section 6.10.7).

6.10.10. *Loading parameters, load vector*

Consider a body with volume Ω and boundary $\partial\Omega$ submitted to a field of body forces $\underline{F}(\underline{x},t) = Q_1(t)\underline{F}(\underline{x})$ proportional to a scalar parameter $Q_1(t)$. The boundary is divided into two complementary parts $\partial\Omega_1$ and $\partial\Omega_2$ that do not overlap and are such that $\partial\Omega = \partial\Omega_1 \cup \partial\Omega_2$. On $\partial\Omega_1$, the boundary data consist of a field of surface forces $\underline{T}_\Omega(\underline{x},t) = Q_2(t)\underline{T}_\Omega(\underline{x})$ proportional to a scalar parameter $Q_2(t)$. On $\partial\Omega_2$, the body is assumed to be fixed to a rigid base without any condition being imposed on the stress field. Given the fields $\underline{F}(\underline{x})$ and $\underline{T}_\Omega(\underline{x})$, $Q_1(t)$ and $Q_2(t)$ are the *loading parameters* of the system, the components of the *load vector* $\underline{Q}(t) = (Q_1(t), Q_2(t)) \in \mathbb{R}^2$.

– Consider now $\underline{Q}^\alpha(t) \in \mathbb{R}^2$ and $\underline{Q}^\beta(t) \in \mathbb{R}^2$ two values of the load vector with $\underline{\underline{\sigma}}^\alpha(\underline{x},t)$ and $\underline{\underline{\sigma}}^\beta(\underline{x},t)$ two stress fields in Ω in equilibrium with the data specified by $\underline{Q}^\alpha(t)$ and $\underline{Q}^\beta(t)$ respectively. Prove that a stress field $\underline{\underline{\sigma}}(\underline{x},t)$ in equilibrium with the data specified through $\underline{Q}(t) = \lambda^\alpha \underline{Q}^\alpha(t) + \lambda^\beta \underline{Q}^\beta(t)$, $\lambda^\alpha \in \mathbb{R}$, $\lambda^\beta \in \mathbb{R}$ can easily be obtained from $\underline{\underline{\sigma}}^\alpha(\underline{x},t)$ and $\underline{\underline{\sigma}}^\beta(\underline{x},t)$.

– It is supposed now that the considered body is made of an isotropic material whose strength condition is defined by the Tresca condition $\sigma_I(\underline{x},t) - \sigma_{III}(\underline{x},t) \le \sigma_0(\underline{x})$ where the physical constant $\sigma_0(\underline{x})$ may vary from one point to another if the material is not homogeneous. We define $K \subset \mathbb{R}^2$ as the set generated by the load vectors $\underline{Q}(t)$ such that it is possible to exhibit a stress field $\underline{\underline{\sigma}}(\underline{x},t)$ that satisfies the corresponding equilibrium equations and complies with the Tresca strength condition all over Ω.

Let $\underline{Q}^\alpha(t)$ and $\underline{Q}^\beta(t)$ be two such load vectors. Prove that any load vector that is a convex combination of $\underline{Q}^\alpha(t)$ and $\underline{Q}^\beta(t)$,

$$\underline{Q}(t) = \lambda \underline{Q}^{\alpha}(t) + (1-\lambda)\underline{Q}^{\beta}(t),\ 0 \le \lambda \le 1,$$

is an element of K.

– Let $\hat{U}(x)$ be any piecewise continuous and continuously differentiable virtual velocity field defined in Ω. Prove that for any given value of the load vector $\underline{Q}(t)$, the virtual rate of work by external forces can be written in the form

$$P_e(\underline{Q}(t), \hat{U}) = Q_1(t)q_1(\hat{U}) + Q_2(t)q_2(\hat{U})$$

where $q_1(\hat{U})$ and $q_2(\hat{U})$ are linear functions of the field $\hat{U}(x)$.

– Prove that for any load vector $\underline{Q}(t)$ in K, the virtual rate of work by external forces in any piecewise continuous and continuously differentiable virtual velocity field $\hat{U}(x)$ admits an upper bound that is a function of $\hat{U}(x)$, the geometrical data and the physical characteristics of the constituent material.

Solution

– Making $\underline{Q}(t) = \lambda^{\alpha}\underline{Q}^{\alpha}(t) + \lambda^{\beta}\underline{Q}^{\beta}(t),\ \lambda^{\alpha} \in \mathbb{R},\ \lambda^{\beta} \in \mathbb{R}$, the data of the problem become

$$\underline{F}(x,t) = (\lambda^{\alpha}Q_1^{\alpha}(t) + \lambda^{\beta}Q_1^{\beta}(t))\underline{F}(x) \text{ and } \underline{T}_{\Omega}(x,t) = (\lambda^{\alpha}Q_2^{\alpha}(t) + \lambda^{\beta}Q_2^{\beta}(t))\underline{T}_{\Omega}(x).$$

The equilibrium field and boundary equations are linear.

It follows that, as $\underline{\underline{\sigma}}^{\alpha}(x,t)$ and $\underline{\underline{\sigma}}^{\beta}(x,t)$ satisfy these equations with $\underline{Q}^{\alpha}(t)$ and $\underline{Q}^{\beta}(t)$ respectively, the field

$$\underline{\underline{\sigma}}(x,t) = \lambda^{\alpha}\underline{\underline{\sigma}}^{\alpha}(x,t) + \lambda^{\beta}\underline{\underline{\sigma}}^{\beta}(x,t) \text{ complies with the equilibrium equations for}$$

the value $\underline{Q}(t) = \lambda^{\alpha}\underline{Q}^{\alpha}(t) + \lambda^{\beta}\underline{Q}^{\beta}(t)$ of the load vector.

– The stress field $\underline{\underline{\sigma}}(x,t) = \lambda\underline{\underline{\sigma}}^{\alpha}(x,t) + (1-\lambda)\underline{\underline{\sigma}}^{\beta}(x,t),\ 0 \le \lambda \le 1$ is in equilibrium with $\underline{Q}(t) = \lambda\underline{Q}^{\alpha}(t) + (1-\lambda)\underline{Q}^{\beta}(t)$ (previous result).

From section 6.10.6, we know that the Tresca strength condition is convex.

Thence, at any point in Ω,

$\sigma_I^{\alpha}(\underline{x},t) - \sigma_{III}^{\alpha}(\underline{x},t) \le \sigma_0(\underline{x})$ and $\sigma_I^{\beta}(\underline{x},t) - \sigma_{III}^{\beta}(\underline{x},t) \le \sigma_0(\underline{x})$ imply,

$$\sigma_I(\underline{x},t) - \sigma_{III}(\underline{x},t) \le \sigma_0(\underline{x})$$

for any stress field in the form $\underline{\underline{\sigma}}(\underline{x},t) = \lambda \underline{\underline{\sigma}}^{\alpha}(\underline{x},t) + (1-\lambda)\underline{\underline{\sigma}}^{\beta}(\underline{x},t),\ 0 \le \lambda \le 1$.

If we consider the set K generated in $\mathbb{R}^2$ by the load vectors for which it is possible to exhibit a stress field that satisfies the equilibrium equations and which complies with Tresca's strength condition all over Ω, this set has thus been proven to be convex in $\mathbb{R}^2$ (Figure 6.22).

– With the data specified by a load vector $\underline{Q}(t) = (Q_1(t), Q_2(t))$, the virtual rate of work by external forces in a virtual velocity field $\underline{\hat{U}}(\underline{x})$ is simply

$$P_e(\underline{Q}, \underline{\hat{U}}) = Q_1(t) \int_{\Omega} \underline{F}(\underline{x}).\underline{\hat{U}}(\underline{x}) \, d\Omega + Q_2(t) \int_{\partial \Omega_1} \underline{T}_{\Omega}(\underline{x}).\underline{\hat{U}}(\underline{x}) \, da,$$

where the cofactors of $Q_1(t)$ and $Q_2(t)$ are the linear functions of the field $\underline{\hat{U}}(\underline{x})$ defined by $q_1(\underline{\hat{U}}) = \int_{\Omega} \underline{F}(\underline{x}).\underline{\hat{U}}(\underline{x}) \, d\Omega$ and $q_2(\underline{\hat{U}}) = \int_{\partial \Omega_1} \underline{T}_{\Omega}(\underline{x}).\underline{\hat{U}}(\underline{x}) \, da$. Symbolically, with $\underline{q}(\underline{\hat{U}}) = (q_1(\underline{\hat{U}}), q_2(\underline{\hat{U}})) \in \mathbb{R}^2$ we may also write $P_e(\underline{Q}(t), \underline{\hat{U}}) = \underline{Q}(t).\underline{q}(\underline{\hat{U}})$.

– As $\underline{Q}(t) \in K$, it follows that we can exhibit a stress field $\underline{\underline{\sigma}}(\underline{x},t)$ that satisfies the equilibrium equations with $\underline{Q}(t)$ and complies with the Tresca strength condition all over Ω. Writing the principle of virtual work with that stress field and a virtual velocity field $\underline{\hat{U}}(\underline{x})$, we obtain

$$P_e(\underline{Q}(t), \underline{\hat{U}}) = \underline{Q}(t).\underline{q}(\underline{\hat{U}}) = -P_i(\underline{\hat{U}}) = \int_{\Omega} \underline{\underline{\sigma}}(\underline{x}):\underline{\underline{\hat{d}}}(\underline{x}) \, d\Omega + \int_{\Sigma_{\hat{U}}} [\![\underline{\hat{U}}(\underline{x})]\!].\underline{\underline{\sigma}}(\underline{x}).\underline{n}(\underline{x}) \, d\Sigma$$

where, as a consequence of section 6.10.7, the right-hand side is bounded by the maximum resisting rate of work

$$P_{mr}(\underline{\hat{U}}) = \int_{\Omega} \pi(\underline{\underline{\hat{d}}}(\underline{x})) \, d\Omega + \int_{\Sigma_{\hat{U}}} \pi([\![\underline{\hat{U}}(\underline{x})]\!], \underline{n}(\underline{x})) \, d\Sigma.$$

Thence, the final equation (Figure 6.22)

$$\underline{Q}(t).\underline{q}(\hat{\underline{U}}) \le \mathcal{P}_{\mathrm{mr}}(\hat{\underline{U}}) = \int_{\Omega} \pi(\hat{\underline{\underline{d}}}(\underline{x}))\,\mathrm{d}\Omega + \int_{\Sigma_{\hat{U}}} \pi([\![\hat{\underline{U}}(\underline{x})]\!], \underline{n}(\underline{x}))\,\mathrm{d}\Sigma.$$

Comments

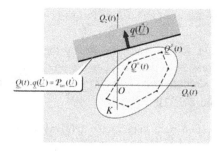

Figure 6.22. *Convex* $K \subset \mathbb{R}^2$

As shown in Figure 6.22, the results obtained in the preceding section prove first that the convex hull of a collection of load vectors $\underline{Q}^{\alpha}(t)$, $\underline{Q}^{\beta}(t)$, $\underline{Q}^{\gamma}(t)$... provide an internal approach for the boundary of K. Then, any virtual velocity field $\hat{\underline{U}}(\underline{x})$ such that $\mathcal{P}_{\mathrm{mr}}(\hat{\underline{U}}) = \int_{\Omega} \pi(\hat{\underline{\underline{d}}}(\underline{x}))\,\mathrm{d}\Omega + \int_{\Sigma_{\hat{U}}} \pi([\![\hat{\underline{U}}(\underline{x})]\!], \underline{n}(\underline{x}))\,\mathrm{d}\Sigma < +\infty$ (isochoric virtual velocity field, see section 6.10.7) provides an external approach for this boundary.

These two mathematically dual approaches are a simple example of an implementation of the general theory of yield design (see [SAL 13]).

7

The Curvilinear
One-dimensional Continuum

7.1. The problem of one-dimensional modeling

Many structures used in civil engineering or industry in general are made from slender elements (Figure 7.1). This geometrical characteristic suggests that we might attempt to describe them by means of a one-dimensional mechanical model where a system would be defined as made of "particles" situated along a *director curve*.

Figure 7.1. *Marble Canyon (Arizona). Solferino footbridge (Paris) (© Jean Salençon)*

At first glance, such a modeling process could appear as less refined than that of the same structure within the formalism of the three-dimensional continuum, but it must be noted that the relevance of a model can only be assessed by the extent to which this model actually achieves relevant results when applied to the problems it has been designed to solve. Moreover, practical applications quite often take advantage of a multi-stage modeling procedure, which means that in the present case they refer to the one-dimensional continuum model as a first step and then shift locally to the three-dimensional continuum when it proves necessary and relevant.

As the intuitive idea of the one-dimensional medium is related to that of a slender three-dimensional solid, a geometrical definition of slenderness will be stated as follows: a slender solid is such that a curve can be traced, with a continuous or piecewise continuous tangent, so that the maximal cross-section diameter of the solid normal to this curve is, all along, small compared with the length of the solid. As a matter of fact, we may attempt to build up a one-dimensional model for any three-dimensional solid without particular geometrical characteristics, but it is easily anticipated that the practical relevance of such a construction is highly dependent on the slenderness hypothesis.

A key problem arises when applying the model to real situations that locate the director curve within the three-dimensional geometry of the considered solid, even a slender one. Clearly enough, the curve introduced in the above definition of slenderness is not unique, although this is sometimes obscured by the way such structural elements are geometrically described. The problem of where to situate the director curve is part of the relevance assessment in each practical case and cannot be discussed independently of the constitutive equation adopted for the one-dimensional continuum *material*. For this reason, the misleading terminology "*mean fibre*" will be avoided here and the approach adopted in the present construction of the one-dimensional continuum *model* simply takes this curve as *given* in the current configuration.

7.2. One-dimensional modeling without an oriented microstructure

7.2.1. *Geometrical modeling*

As in Chapter 4, the first step of the virtuous circle consists of building up the geometrical modeling of a system following the intuitive idea described in the preceding section. Considering the director curve (C) embedded in $\mathbb{R}^3$ in the current configuration κ as a *data* (Figure 7.2), we may say that the geometrical modeling process aims at squeezing a three-dimensional slender system onto a one-dimensional description made on an arc $\overset{\frown}{AB}$ of (C). Thence, a particle of the one-dimensional system S is no longer a kind of "diluted material point" with volume $d\Omega$, as it used to be in the case of the three-dimensional continuum, but will be represented as a "diluted material point" with length ds on (C).

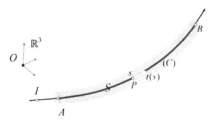

Figure 7.2. *One-dimensional geometrical modeling*

A system S is described on the oriented director curve, choosing an origin I for measurement of the curvilinear abscissa s of the field point P. It is made of the particles that occupy the arc $\overset{\frown}{AB}$ with, by convention, $s_A < s_B$. A subsystem S' is described in the same way on an arc $\overset{\frown}{A'B'}$ part of $\overset{\frown}{AB}$. The particle at point P has a mass $dm = \rho(s,t)\,ds$ with $\rho(s,t)$ the mass per unit length. This description of the geometrical configuration of the one-dimensional system S is quite similar to the one given in Chapter 4 for the three-dimensional medium. It is worth noting that, due to the specificity of the one-dimensional geometry, the arc $\overset{\frown}{AB}$ and ds here play the roles of Ω and $d\Omega$ in the three-dimensional model, while the endpoints B and A stand for $\partial\Omega$ with outward normal represented by the tangent to (C) in the positive and negative directions respectively.

7.2.2. Kinematics: actual and virtual motions

The geometrical state of the system at time t is defined by the vector field $\underline{x}(s) = \underline{OP}$ which describes the arc $\overset{\frown}{AB}$ on (C) in the current configuration κ^1, with O the origin of a reference frame in $\mathbb{R}^3$. The vector field $\underline{x}(s)$ is continuous and piecewise continuously differentiable with respect to s.

Actual motions of S at time t are defined on $\overset{\frown}{AB}$ by $\underline{U}(s,t)$, the velocity field of the constituent particles in $\mathbb{R}^3$, continuous and continuously differentiable with respect to s. The definition is similar for any subsystem.

Complying with the condition that they must include rigid body motions and encompass the actual motions, *virtual motions* of the system S or any subsystem S', denoted by $\underline{\hat{U}}$, will be defined on $\overset{\frown}{AB}$ (respectively $\overset{\frown}{A'B'}$) by vector fields

---

1 As in Chapter 4, the subscript t is unambiguously deleted for simplicity.

denoted by $\hat{\underline{U}}(s)$ that are first assumed continuous and continuously differentiable and will ultimately be allowed piecewise continuity (section 7.2.8). In each case, they do generate a vector space with infinite dimension.

7.2.3. *Virtual rate of work by quantities of acceleration*

Let $\underline{a}(s,t)$ denote the *acceleration* of the particle at the point P. The quantity of acceleration of the element with mass $dm = \rho(s,t)\,ds$ is just

$$\underline{a}(s,t)\,dm = \rho(s,t)\underline{a}(s,t)\,ds \qquad [7.1]$$

and consequently the linear forms $\mathcal{A}(\hat{\underline{U}})$ and $\mathcal{A}'(\hat{\underline{U}})$ are the integrals of the density $\rho(s,t)\underline{a}(s,t).\hat{\underline{U}}(s)\,ds$ on $\widehat{AB}$ (respectively $\widehat{A'B'}$):

$$\mathcal{A}(\hat{\underline{U}}) = \int_{\widehat{AB}} \rho(s,t)\underline{a}(s,t).\hat{\underline{U}}(s)\,ds \qquad [7.2]$$

$$\mathcal{A}'(\hat{\underline{U}}) = \int_{\widehat{A'B'}} \rho(s,t)\underline{a}(s,t).\hat{\underline{U}}(s)\,ds. \qquad [7.3]$$

7.2.4. *Virtual rate of work by external forces for the system*

External forces on the system S consist of *body forces* exerted on $\widehat{AB}$ and *boundary forces* exerted on the endpoints B and A. Thence, the expression for $\mathcal{P}_e(\hat{\underline{U}})$ is the sum of the integral over $\widehat{AB}$ of a line density $\underline{f}(s,t).\hat{\underline{U}}(s)$ and two point contributions at B and A denoted by $\underline{R}_B^S(t).\hat{\underline{U}}(s_B)$ and $\underline{R}_A^S(t).\hat{\underline{U}}(s_A)$ respectively:

$$\mathcal{P}_e(\hat{\underline{U}}) = \int_{\widehat{AB}} \underline{f}(s,t).\hat{\underline{U}}(s)\,ds + \underline{R}_B^S(t).\hat{\underline{U}}(s_B) + \underline{R}_A^S(t).\hat{\underline{U}}(s_A). \qquad [7.4]$$

A physical meaning can be given to cofactors $\underline{f}(s,t)$, $\underline{R}_B^S(t)$ and $\underline{R}_A^S(t)$ from the intuitive concept of the one-dimensional modeling process, as shown in Figure 7.3: "squeezing" of the three-dimensional system onto the arc $\widehat{AB}$. The infinitesimal distributed force $\underline{f}(s,t)\,ds$ on the length element ds accounts for the body forces in the volume Ω and surface forces on the boundary $\partial\Omega_L$ exerted on the three-dimensional original system between the cross-sections at s and $(s+ds)$. The point forces $\underline{R}_B^S(t)$ and $\underline{R}_A^S(t)$ account for the forces exerted by the world external to the three-dimensional system on the boundary sections S_B and S_A respectively. Note that, due to the very definition of the considered virtual motions, external forces are modeled by *vectors exclusively* as for the classical three-dimensional continuum.

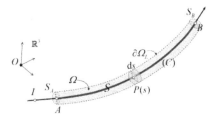

Figure 7.3. *"Squeezing" the three-dimensional system*

7.2.5. *Virtual rate of work by external forces for subsystems*

Let $S' = \widehat{A'B'} \subset S = \widehat{AB}$ be a subsystem of S *stricto sensu* (Figure 7.4).

The expression for $\mathcal{P}_e'(\hat{U})$ is similar to [7.4]:

$$\mathcal{P}_e'(\hat{U}) = \int_{\widehat{A'B'}} \underline{f}^{S'}(s,t).\hat{\underline{U}}(s)\,ds + \underline{R}_{B'}^{S'}(t).\hat{\underline{U}}(s_{B'}) + \underline{R}_{A'}^{S'}(t).\hat{\underline{U}}(s_{A'}), \qquad [7.5]$$

where the line density $\underline{f}^{S'}(s,t)\,ds$ stands for the forces exerted on the particle of $S' = \widehat{A'B'}$ by the particles of $(S-S')$ and the world external to S. Consistently with the intuitive concept of the one-dimensional model being derived from the classical model of a three-dimensional continuum as recalled in section 7.2.4, we assume that the particles of S *do not exert any action at a distance* on each other as in Chapter 6 (section 6.3.4). Thence, $\underline{f}^{S'}(s,t)$ is only exerted by the world external to S and

$$\forall S' = \widehat{A'B'} \subset S = \widehat{AB}, \ \underline{f}^{S'}(s,t) \equiv \underline{f}(s,t). \qquad [7.6]$$

It being recalled that the concept of a subsystem is just the result of a thought experiment, the concentrated forces $\underline{R}_{B'}^{S'}(t)$ and $\underline{R}_{A'}^{S'}(t)$ stand for the external forces exerted, by the world external to S', on the boundary of $S' = \widehat{A'B'}$. As in the case of the three-dimensional continuum (Chapter 6, section 6.2.3) and consistently with [7.4] where no point contribution inside $\widehat{AB}$ appears, we observe that they are only exerted by the *constituent particles of S external* to S'.

Finally, the virtual rate of work by external forces for a subsystem is written as:

$$\mathcal{P}'_e(\underline{\hat{U}}) = \int_{\widehat{A'B'}} \underline{f}(s,t).\underline{\hat{U}}(s)\,ds + \underline{R}^{S'}_{B'}(t).\underline{\hat{U}}(s_{B'}) + \underline{R}^{S'}_{A'}(t).\underline{\hat{U}}(s_{A'}). \qquad [7.7]$$

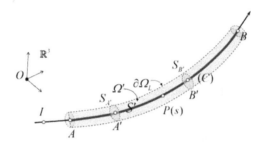

Figure 7.4. *Subsystem $S' = \widehat{A'B'} \subset S = \widehat{AB}$*

7.2.6. Virtual rate of work by internal forces

For the continuous linear forms $\mathcal{P}_i(\underline{\hat{U}})$ and $\mathcal{P}'_i(\underline{\hat{U}})$ that express the virtual rate of work by internal forces for the system or any subsystem, we make the fundamental assumption that

The virtual rate of work by internal forces for the system or any subsystem can be written as the integral of a line density $p_i(\underline{\hat{U}})$ independent of the considered subsystem, which is a linear form in the local values of the virtual velocity field and its first derivative along (C) :

$$\forall \underline{\hat{U}} \text{ m. v., } \forall S' \subseteq S, \; \mathcal{P}'_i(\underline{\hat{U}}) = \int_{\widehat{A'B'}} p_i(\underline{\hat{U}})\,ds \qquad [7.8]$$

with

$$p_i(\underline{\hat{U}}) = -\underline{A}(s,t).\underline{\hat{U}}(s) - \underline{X}(s,t).\frac{d\underline{\hat{U}}(s)}{ds}. \qquad [7.9]$$

This assumption is obviously analogous to the hypothesis underlying the classical three-dimensional continuum model. From the experience gained in Chapter 6 (section 6.3.4), we can already guess that its validity depends on the assumption [7.6] related to interactions between particles of the system.

7.2.7. Implementation of the principle of virtual work

The implementation of the principle of virtual work follows the same path as in Chapter 6, section 6.3.

7.2.7.1. Specifying the virtual rate of work by internal forces

We first refer to the dual statement of the law of mutual actions (Chapter 3, section 3.2.4)

$$\begin{cases} \forall S' \subseteq S \\ \forall \hat{\underline{U}} \text{ r.b.v.m. of } S', \ \mathcal{P}'_i(\hat{\underline{U}}) = 0 \end{cases} \qquad [7.10]$$

where $\hat{\underline{U}}$, arbitrary rigid body virtual motion (r.b.v.m.) of S', is generated by *arbitrary vectors* $\hat{\underline{U}}_0$ and $\hat{\underline{\omega}}_0$ and written as

$$\hat{\underline{U}}(\underline{x}) = \hat{\underline{U}}_0 + \hat{\underline{\omega}}_0 \wedge \underline{OM}. \qquad [7.11]$$

Making $\hat{\underline{\omega}}_0 = 0$, we prove that $\underline{A}(s,t) = 0$ identically over $\widehat{AB}$. From [7.11], it comes out that $\dfrac{d\hat{\underline{U}}(s)}{ds} = \hat{\underline{\omega}}_0 \wedge \underline{t}(s)$ where $\underline{t}(s)$ is the unit vector tangent to (C) at the point P (Figure 7.2). Then [7.10] implies:

$$\begin{cases} \forall \widehat{A'B'} \subseteq \widehat{AB} \\ \forall \hat{\underline{\omega}}_0 \in \mathbb{R}^3, \\ \displaystyle\int_{\widehat{A'B'}} \underline{X}(s,t).(\hat{\underline{\omega}}_0 \wedge \underline{t}(s))\,ds = 0. \end{cases} \qquad [7.12]$$

It follows that $\underline{X}(s,t)$ must be collinear with $\underline{t}(s)$ at all points of $\widehat{AB}$

$$\underline{X}(s,t) = X(s,t)\underline{t}(s) \qquad [7.13]$$

with $X(s,t)$ a scalar function. The most general form for $p_i(\hat{\underline{U}})$ according to the initial fundamental assumption reduces to

$$p_i(\hat{\underline{U}}) = -X(s,t)\frac{\mathrm{d}\hat{\underline{U}}(s)}{\mathrm{d}s} \cdot \underline{t}(s) = -X(s,t)\hat{D}(s) \qquad [7.14]$$

and

$$\forall S' \subseteq S, \ \mathcal{P}'_i(\hat{\underline{U}}) = \int_{\widehat{A'B'}} -X(s,t)\hat{D}(s)\,\mathrm{d}s, \qquad [7.15]$$

where $\hat{D}(s) = \dfrac{\mathrm{d}\hat{\underline{U}}(s)}{\mathrm{d}s} \cdot \underline{t}(s)$ represents the *virtual rate of stretch* of the director curve at the point P.

7.2.7.2. Equations of motion for the system

With this expression for the virtual rate of work by internal forces, we apply the second statement of the principle of virtual work to the system, which takes the form:

$$\begin{cases} \text{in a Galilean frame,} \\ \forall \hat{\underline{U}} \text{ v.m.,} \\ \displaystyle\int_{\widehat{AB}} \underline{f}(s,t).\hat{\underline{U}}(s)\,\mathrm{d}s + \underline{R}^S_B(t).\hat{\underline{U}}(s_B) + \underline{R}^S_A(t).\hat{\underline{U}}(s_A) \\ + \displaystyle\int_{\widehat{AB}} -X(s,t)\hat{D}(s)\,\mathrm{d}s = \int_{\widehat{AB}} \rho(s,t)\underline{a}(s,t).\hat{\underline{U}}(s)\,\mathrm{d}s \end{cases} \qquad [7.16]$$

Assuming that the field $\underline{X}(s,t) = X(s,t)\underline{t}(s)$ is continuous and continuously differentiable, we can transform the integrand in $\mathcal{P}_i(\hat{\underline{U}})$ and [7.16] becomes:

$$\begin{cases} \text{in a Galilean frame,} \\ \forall \hat{\underline{U}} \text{ v.m.,} \\ \displaystyle\int_{\widehat{AB}} \left(\frac{\mathrm{d}X(s,t)\underline{t}(s)}{\mathrm{d}s} + \underline{f}(s,t)\right).\hat{\underline{U}}(s)\,\mathrm{d}s - \int_{\widehat{AB}} \rho(s,t)\underline{a}(s,t).\hat{\underline{U}}(s)\,\mathrm{d}s \\ + (\underline{X}(s_A,t) + \underline{R}^S_A(t)).\hat{\underline{U}}(s_A) - (\underline{X}(s_B,t) - \underline{R}^S_B(t)).\hat{\underline{U}}(s_B) = 0. \end{cases} \qquad [7.17]$$

With the same arguments as in Chapter 6 (section 6.3.2), we derive *the field equation of motion*:

$$\begin{cases} \text{in a Galilean frame, } \forall P \in \widehat{AB}, \\ \dfrac{\mathrm{d}X(s,t)\underline{t}(s)}{\mathrm{d}s} + \underline{f}(s,t) - \boldsymbol{\rho}(s,t)\underline{a}(s,t) = 0 \end{cases} \qquad [7.18]$$

and the *boundary conditions* at the endpoints

$$\begin{cases} X(s_A,t)\underline{t}(s_A) = -\underline{R}_A^S(t) \\ X(s_B,t)\underline{t}(s_B) = \underline{R}_B^S(t). \end{cases} \qquad [7.19]$$

7.2.7.3. *Equations of motion for the subsystem*

For an arbitrary subsystem *stricto sensu* $S' \subset S$, the same reasoning can be performed with corresponding expressions for $\mathcal{P}'_e(\hat{U})$, $\mathcal{P}'_i(\hat{U})$ and $\mathcal{A}'(\hat{U})$. We obtain the *field equation* for $\underline{X}(s,t) = X(s,t)\underline{t}(s)$ in exactly the same form as [7.18], as a consequence of [7.6].

The *boundary equations* at the endpoints of the subsystem

$$\begin{cases} \underline{R}_{A'}^{S'}(t) = -X(s_{A'},t)\underline{t}(s_{A'}) \\ \underline{R}_{B'}^{S'}(t) = X(s_{B'},t)\underline{t}(s_{B'}) \end{cases} \qquad [7.20]$$

determine the external forces at the boundary of the subsystem and deserve some comments:

– The value of the external force exerted on the subsystem at the endpoint A' (respectively B') proceeds from the knowledge of the field of internal forces $\underline{X}(s,t) = X(s,t)\underline{t}(s)$ over $\widehat{AB}$ and does not depend on the considered subsystem $\widehat{A'B'}$. As in the case of the three-dimensional continuum, it is just the result of *contact actions* at point A' (respectively B') exerted by $(S\text{-}S')$. As a result, the superscript S' in $\underline{R}_{B'}^{S'}(t)$ and $\underline{R}_{A'}^{S'}(t)$ is no longer necessary.

– It is worth noting the sign difference between the two lines in [7.20]; it is related to the fact that the outward normal is directed along $-\underline{t}(s_{A'})$ at the point A' and along. $\underline{t}(s_{B'})$ at the point B'.

7.2.8. *Piecewise continuous fields*

7.2.8.1. *Piecewise continuous field of internal forces*

Returning to equation [7.16], we now assume that the field $\underline{X}(s,t) = X(s,t)\underline{t}(s)$ is piecewise continuous and continuously differentiable with discontinuities denoted by $[\![\underline{X}(s_i,t)]\!] = \underline{X}(s_i^+,t) - \underline{X}(s_i^-,t)$ at the points P_i. We also assume that, in addition to the line density $\underline{f}(s,t).\underline{\hat{U}}(s)$, the virtual rate of work by external forces includes *concentrated terms* $\underline{F}(s_k,t).\underline{\hat{U}}(s_k)$. The integrand in $\mathcal{P}_i(\underline{\hat{U}})$ can still be integrated by parts and we obtain for [7.16]:

$$
\left\{
\begin{aligned}
&\text{in a Galilean frame,}\\
&\forall \underline{\hat{U}} \text{ v.m.,}\\
&\int_{\widehat{AB}} (\frac{d\underline{X}(s,t)}{ds} + \underline{f}(s,t)).\underline{\hat{U}}(s)\,ds + \sum_i [\![\underline{X}(s_i,t)]\!].\underline{\hat{U}}(s_i) + \sum_k \underline{F}(s_k,t).\underline{\hat{U}}(s_k) \quad [7.21]\\
&- \int_{\widehat{AB}} \rho(s,t)\underline{a}(s,t).\underline{\hat{U}}(s)\,ds\\
&+ (\underline{X}(s_A,t) + \underline{R}_A^S(t)).\underline{\hat{U}}(s_A) - (\underline{X}(s_B,t) - \underline{R}_B^S(t)).\underline{\hat{U}}(s_B) = 0.
\end{aligned}
\right.
$$

Thence, in addition to [7.18] and [7.19], we obtain the jump condition for the field $\underline{X}(s,t) = X(s,t)\underline{t}(s)$

$$[\![X(s_i,t)\underline{t}(s_i)]\!] + \underline{F}(s_k,t) = 0, \qquad\qquad [7.22]$$

which means that $\underline{X}(s,t) = X(s,t)\underline{t}(s)$ is *discontinuous* at a point where a *concentrated force* is exerted on $\widehat{AB}$. (In the same way, as for the three-dimensional continuum (Chapter 6, section 6.4.2), *shock waves* would also induce discontinuities.)

7.2.8.2. *Piecewise continuous virtual velocity field*

Having completed the force modeling for this one-dimensional continuum using continuous and continuously differentiable virtual velocity fields as mathematical test functions, we examine the changes to be made in the expressions of $\mathcal{P}_e(\underline{\hat{U}})$, $\mathcal{P}_e'(\underline{\hat{U}})$, $\mathcal{P}_i(\underline{\hat{U}})$, $\mathcal{P}_i'(\underline{\hat{U}})$, $\mathcal{A}(\underline{\hat{U}})$ and $\mathcal{A}'(\underline{\hat{U}})$ in order to express the principle of virtual work with this model on the vector space of *piecewise continuous and continuously differentiable virtual velocity fields*.

Only $\mathcal{P}_i(\hat{U})$ and $\mathcal{P}'_i(\hat{U})$ need to be modified. Following the same reasoning path as in Chapter 6 (section 6.4.3), we consider a piecewise continuous and continuously differentiable virtual velocity field $\hat{U}$ with discontinuities $[\![\hat{U}(s_j)]\!] = \hat{U}(s_j^+) - \hat{U}(s_j^-)$ at points P_j on $\widehat{AB}$ where no concentrated force (nor shock wave) is exerted (Figure 7.5).

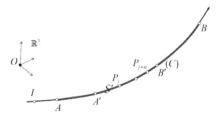

Figure 7.5. *Discontinuous virtual velocity field on* $S' = \widehat{A'B'}$

For an arbitrary subsystem $\widehat{A'B'} \subseteq \widehat{AB}$, we write [7.16] separately for each subsystem $\widehat{A'P_j}$, $\widehat{P_jP_{j+1}},\dots \widehat{P_{j+n}B'}$. From [7.22], we know that $\underline{X}(s_j,t)$ is continuous at the points P_j and equal to $\underline{X}(s_j,t) = X(s_j,t)\underline{t}(s_j)$. Summing up the obtained equations, we arrive at the expression of the principle of virtual work for the whole subsystem $\widehat{A'B'}$

$$
\left|
\begin{aligned}
&\text{in a Galilean frame, } \forall S' \subseteq S, \\
&\forall \hat{U} \text{ v.m.,} \\
&\int_{A'B'} f(s,t).\hat{U}(s)\,ds + \underline{R}_{B'}^{S'}(t).\hat{U}(s_{B'}) + \underline{R}_{A'}^{S'}(t).\hat{U}(s_{A'}) \\
&-\int_{A'B'} X(s,t)\hat{D}(s)\,ds - \sum_{s_{A'}<s_j<s_{B'}} X(s_j,t)[\![\hat{U}(s_j)]\!].\underline{t}(s_j) \\
&= \int_{A'B'} \rho(s,t)\underline{a}(s,t).\hat{U}(s)\,ds
\end{aligned}
\right. \qquad [7.23]
$$

that shows that the expression to be adopted for $\mathcal{P}'_i(\hat{U})$ is[2]

---

2 This formula can easily be interpreted within the framework of the theory of distributions.

$$\mathcal{T}_i'(\hat{\underline{U}}) = \int_{\widehat{A'B'}} -X(s,t)\hat{D}(s)\,\mathrm{d}s - \sum_{s_{A'} < s_j < s_{B'}} X(s_j,t)[\![\hat{\underline{U}}(s_j)]\!].\underline{t}(s_j). \tag{7.24}$$

7.2.9. Consistency and validation of the model

7.2.9.1. Mathematical consistency

The analysis of the mathematical consistency of the model amounts to finding out under what conditions the field and boundary equations [7.18] and [7.19] are compatible with each other and with the data.

From [7.19], it comes out straightforwardly that a first condition that must be satisfied is that the concentrated external forces $\underline{R}_B^S(t)$ and $\underline{R}_A^S(t)$ applied at the endpoints must be *directed along the tangents* to the director curve at points A and B.

Regarding the line density of external forces, the field equation [7.18] can be written explicitly in the local basis of the right-handed orthonormal triad $\underline{t}(s), \underline{n}(s), \underline{b}(s)$ defined by the Frenet–Serret equations (Figure 7.6) with $r(s)$ and $\tau(s)$ respectively the radii of curvature and torsion of (C) at point P.

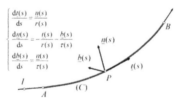

Figure 7.6. *Frenet–Serret equations and local orthonormal basis*

Equation [7.18] yields three scalar equations; as an example, in the case of equilibrium, i.e. $\underline{a}(s,t) = 0$:

$$\begin{cases} \forall P \in \widehat{AB}, \\[1mm] \dfrac{\mathrm{d}X(s,t)}{\mathrm{d}s} + f_t(s,t) = 0 \\[2mm] \dfrac{X(s,t)}{r(s)} + f_n(s,t) = 0 \\[2mm] f_b(s,t) = 0. \end{cases} \tag{7.25}$$

In order to be mathematically compatible, these equations impose that the data on the line density of external forces satisfy the following conditions along $\overset{\frown}{AB}$:

$$\begin{cases} \forall P \in \overset{\frown}{AB}, \\ f_t(s,t) = \dfrac{d}{ds}(r(s)f_n(s,t)) \\ f_b(s,t) = 0. \end{cases}$$ [7.26]

In the case of concentrated forces being exerted on $\overset{\frown}{AB}$, equation [7.22] must also be satisfied, which implies that the director curve (C) will but exceptionally be regular at a point where such a force is acting (Figure 7.7).

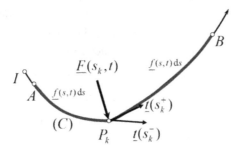

Figure 7.7. *Concentrated external force acting on the system*

The compatibility conditions on the data stated here above come in addition to the general necessary condition that they should be compatible with the *fundamental law of dynamics* as stated in Chapter 3 (sections 3.2.5 and 3.3.3).

7.2.9.2. *Integration of the field equations of motion*

The field equations of motion [7.18] and [7.22] can be integrated explicitly along $\overset{\frown}{AB}$, with the boundary conditions [7.19]. It follows that $\underline{X}(s,t) = X(s,t)\underline{t}(s)$ is given by either equation

$$\underline{X}(s,t) = \underline{R}_B^S(t) + \int_{\overset{\frown}{PB}}(\underline{f}(\sigma,t) - \rho(\sigma,t)\underline{a}(\sigma,t))\,d\sigma + \sum_{s<s_k<s_B} \underline{F}(s_k,t)$$ [7.27]

$$\underline{X}(s,t) = -\underline{R}_A^S(t) - \int_{\overset{\frown}{AP}}(\underline{f}(\sigma,t) - \rho(\sigma,t)\underline{a}(\sigma,t))\,d\sigma - \sum_{s_A<s_k<s} \underline{F}(s_k,t)$$ [7.28]

which are equivalent through the *fundamental law of dynamics*.

In the case of statics, i.e. $\underline{a}(\sigma,t) \equiv 0$, this result is often stated in the equivalent following forms:

– the internal force at the field point P is *equal to the sum* of all external forces acting on the *downstream* subsystem $S' = \overset{\frown}{PB}$;

– the internal force at the field point P is *the opposite of the sum* of all external forces acting on the *upstream* subsystem $S' = \overset{\frown}{AP}$.

7.2.9.3. *Physical interpretation*

The physical interpretation of the model comes out from equation [7.20] which expresses the fundamental assumption/intuition underlying the modeling process: the particles of S only exert contact actions on each other. Thence, at the endpoint B' of a subsystem $S' = \overset{\frown}{A'B'}$, the force $\underline{X}(s_{B'},t) = X(s_{B'},t)\underline{t}(s_{B'})$ represents the contact action of the "downstream" particle $P(s_{B'}^+)$ on the "upstream" particle $P(s_{B'}^-)$. This force is tangent to (C) and its magnitude $X(s_{B'},t)$ is positive when it corresponds to a *traction* exerted by $P(s_{B'}^+)$ on $P(s_{B'}^-)$. It is therefore named as the *tension* in the one-dimensional medium at the considered point consistently with [7.24].

7.2.9.4. *Relevance of the model. Validation*

In order to make the discussion simpler (without loss of generality regarding the conclusions), we now restrict the analysis to the case of equilibrium.

Since one-dimensional modeling is supposed to be the result of the "squeezing process" of a slender three-dimensional system referred to in section 7.2.1, the first condition for its practical relevance is that, through this process, external forces acting on the three-dimensional system may actually and conveniently be reduced to a *line density* $\underline{f}(s,t)$ and *concentrated forces* on (C), and *concentrated forces* $\underline{R}_B^S(t)$ and $\underline{R}_A^S(t)$ in B and A for the one-dimensional system.

This first condition having been fulfilled, practical relevance depends on whether these external forces comply with the mathematical compatibility conditions already listed. These conditions explicitly refer to the geometry of the director curve in the configuration which it takes when loaded with the given external forces.

Without pretending to be exhaustive, we may examine two cases that can be considered iconic.

The first concerns a slender three-dimensional solid that may be considered *stiff*, meaning that the shape of the director curve under loading only differs slightly from its known shape in the initial configuration. The model is then relevant if the shape of the director curve specified from the three-dimensional solid is suitably related to the given external forces regarding the mathematical compatibility conditions. This is precisely what is sought in *trusses* with ball-and-sockets connected rods, in *funicular* arches and even in some *vaults*.

The other one is, so to speak, the opposite and relates to slender three-dimensional solids *without stiffness*. Physically this means that, due to their extreme slenderness (wires) or their material or structural constitution (such as a bundle of loose wires and cables without stiffness), these solids *offer no resistance to any external moments applied on the boundary sections* S_B and S_A (Figure 7.3). In these cases, the shape of the director curve of the corresponding one-dimensional solid under loading is determined as part of the solution to the equilibrium problem in order that the mathematical compatibility conditions are satisfied. This practical and most frequent domain of application for the model is called the *Statics (and Dynamics) of wires* (and *cables without stiffness*). Moreover, due to the same extreme slenderness, these solids offer no resistance to negative values of the tension $X(s,t)$ (i.e. compression), which implies that $X(s,t) \geq 0$ must be added to the conditions imposed on the solution to the equilibrium problem.

Clearly enough, it turns out that the domain of relevance of the one-dimensional model that has just been constructed is not wide enough to cover what was initially expected when referring to the examples of slender structures encountered in practice as those presented in Figure 7.1. On this occasion, we may find some similarity with the discussion in Chapter 6 (section 6.6) about the relevance of the hydrostatic pressure force modeling where mathematical compatibility conditions were also a strong constraint and called for "freedom" to be given to the internal force field.

7.2.10. An example

To illustrate the preceding section, we consider the equilibrium of an arc made of a perfectly flexible wire, with no resistance to compression under the action of gravity. Let $\overparen{AB}$ be such an arc with constant mass per unit length $\rho(s) = \rho$, gravity being the only active external force. The arc is connected by means of ball and socket assembly joints to fixed supports at the points A and B

(see section 7.6.2). The length of the arc is $\ell = \left| \widehat{AB} \right| \geq AB$. The goal of the analysis is to determine the shape of the arc and the internal force field in the equilibrium state.

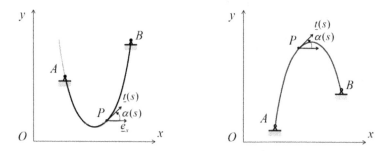

Figure 7.8. *a) The catenary problem. b) Funicular arch*

As presented in Figure 7.8, we denote by Ox and Oy the horizontal and vertical axes on the vertical plane defined by AB. Let H_A, H_B, V_A, V_B denote the horizontal and vertical components of $\underline{R}_A$ and $\underline{R}_B$:

$$\underline{R}_A = H_A \underline{e}_x + V_A \underline{e}_y, \; \underline{R}_B = H_B \underline{e}_x + V_B \underline{e}_y. \tag{7.29}$$

Without any loss of generality, we assume, as in Figure 7.8, that $x_A \leq x_B$ and $y_A \leq y_B$.

Global equilibrium of the arc implies

$$\begin{cases} H_A + H_B = 0 \\ V_A + V_B - \rho\ell = 0 \\ V_B(x_B - x_A) - H_B(y_B - y_A) - \int_{s_A}^{s_B} \rho(x - x_A)ds = 0. \end{cases} \tag{7.30}$$

With $\underline{X}(s) = X(s)\underline{t}(s) = X_x(s)\underline{e}_x + X_y(s)\underline{e}_y$, the equilibrium equations [7.18] and [7.19] are written explicitly as

$$\begin{cases} \dfrac{dX_x(s)}{ds} = 0 \\[2mm] \dfrac{dX_y(s)}{ds} = \rho \end{cases} \quad \text{and} \quad \begin{cases} X_x(s_A) = -H_A\,,\ X_y(s_A) = -V_A \\[2mm] X_x(s_B) = H_B\,,\ X_y(s_B) = V_B \end{cases}. \qquad [7.31]$$

It follows that $X_x(s)$ is equal to an arbitrary constant a along $\widehat{AB}$:

$$X_x(s) = a = -H_A = H_B.. \qquad [7.32]$$

Then, denoting by $\alpha(s)$ the angle between $\underline{t}(s)$ and $\underline{e}_x$, we may write

$$X_y(s) = a \tan \alpha(s) = -V_A + \rho(s - s_A), \qquad [7.33]$$

from which we derive the value of the radius of curvature of the equilibrium curve at the field point

$$r(s) = \frac{a}{\rho}(1 + \tan^2 \alpha(s)) \qquad [7.34]$$

and the tension in the wire

$$X(s) = a / \cos \alpha(s). \qquad [7.35]$$

7.2.10.1. Vertical equilibrium

$a = 0$ corresponds to the specific case when $\alpha(s) = \dfrac{\pi}{2}$ along $\widehat{AB}$, i.e. the vertical equilibrium position of the wire: $x_A = x_B$. No support is needed at the point A and the internal force field is just

$$\underline{X}(s) = X_y(y)\underline{e}_y = \rho(y - y_A)\underline{e}_y \qquad [7.36]$$

which is always tensile.

7.2.10.2. General equilibrium

In the case when $-\dfrac{\pi}{2} < \alpha(s) < \dfrac{\pi}{2}$ along $\widehat{AB}$, we derive the second order differential equation for the equilibrium curve in Cartesian coordinates from [7.34]

$$\frac{d^2y}{dx^2} = \frac{\rho}{a}(1+(\frac{dy}{dx})^2)^{1/2},$$

[7.37]

whose general solution is written as

$$y = \frac{a}{\rho}\cosh\frac{\rho}{a}(x-k_1)+k_2.$$

[7.38]

It is the equation of a *catenary*, with k_1 and k_2 as two arbitrary constants, to comply with the geometrical boundary conditions

$$\begin{cases} y_A = \frac{a}{\rho}\cosh\frac{\rho}{a}(x_A - k_1)+k_2 \\ \\ y_B = \frac{a}{\rho}\cosh\frac{\rho}{a}(x_B - k_1)+k_2 \end{cases}$$

[7.39]

while

$$\ell = \int_A^B ds = \frac{a}{\rho}\left[\sinh\frac{\rho}{a}(x_B - k_1) - \sinh\frac{\rho}{a}(x_A - k_1)\right].$$

[7.40]

These three conditions determine a, k_1, k_2. Provided that $\ell = \left|\widehat{AB}\right| > AB$, $\frac{a}{\rho} = L$ is such that

$$L\sinh\frac{x_B - x_A}{2L} = \frac{1}{2}(\ell^2 - (y_B - y_A)^2)^{1/2}.$$

[7.41]

This equation admits two roots with opposite signs, which depend solely on the geometrical data. In order that $X(s) = \rho L/\cos\alpha(s)$ be tensile, the *positive* root is chosen. Then, k_1, k_2 can be expressed conveniently as functions of the coordinates (x_0, y_0) of the turning point where $\alpha(s) = 0$ on the catenary so that [7.38] becomes

$$y - y_0 = L(\cosh\frac{x - x_0}{L} - 1)$$

[7.42]

and completely defines the arc $\widehat{AB}$ geometrically. Depending on the value of $\ell = \left| \widehat{AB} \right| > AB$, the turning point will actually be located on $\widehat{AB}$ as it shown in Figure 7.8(a), or upstream of A ($x_0 < x_A$) on the geometrical curve.

The internal force field is

$$\underline{X}(s) = X(s)\underline{t}(s) = \frac{\rho L}{\cos \alpha(s)} \underline{t}(s) \tag{7.43}$$

that identifies $a = \rho L$ as the tension in the arc at the turning point. Thence, as from [7.34] we derive $s - s_0 = L \tan \alpha(s)$, we may also write

$$\underline{X}(s) = \frac{\rho(s - s_0)}{\sin \alpha(s)} \underline{t}(s). \tag{7.44}$$

At the endpoints A and B

$$\begin{cases} H_A = -\rho L, V_A = -\rho L \tan \alpha(s_A) \\ H_B = \rho L, V_B = \rho L \tan \alpha(s_B) \end{cases} \tag{7.45}$$

and along $\widehat{AB}$

$$X(s) = \rho(L + y - y_0). \tag{7.46}$$

The negative root in [7.41] leads to the equilibrium curve presented in Figure 7.8(b), with $X(s)$ being compressive all along $\widehat{AB}$. This would be the solution for a *funicular arch*.

7.3. One-dimensional model with an oriented microstructure

7.3.1. *Guiding ideas*

From the relevance and validation analyses presented in the preceding section, it clearly appears that, apart from the specific case of slender solids when the geometry of the director curve and the applied external forces comply with the mathematical compatibility conditions, the true relevance of the model is to the Statics (and Dynamics) of slender solids with very little stiffness due to the very small diameter of their transverse cross-section such as wires or loose bundles of wires.

In order to cover the case of stiff slender solids such as beams or arcs that can sustain non-tangential external forces, external moments at the endpoints, etc. and model them as *one-dimensional continua*, it is necessary to *enrich the geometrical description* of the particles on the director curve by means of *additional geometrical parameters* that reflect the structure of the three-dimensional solid in a significant way.

As it is commonly confirmed by everyday practice that the stiffness of slender solids is directly related to the dimensions of their *transverse cross-section* (see Figure 7.1), the physical idea governing this more sophisticated approach is that a suitable one-dimensional model for stiff slender solids could be obtained by introducing additional geometrical parameters that would manifest the existence of a *significant transverse cross-section* as an *underlying oriented microstructure* of the one-dimensional continuum.

7.3.2. Geometrical modeling

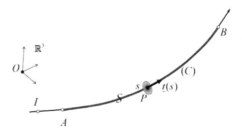

Figure 7.9. *The particle at the point* P

A system S is still described geometrically as on an arc $\overarc{AB}$ ($s_A < s_B$) of the director curve (C) and made of particles, "diluted material points", with length ds and mass $dm = \rho(s,t)\,ds$. At the field point P, the particle is geometrically characterized by its position $\underline{x}(s) = \underline{OP}$ and the orientation of an *attached transverse microstructure* with respect to the reference frame in $\mathbb{R}^3$, which accounts for the transverse cross-section. To make this concept more concrete, the particle can be interpreted as a plane surface element orthogonal to (C) at the field point P at time t (Figure 7.9). The definition is similar for a subsystem S' on an arc $\overarc{A'B'}$.

7.3.3. *Kinematics: actual and virtual motions*

7.3.3.1. *Actual motions*

Actual motions of S at time t, denoted by $\mathbf{U}(t)$, are defined on $\overset{\frown}{AB}$ by the evolution of the geometrical parameters of the particles, namely the velocity $\underline{U}(s,t)$ of the generic point P and the angular velocity $\underline{\Omega}(s,t)$ of the attached microstructure. Note that at point P, the rotation rate of the tangent $\underline{t}(s)$ to (C) is determined from $\underline{U}(s,t)$ only, *independently* of the rotation rate of the microstructure $\underline{\Omega}(s,t)$.

The fields $\underline{U}(s,t)$ and $\underline{\Omega}(s,t)$ are *piecewise continuous and continuously differentiable* on $\overset{\frown}{AB}$.

At each point P on $\overset{\frown}{AB}$, the velocity distributor (Appendix 3) denoted by $\{\mathbb{U}(s,t)\} = \{ P, \underline{U}(s,t), \underline{\Omega}(s,t)\}$ defines the rigid body motion of the particle interpreted as in Figure 7.9. The general expression for the corresponding velocity field in $\mathbb{R}^3$ is

$$\forall M \in \mathbb{R}^3, \ \underline{U}(M,t) = \underline{U}(s,t) + \underline{\Omega}(s,t) \wedge \underline{PM}. \tag{7.47}$$

Any rigid body motion of the system S (respectively subsystem S') is characterized by the rigid body motion of the current particle P being constant on $\overset{\frown}{AB}$ (respectively $\overset{\frown}{A'B'}$). In such a case, the system (respectively subsystem) is not deformed.

Deformation of the system is the result of the *variation* of the rigid body motion of the particles along the director curve. The *strain rate* of the one-dimensional continuum at the point P is the derivative, with respect to s, of the rigid body motion attached to the particle P, i.e. the derivative of the velocity distributor $\{\mathbb{U}(s,t)\}$ with respect to s (Appendix 3, section A3.4.1)

$$\frac{d}{ds}\{\mathbb{U}(s,t)\} = \{\mathbb{D}(s,t)\} = \left\{ P, \frac{d\underline{U}(s,t)}{ds} - \underline{\Omega}(s,t) \wedge \underline{t}(s), \frac{d\underline{\Omega}(s,t)}{ds} \right\}. \tag{7.48}$$

This means that the rigid body motion rate of the particle $P(s+ds)$ with respect to the particle $P(s)$ consists of the *translation rate* with vector

$$(\frac{\mathrm{d}\underline{U}(s,t)}{\mathrm{d}s} - \underline{\Omega}(s,t) \wedge \underline{t}(s))\, \mathrm{d}s \qquad\qquad [7.49]$$

and the rotation rate

$$\frac{\mathrm{d}\underline{\Omega}(s,t)}{\mathrm{d}s}\, \mathrm{d}s. \qquad\qquad [7.50]$$

From [7.49], we easily identify the *rate of stretch of the director curve* at the point P from the component of the translation rate along $\underline{t}(s)$

$$D(s,t) = \frac{\mathrm{d}U(s,t)}{\mathrm{d}s}.\underline{t}(s). \qquad\qquad [7.51]$$

The component of [7.49] orthogonal to the director curve

$$(\frac{\mathrm{d}U(s,t)}{\mathrm{d}s} - D(s,t)\underline{t}(s) - \underline{\Omega}(s,t) \wedge \underline{t}(s))\, \mathrm{d}s \qquad\qquad [7.52]$$

will be interpreted in section 7.4.4.

7.3.3.2. *Virtual motions*

Virtual motions of the system S or any subsystem S', denoted by $\hat{U}$, will be defined on $\widehat{AB}$ (respectively $\widehat{A'B'}$) by vector fields $\underline{\hat{U}}(s)$ and $\underline{\hat{\Omega}}(s)$ in the same way as actual motions, obviously including rigid body motions and encompassing actual motions. As in the preceding examples of the implementation of the virtual work method, these fields will first be assumed continuous and continuously differentiable and will ultimately (section 7.3.9) be allowed piecewise continuity. In each case, they generate a vector space with infinite dimension.

7.3.4. *Virtual rate of work by quantities of acceleration*

For any subsystem $S' \subseteq S$, the virtual rate of work by quantities of acceleration in the virtual motions defined in section 7.3.3 is the integral on $S' \subseteq S$ of a line density $[\, \boldsymbol{pa}(s,t)) \,].\{\hat{U}(s)\}$ where $[\, \boldsymbol{pa}(s,t) \,]$ stands for the wrench of quantities of acceleration:

$$\mathcal{A}(\hat{U}) = \int_{\widehat{AB}} [\, \boldsymbol{pa}(s,t) \,].\{\hat{U}(s)\}\, \mathrm{d}s \qquad\qquad [7.53]$$

$$\mathcal{A}'(\hat{\mathbf{U}}) = \int_{\overline{A'B'}} [\,\rho \boldsymbol{a}(s,t)\,].\{\hat{U}(s)\}\,ds.$$

[7.54]

7.3.5. *Virtual rate of work by external forces for the system*

The continuous linear form $\mathcal{P}_{\mathrm{e}}(\hat{\mathbf{U}})$ which expresses the virtual rate of work by external forces for the system consists of a contribution in the form of the integral over $\overline{AB}$ of a line density, playing the role of a "volume" term, and two point contributions at B and A as boundary terms.

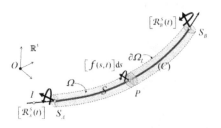

Figure 7.10. *External forces exerted on the system*

The line density on $\overline{AB}$ involves two vector fields $\underline{f}(s,t)$ and $\underline{m}(s,t)$ as cofactors of the fields $\hat{U}(s)$ and $\hat{\underline{\Omega}}(s)$. It is written as $\underline{f}(s,t).\hat{U}(s) + \underline{m}(s,t).\hat{\underline{\Omega}}(s)$. In addition, the boundary terms naturally introduce the vectors $\underline{R}_A^S(t)$, $\underline{H}_A^S(t)$, and $\underline{R}_B^S(t)$, $\underline{H}_B^S(t)$, as cofactors of $\hat{U}(s_A)$ and $\hat{\underline{\Omega}}(s_A)$, $\hat{U}(s_B)$ and $\hat{\underline{\Omega}}(s_B)$. Thence, for $\mathcal{P}_{\mathrm{e}}(\hat{\mathbf{U}})$:

$$\mathcal{P}_{\mathrm{e}}(\hat{\mathbf{U}}) = \int_{\overline{AB}} \underline{f}(s,t).\hat{U}(s)\,ds + \underline{R}_B^S(t).\hat{U}(s_B) + \underline{R}_A^S(t).\hat{U}(s_A)$$
$$+ \int_{\overline{AB}} \underline{m}(s,t).\hat{\underline{\Omega}}(s)\,ds + \underline{H}_B^S(t).\hat{\underline{\Omega}}(s_B) + \underline{H}_A^S(t).\hat{\underline{\Omega}}(s_A).$$

[7.55]

Introducing the wrench field $\left[\,\boldsymbol{f}(s,t)\,\right]$ (Appendix 3, section A3.4.2) defined from $\underline{f}(s,t)$ and $\underline{m}(s,t)$ by

$$\forall P \in \overline{AB},\, \left[\,\boldsymbol{f}(s,t)\,\right] = \left[\,P, \underline{f}(s,t), \underline{m}(s,t)\,\right]$$

[7.56]

makes it possible to write the line density in the compact form

$$\underline{f}(s,t).\hat{\underline{U}}(s) + \underline{m}(s,t).\hat{\underline{\Omega}}(s) = \left[\boldsymbol{f}(s,t)\right].\left\{\hat{\mathbb{U}}(s)\right\} \qquad [7.57]$$

with $\left\{\hat{\mathbb{U}}(s)\right\} = \left\{P, \hat{\underline{U}}(s), \hat{\underline{\Omega}}(s)\right\}$. External force wrenches can be defined in the same way at the endpoints:

$$\begin{cases} \left[\mathcal{R}_A^S(t)\right] = \left[A, \underline{R}_A^S(t), \underline{H}_A^S(t)\right] \\ \left[\mathcal{R}_B^S(t)\right] = \left[B, \underline{R}_B^S(t), \underline{H}_B^S(t)\right]. \end{cases} \qquad [7.58]$$

Finally:

$$\mathcal{P}_e(\hat{\mathbb{U}}) = \int_{\widehat{AB}}\left[\boldsymbol{f}(s,t)\right].\left\{\hat{\mathbb{U}}(s)\right\}ds + \left[\mathcal{R}_B^S(t)\right].\left\{\hat{\mathbb{U}}(s_B)\right\} + \left[\mathcal{R}_A^S(t)\right].\left\{\hat{\mathbb{U}}(s_A)\right\}. \qquad [7.59]$$

The physical meaning of cofactors $\left[\boldsymbol{f}(s,t)\right]$, $\left[\mathcal{R}_B^S(t)\right]$ and $\left[\mathcal{R}_A^S(t)\right]$, which represent the external force modeling on the one-dimensional continuum, follows the same track as in section 7.2.4 and is shown in Figure 7.10. The infinitesimal force wrench $\left[\boldsymbol{f}(s,t)\right]ds$ on the length element ds accounts for the body forces in the volume Ω and surface forces on the boundary $\partial\Omega_L$ exerted on the three-dimensional original system between the cross-sections at s and $(s+ds)$. Thence, for its resultant force $\underline{f}(s,t)ds$, the physical meaning is clearly identical to what was said in section 7.2.4 while the possibility of a line density of moment $\underline{m}(s,t)$ is illustrated by the example in Figure 7.11 where a plate (or a shell) is embedded in a beam as a built-in support along a part of its boundary.

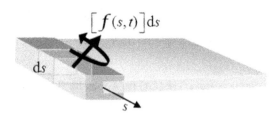

Figure 7.11. *Beam as the built-in support of a plate element*

The force wrenches $\left[\mathcal{R}_B^S(t)\right]$ and $\left[\mathcal{R}_A^S(t)\right]$ account for the forces exerted by the world external to the three-dimensional system on the boundary sections S_B and S_A respectively. Note that, with the definition of the considered virtual motions being wider than in section 7.2, it is now possible, in the "squeezing" process, to model the external forces exerted on S_B and S_A by their resulting forces $\underline{R}_B^S(t)$, $\underline{R}_A^S(t)$ and moments $\underline{H}_B^S(t)$, $\underline{H}_A^S(t)$ at points B and A respectively.

7.3.6. Virtual rate of work by external forces for subsystems

For $S' = \widehat{A'B'} \subset S = \widehat{AB}$, a subsystem of S *stricto sensu* (Figure 7.4), the expression for $\mathcal{P}_e'(\hat{\mathbf{U}})$ is similar to [7.59] and can be written in the compact form, homologous to [7.5]:

$$\mathcal{P}_e'(\hat{\mathbf{U}}) = \int_{\widehat{A'B'}}\left[f^{S'}(s,t)\right].\left\{\hat{\mathbb{U}}(s)\right\}ds + \left[\mathcal{R}_{B'}^{S'}(t)\right].\left\{\hat{\mathbb{U}}(s_{B'})\right\} + \left[\mathcal{R}_{A'}^{S'}(t)\right].\left\{\hat{\mathbb{U}}(s_{A'})\right\}. \quad [7.60]$$

With the assumption that the particles of S *do not exert any action at a distance* on each other, the force wrench line density $\left[f^{S'}(s,t)\right]$ does not depend on S'

$$\left[f^{S'}(s,t)\right] = \left[f(s,t)\right]. \qquad [7.61]$$

$\mathcal{P}_e'(\hat{\mathbf{U}})$ is finally written in the form

$$\mathcal{P}_e'(\hat{\mathbf{U}}) = \int_{\widehat{A'B'}}\left[f(s,t)\right].\left\{\hat{\mathbb{U}}(s)\right\}ds + \left[\mathcal{R}_{B'}^{S'}(t)\right].\left\{\hat{\mathbb{U}}(s_{B'})\right\} + \left[\mathcal{R}_{A'}^{S'}(t)\right].\left\{\hat{\mathbb{U}}(s_{A'})\right\}, \quad [7.62]$$

where the force wrenches $\left[\mathcal{R}_{B'}^{S'}(t)\right]$ and $\left[\mathcal{R}_{A'}^{S'}(t)\right]$ model the external forces exerted by the world external to S' on the boundary of $S' = \widehat{A'B'}$. Upon the same arguments as before, they are only exerted by the *constituent particles of S external to S'*.

7.3.7. Virtual rate of work by internal forces

As in the preceding modeling processes, we postulate that, for any subsystem $S' = \widehat{A'B'} \subseteq S = \widehat{AB}$, the continuous linear form $\mathcal{P}_i'(\hat{\mathbf{U}})$ is the integral on $\widehat{A'B'}$ of a line density $p_i(\hat{\mathbf{U}})$, independent of S', that *is a linear form in the local values of the virtual fields* $\hat{\underline{U}}(s)$, $\hat{\underline{\Omega}}(s)$ *and their first derivatives along* (C):

$$\forall \hat{\mathbf{U}} \text{ m. v.}, \forall S' \subseteq S, \; \mathcal{P}_i'(\hat{\mathbf{U}}) = \int_{\widehat{A'B'}} p_i(\hat{\mathbf{U}}) \, \mathrm{d}s. \qquad [7.63]$$

Equivalently, we may say that $p_i(\hat{\mathbf{U}})$ at the field point P on (C) is a linear form in $\left\{ \hat{U}(s) \right\}$ and $\dfrac{\mathrm{d}}{\mathrm{d}s}\left\{ \hat{U}(s) \right\}$. Referring to Appendix 3 (section A3.4.1), we define

$$\left\{ \hat{\mathbb{D}}(s) \right\} = \frac{\mathrm{d}}{\mathrm{d}s}\left\{ \hat{U}(s) \right\} = \left\{ P, \; \frac{\mathrm{d}\hat{\underline{U}}(s)}{\mathrm{d}s} - \hat{\underline{\Omega}}(s) \wedge \underline{t}(s), \; \frac{\mathrm{d}\hat{\underline{\Omega}}(s)}{\mathrm{d}s} \right\} \qquad [7.64]$$

and we can write

$$p_i(\hat{\mathbf{U}}) = -\left[A(s,t) \right].\left\{ \hat{U}(s) \right\} - \left[\boldsymbol{X}(s,t) \right].\frac{\mathrm{d}}{\mathrm{d}s}\left\{ \hat{U}(s) \right\} \qquad [7.65]$$

where $\left[A(s,t) \right]$ and $\left[\boldsymbol{X}(s,t) \right]$ are two wrenches.

7.3.8. Implementation of the principle of virtual work

7.3.8.1. Specifying the virtual rate of work by internal forces

The dual statement of the law of mutual actions is now written as

$$\left\{ \begin{array}{l} \forall S' \subseteq S, \; \forall \hat{\mathbf{U}} \text{ r.b.v.m. of } S', \\[2mm] \mathcal{P}_i'(\hat{\mathbf{U}}) = -\int_{\widehat{A'B'}} \left(\left[A(s,t) \right].\left\{ \hat{U}(s) \right\} + \left[\boldsymbol{X}(s,t) \right].\dfrac{\mathrm{d}}{\mathrm{d}s}\left\{ \hat{U}(s) \right\} \right) \mathrm{d}s = 0. \end{array} \right. \qquad [7.66]$$

As stated in section 7.3.3, any rigid body motion of the system $S' \subseteq S$ is characterized by the rigid body motion $\left\{ \hat{U}(s) \right\}$ of the current particle $P(s)$ being constant over $\widehat{A'B'}$

$$\left\{ \hat{\mathbb{U}}(s) \right\} = \left\{ P, \; \hat{\underline{U}}(s), \; \hat{\underline{\Omega}}(s) \right\} = \left\{ O, \; \hat{\underline{U}}_0, \; \hat{\underline{\Omega}}_0 \right\}, \qquad [7.67]$$

with $\hat{\underline{U}}_0$ and $\hat{\underline{\Omega}}_0$ arbitrary vectors in $\mathbb{R}^3$. It then follows from [7.66] that $[A(s,t)]$ is zero on $\widehat{AB}$ and $p_i\,(\hat{\mathbf{U}})$ reduces to

$$p_i\,(\hat{\mathbf{U}}) = -[\boldsymbol{X}(s,t)]. \frac{d}{ds}\{\hat{\mathbb{U}}(s)\} = -[\boldsymbol{X}(s,t)].\{\hat{\mathbb{D}}(s)\}. \qquad [7.68]$$

Internal forces at the field point P are represented by the *wrench* $[\boldsymbol{X}(s,t)]$ which is associated in the expression of $p_i\,(\hat{\mathbf{U}})$ with the *virtual strain rate* $\{\hat{\mathbb{D}}(s)\}$ of the one-dimensional medium, the derivative of the virtual distributor field $\{\hat{\mathbb{U}}(s)\}$.

With $\underline{X}(s,t)$ and $\underline{\Gamma}(s,t)$ denoting the reduced elements of the *internal force wrench* at the field point, we write $[\boldsymbol{X}(s,t)] = [P, \underline{X}(s,t), \underline{\Gamma}(s,t)]$ and, taking [7.64] into account, $p_i\,(\hat{\mathbf{U}})$ can be written in the expanded form

$$p_i\,(\hat{\mathbf{U}}) = -\underline{X}(s,t).(\frac{d\hat{\underline{U}}(s)}{ds} + \underline{t}(s) \wedge \hat{\underline{\Omega}}(s)) - \underline{\Gamma}(s,t). \frac{d\hat{\underline{\Omega}}(s)}{ds} \qquad [7.69]$$

It is worth noting the similarity between [7.14] and the compact expression [7.68] in terms of wrench and distributor that favors reflection and simplifies calculations as it will appear in the following section.

7.3.8.2. Equations of motion for the system

With [7.68] as an expression for the virtual rate of work by internal forces, we write the second statement of the principle of virtual work applied to the system $S = \widehat{AB}$

$$\left\{ \begin{array}{l} \text{in a Galilean frame,} \\[4pt] \forall \hat{\mathbf{U}} \text{ v.m.,} \\[4pt] \int_{\widehat{AB}}[\boldsymbol{f}(s,t)].\{\hat{\mathbb{U}}(s)\}ds + [\mathcal{R}_B^S(t)].\{\hat{\mathbb{U}}(s_B)\} + [\mathcal{R}_A^S(t)].\{\hat{\mathbb{U}}(s_A)\} \\[4pt] -\int_{\widehat{AB}}[\boldsymbol{X}(s,t)]. \frac{d}{ds}\{\hat{\mathbb{U}}(s)\}ds = \int_{\widehat{AB}}[\boldsymbol{\rho a}(s,t)].\{\hat{\mathbb{U}}(s)\}ds. \end{array} \right. \qquad [7.70]$$

As a matter of fact, in consideration of the slenderness of the three-dimensional solid which is at the origin of the model, the wrench of quantities of acceleration at point P usually reduces to $[\boldsymbol{\rho a}(s,t)]=[P, \rho(s,t)\underline{a}(s,t), 0]$ with $\underline{a}(s,t)$, the *acceleration* of the particle at the point P and $\rho(s,t)$ the mass line density (see section 7.4.2).

Thence, assuming the field $[\boldsymbol{X}(s,t)]=[P,\underline{X}(s,t),\underline{\Gamma}(s,t)]$ to be continuous and continuously differentiable, we can transform the integrand in $\mathcal{P}_i(\hat{\mathbf{U}})$ from the expression of the derivative of the duality product (Appendix 3, section A3.4.3)

$$\frac{d}{ds}\left\langle[\boldsymbol{X}(s,t)].\{\hat{\mathbb{U}}(s)\}\right\rangle=\{\hat{\mathbb{U}}(s)\}.\frac{d}{ds}[\boldsymbol{X}(s,t)]+[\boldsymbol{X}(s,t)].\frac{d}{ds}\{\hat{\mathbb{U}}(s)\} \qquad [7.71]$$

where (Appendix 3, section A3.4.2)

$$\frac{d}{ds}[\boldsymbol{X}(s,t)]=\left[P, \frac{d\underline{X}(s,t)}{ds}, \frac{d\underline{\Gamma}(s,t)}{ds}-\underline{X}(s,t)\wedge \underline{t}(s)\right]. \qquad [7.72]$$

and, integrating by parts we obtain

$$\left\{\begin{array}{l} \text{in a Galilean frame,} \\[4pt] \forall \hat{\mathbf{U}} \text{ v.m.,} \\[4pt] \int_{\widehat{AB}}(\frac{d}{ds}[\boldsymbol{X}(s,t)]+[\boldsymbol{f}(s,t)]).\{\hat{\mathbb{U}}(s)\}\,ds \qquad [7.73] \\[4pt] +([\mathcal{R}_B^S(t)]-[\boldsymbol{X}(s_B,t)]).\{\hat{\mathbb{U}}(s_B)\}+([\mathcal{R}_A^S(t)]+[\boldsymbol{X}(s_A,t)].\{\hat{\mathbb{U}}(s_A)\} \\[4pt] =\int_{\widehat{AB}}[\boldsymbol{\rho a}(s,t)].\{\hat{\mathbb{U}}(s)\}\,ds=\int_{\widehat{AB}}\rho(s,t)\underline{a}(s,t).\hat{\underline{U}}(s). \end{array}\right.$$

From [7.73], we derive the field differential equation of motion on $\widehat{AB}$

$$\left\{\begin{array}{l} \text{in a Galilean frame,} \\[4pt] \forall P\in \widehat{AB}, \qquad\qquad\qquad\qquad\qquad [7.74] \\[4pt] \frac{d}{ds}[\boldsymbol{X}(s,t)]+[\boldsymbol{f}(s,t)]-[\boldsymbol{\rho a}(s,t)]=0 \end{array}\right.$$

and the boundary conditions at the end points of the system

$$\begin{cases} [\boldsymbol{X}(s_A,t)] = -\left[\mathcal{R}_A^S(t)\right] \\ [\boldsymbol{X}(s_B,t)] = \left[\mathcal{R}_B^S(t)\right]. \end{cases} \qquad [7.75]$$

7.3.8.3. *Equations of motion for a subsystem*

For an arbitrary subsystem *stricto sensu* $S' \subset S$, with corresponding expressions for $\mathcal{P}_e'(\hat{\mathbf{U}})$, $\mathcal{P}_i'(\hat{\mathbf{U}})$ and $\mathcal{A}'(\hat{\mathbf{U}})$, we obtain the *field equation* for $[\boldsymbol{X}(s,t)]$ in exactly the same form as [7.74], as a consequence of [7.61].

The *boundary equations* at the endpoints of the subsystem

$$\begin{cases} \left[\mathcal{R}_{A'}^{S'}(t)\right] = -[\boldsymbol{X}(s_{A'},t)] \\ \left[\mathcal{R}_{B'}^{S'}(t)\right] = [\boldsymbol{X}(s_{B'},t)] \end{cases} \qquad [7.76]$$

determine the external forces at the boundary of this subsystem.

7.3.8.4. *Equations of motion in terms of reduced elements*

Equations [7.74]–[7.76] can be expressed in terms of the reduced elements at the field point of $[\boldsymbol{X}(s,t)]$ and its derivative [7.72] in order to obtain the vector equations for $\underline{X}(s,t)$ and $\underline{\Gamma}(s,t)$:

Differential equations

$$\begin{cases} \text{in a Galilean frame,} \\ \forall P \in \widehat{AB}, \\ \dfrac{d\underline{X}(s,t)}{ds} + \underline{f}(s,t) = \rho(s,t)\underline{a}(s,t) \\ \dfrac{d\underline{\Gamma}(s)}{ds} + \underline{t}(s) \wedge \underline{X}(s) + \underline{m}(s) = 0. \end{cases} \qquad [7.77]$$

Boundary conditions at the end points of S

$$\begin{cases} \underline{X}(s_A,t) = -\underline{R}_A^S(t), \quad \underline{\Gamma}(s_A,t) = -\underline{H}_A^S(t) \\ \underline{X}(s_B,t) = \underline{R}_B^S(t), \quad \underline{\Gamma}(s_B,t) = \underline{H}_B^S(t) \end{cases} \qquad [7.78]$$

Boundary conditions at the end points of S'

$$\begin{cases} \underline{R}_{A'}^{S'}(t) = -\underline{X}(s_{A'},t), & \underline{H}_{A'}^{S'}(t) = -\underline{\Gamma}(s_{A'},t) \\ \underline{R}_{B'}^{S'}(t) = \underline{X}(s_{B'},t), & \underline{H}_{B'}^{S'}(t) = \underline{\Gamma}(s_{B'},t). \end{cases}$$

[7.79]

7.3.9. Comments

The equations of motion [7.74]–[7.76] in terms of wrenches are quite similar to [7.18]–[7.20] and deserve the same comments.

– The value of the external force wrench exerted on the subsystem at the endpoint A' (respectively B') proceeds from the knowledge of the field of internal force wrench $[\mathcal{X}(s,t)]$ over $\widehat{AB}$ and does not depend on the considered subsystem $\widehat{A'B'}$. This implies that it is only the result of *contact actions* at point A' (respectively B') exerted by $(S-S')$. As a result, the superscript S' in $\left[\mathcal{R}_{B'}^{S'}(t)\right]$ and $\left[\mathcal{R}_{A'}^{S'}(t)\right]$ is no longer necessary.

– Again, the sign difference between the two lines in [7.76] is due to the fact that the outward normal at points A' and B' are opposite to each other.

But, contrary to section 7.2.9, the only *mathematical compatibility condition* to be satisfied by the data is the general necessary condition that they should comply with the *fundamental law of dynamics* (Chapter 3, section 3.2.5, 3.3.3). In particular, forces arbitrarily orientated with respect to the director curve and moments can be applied at the end points of the system, while the line density of external forces, like the concentrated forces, suffers no restriction.

In other words, *this more elaborate model fulfills the objectives* set out in section 7.3.1. It actually provides a one-dimensional representation for stiff slender solids such as beams, arches or cables with larger cross-sections than wires. For this reason, it is often referred to as the *Statics or Dynamics of beams*.

7.3.10. Piecewise continuous fields

7.3.10.1. Piecewise continuous field of internal forces

As in section 7.2.8, we now assume that the field $[\mathcal{X}(s,t)]$ is piecewise continuous and continuously differentiable with discontinuities $\left[\!\left[[\mathcal{X}(s_i,t)] \right]\!\right]$ at the points P_i. We also assume that, in addition to the line density $\left[\boldsymbol{f}(s,t)\right].\left\{\hat{\mathbb{U}}(s)\right\}$, the virtual rate of work by external forces includes concentrated terms $\left[\boldsymbol{\mathcal{F}}(s_k,t)\right].\left\{\hat{\mathbb{U}}(s_k)\right\}$.

Thanks to its compact expression, we can proceed with [7.70] as in section 7.2.8 with [7.16] and obtain, in addition to [7.74] and [7.75], the jump condition on $\left[\boldsymbol{X}(s,t)\right]$

$$\left[\!\left[\left[\boldsymbol{X}(s_i,t)\right]\right]\!\right]+\left[\boldsymbol{\mathcal{F}}(s_k,t)\right]=0, \tag{7.80}$$

which means that $\left[\boldsymbol{X}(s,t)\right]$ is *discontinuous* when passing a point where a *concentrated wrench* is exerted on $\overset{\frown}{AB}$. (Shock waves would also induce discontinuities.)

In such a point P_k, [7.80] can be written in terms of reduced elements

$$\begin{cases} \left[\!\left[\underline{X}(s_k,t)\right]\!\right]+\underline{F}(s_k,t)=0 \\ \left[\!\left[\underline{\Gamma}(s_k,t)\right]\!\right]+\underline{M}(s_k,t)=0 \end{cases} \tag{7.81}$$

with $\underline{F}(s_k,t)$ and $\underline{M}(s_k,t)$ the reduced elements of $\left[\boldsymbol{\mathcal{F}}(s_k,t)\right]$ at that point

$$\left[\boldsymbol{\mathcal{F}}(s_k,t)\right]=\left[P_k,\ \underline{F}(s_k,t),\ \underline{M}(s_k,t)\right]. \tag{7.82}$$

As for $\underline{m}(s,t)$ in section 7.2.4, the physical meaning of $\underline{M}(s_k,t)$ is shown in Figure 7.12 where two beams are connected to each other by a rigid joint.

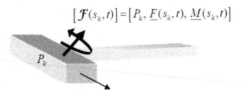

Figure 7.12. *Beams connected by a rigid joint*

7.3.10.2. *Piecewise continuous virtual motions*

With the model that has just been constructed, we now examine, as in section 7.2.8, the expression to be adopted for the principle of virtual work when piecewise continuous virtual motions are considered.

Let $\hat{\mathbf{U}}$ be such a virtual motion defined by $\left\{\hat{\mathbb{U}}(s)\right\} = \left\{ P, \underline{\hat{U}}(s), \underline{\hat{\Omega}}(s) \right\}$ with discontinuities $\left[\!\left[\left\{\hat{\mathbb{U}}(s_j)\right\} \right]\!\right] = \left\{ P_j, [\![\underline{\hat{U}}(s_j)]\!], [\![\underline{\hat{\Omega}}(s_j)]\!] \right\}$ at points P_j on $\overparen{AB}$ where no concentrated external force wrench is exerted. Referring again to Figure 7.5, we write [7.70] separately for each subsystem $\overparen{A'P_j}$, $\overparen{P_jP_{j+1}}, \dots \overparen{P_{j+n}B'}$ of any subsystem $\overparen{A'B'} \subseteq \overparen{AB}$. Taking [7.80] into account and summing up the equations so obtained, we arrive at the expression of the principle of virtual work for the whole subsystem $\overparen{A'B'}$ in the form

$$
\left\{
\begin{aligned}
&\text{in a Galilean frame, } \forall S' \subseteq S, \\
&\forall \hat{\mathbf{U}} \text{ v.m.,} \\
&\int_{\overparen{A'B'}} \left[\boldsymbol{f}(s,t) \right].\left\{\hat{\mathbb{U}}(s)\right\} ds + \left[\mathcal{R}_{B'}^{S'}(t) \right].\left\{\hat{\mathbb{U}}(s_{B'})\right\} + \left[\mathcal{R}_{A'}^{S'}(t) \right].\left\{\hat{\mathbb{U}}(s_{A'})\right\} \\
&- \int_{\overparen{A'B'}} \left[\boldsymbol{X}(s,t) \right].\frac{d}{ds}\left\{\hat{\mathbb{U}}(s)\right\} ds - \sum_{s_{A'} < s_j < s_{B'}} \left[\boldsymbol{X}(s_j,t) \right].\left[\!\left[\left\{\hat{\mathbb{U}}(s_j)\right\} \right]\!\right] \\
&= \int_{\overparen{A'B'}} \rho(s,t)\underline{a}(s,t).\underline{\hat{U}}(s)\, ds
\end{aligned}
\right.
\qquad [7.83]
$$

that yields the final expression to be retained for $\mathcal{P}_i'(\hat{\mathbf{U}})$:

$$
\mathcal{P}_i'(\hat{\mathbf{U}}) = -\int_{\overparen{A'B'}} \left[\boldsymbol{X}(s,t) \right].\frac{d}{ds}\left\{\hat{\mathbb{U}}(s)\right\} ds - \sum_{s_{A'} < s_j < s_{B'}} \left[\boldsymbol{X}(s,t) \right].\left[\!\left[\left\{\hat{\mathbb{U}}(s_j)\right\} \right]\!\right].
\qquad [7.84]
$$

In terms of the reduced elements of $\left[\boldsymbol{X}(s_j,t) \right]$ and $\left[\!\left[\left\{\hat{\mathbb{U}}(s_j)\right\} \right]\!\right]$ at the field point P_j, the last term in [7.84] is written as

$$
-\sum_{s_{A'} < s_j < s_{B'}} \left[\boldsymbol{X}(s_j,t) \right].\left[\!\left[\left\{\hat{\mathbb{U}}(s_j)\right\} \right]\!\right] = -\underline{X}(s_j,t).[\![\underline{\hat{U}}(s_j)]\!] - \underline{\Gamma}(s,t).[\![\underline{\hat{\Omega}}(s_j)]\!]
\qquad [7.85]
$$

7.3.11. *Integration of the field equations of motion*

The field equations of motion [7.74] and [7.80] can now be integrated explicitly along $\overset{\frown}{AB}$ with the boundary conditions [7.75], without any mathematical compatibility condition on the data besides the *fundamental law of dynamics*, and determine $[\boldsymbol{X}(s,t)]$ by either equation

$$[\boldsymbol{X}(s,t)] = \left[\mathcal{R}_B^S(t)\right] + \int_{\overset{\frown}{PB}} \left(\left[\boldsymbol{f}(\sigma,t)\right] - [\boldsymbol{pa}(\sigma,t))]\right)d\sigma + \sum_{s<s_k<s_B} [\boldsymbol{F}(s_k,t)] \qquad [7.86]$$

$$[\boldsymbol{X}(s,t)] = -\left[\mathcal{R}_A^S(t)\right] - \int_{\overset{\frown}{AP}} \left(\left[\boldsymbol{f}(\sigma,t)\right] - [\boldsymbol{pa}(\sigma,t))]\right)d\sigma - \sum_{s_A<s_k<s} [\boldsymbol{F}(s_k,t)], \qquad [7.87]$$

that are equivalent to each other.

In the case of statics ($[\boldsymbol{pa}(s,t))] = 0$), this result is commonly stated in the equivalent following forms (Figure 7.13):

– the internal force wrench at the field point P is equal to the wrench of all external forces acting on the downstream subsystem $S' = \overset{\frown}{PB}$;

– the internal force wrench at the field point P is the opposite of the wrench of all external forces acting on the upstream subsystem $S' = \overset{\frown}{AP}$.

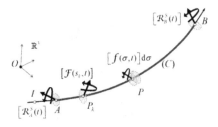

Figure 7.13. *External forces acting on the system*

7.4. Relevance of the model

7.4.1. *Physical interpretation*

As a consequence of section 7.3.9, the internal force wrench $[\boldsymbol{X}(s,t)]$ is the resulting wrench of the *contact actions* exerted by any downstream subsystem $\overset{\frown}{PB'}$

on any upstream subsystem $\widehat{A'P}$ at the field point P, with $\underline{X}(s,t)$ and $\underline{\Gamma}(s,t)$ as reduced elements at that point. Note that $-[X(s,t)]$ represents the contact actions exerted by any upstream subsystem $\widehat{A'P}$ on any downstream subsystem $\widehat{PB'}$ at the same point. Figure 7.14 shows the corresponding thought experiment.

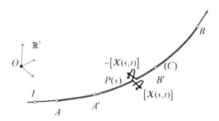

Figure 7.14. *Internal force wrench at the field point*

The field differential equation of motion [7.74] expresses the fundamental law of dynamics applied to the infinitesimal line element ds at the field point P (Figure 7.15).

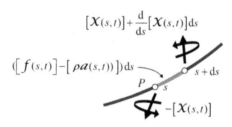

Figure 7.15. *Fundamental law of dynamics applied to the infinitesimal line element*

7.4.2. Matching the one-dimensional model with the Cauchy stress model

Referring to the initial guiding idea of a one-dimensional model being obtained by "squeezing" a three-dimensional slender solid onto a director curve, it is natural to try to establish a connection with the three-dimensional force model presented in the preceding chapter. The guiding idea in this attempt is to seek a way of connecting, for the same slender solid, the wrench field representing the internal forces in the one-dimensional model with the Cauchy tensor stress field that describes internal forces within the formalism of the classical three-dimensional continuum.

Following the virtual work approach, for any virtual motion $\hat{\mathbf{U}}$ of the one-dimensional model, we define a virtual motion for the three-dimensional solid through a virtual velocity field generated by the velocity distributor field $\left\{\hat{\mathbb{U}}(s)\right\}=\left\{P, \hat{\underline{U}}(s), \hat{\underline{\Omega}}(s)\right\}$ that defines $\hat{\mathbf{U}}$. Then, we identify the virtual rates of work by quantities of acceleration, external forces and internal forces in this virtual motion for the three-dimensional volume element presented in Figure 7.16, denoted by $(ds \times S)$ and delimited by the transverse cross-sections $S(s)$ and $S(s+ds)$, with the corresponding quantities for the particle ds in $\hat{\mathbf{U}}$.

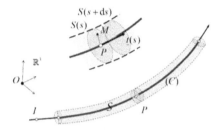

Figure 7.16. *Matching the one-dimensional model with the three-dimensional one*

7.4.2.1. *Three-dimensional virtual velocity field*

We assume that the director curve (C), wisely chosen in a direction along which the solid is slender, is regular. Given a velocity distributor field continuous and continuously differentiable $\left\{\hat{\mathbb{U}}(s)\right\}=\left\{P, \hat{\underline{U}}(s), \hat{\underline{\Omega}}(s)\right\}$, we can construct a virtual motion $\hat{\underline{U}}(x)$ for the three-dimensional solid, *in the neighborhood of* (C), in the following way. Let M be a field point in the three-dimensional solid, in the transverse cross-section of (C) corresponding to P. It is assumed as the *condition associated with slenderness and defining "neighborhood"* here that

$$\underline{PM}.\underline{n}(s) \ll r(s) \tag{7.88}$$

with $r(s)$ the radius of curvature and $\underline{n}(s)$ the normal to (C) at that point (see Figure 7.6). Then, the virtual velocity at point M is generated by the velocity distributor $\left\{\hat{\mathbb{U}}(s)\right\}$

$$\hat{\underline{U}}(\underline{x}) = \hat{\underline{U}}(s) + \hat{\underline{\Omega}}(s) \wedge \underline{PM} = \hat{\underline{U}}(s) + \hat{\underline{\underline{\Omega}}}(s).\underline{PM} \tag{7.89}$$

as a one-to-one definition, and the gradient[3] of this velocity field at point M reduces to

$$\operatorname{grad}\hat{U}(\underline{x}) = (\hat{\underline{\underline{\Omega}}}(s) + \frac{\mathrm{d}\hat{U}(s)}{\mathrm{d}s} - \hat{\underline{\underline{\Omega}}}(s).\underline{t}(s) + \frac{\mathrm{d}\hat{\underline{\underline{\Omega}}}(s)}{\mathrm{d}s}.\underline{PM}) \otimes \underline{t}(s). \qquad [7.90]$$

7.4.2.2. Virtual rate of work by quantities of acceleration

At the field point M in the three-dimensional solid, located in the transverse cross-section of (C) corresponding to P, the quantity of acceleration of the volume element $\mathrm{d}\Omega = \mathrm{d}s\,\mathrm{d}a$ is $\rho(\underline{x},t)\underline{a}(\underline{x},t)\mathrm{d}s\,\mathrm{d}a$. Thence, in the virtual velocity field [7.89], the virtual rate of work by quantities of acceleration for the three-dimensional volume $(\mathrm{d}s \times S)$ is equal to

$$\mathrm{d}s\left[P, \int_{S}\rho(\underline{x},t)\underline{a}(\underline{x},t)\mathrm{d}a, \int_{S}\underline{PM} \wedge \rho(\underline{x},t)\underline{a}(\underline{x},t)\mathrm{d}a\right].\{\hat{\underline{U}}(s)\} \text{ with } \underline{x} = \underline{s} + \underline{PM}. \quad [7.91]$$

In consideration of the slenderness of the three-dimensional solid, the resultant moment term in the wrench is usually considered negligible and the resultant force term reduces to $\boldsymbol{\rho}(s,t)\underline{a}(s,t)$ with $\boldsymbol{\rho}(s,t) = \int_{S}\rho(\underline{x},t)\mathrm{d}a$ the mass line density.

7.4.2.3. Virtual rate of work by external forces

The virtual rate of work by external forces acting in the three-dimensional volume and on its boundary surface is also expressed by a duality product similar to [7.91] with the wrench $\left[\boldsymbol{f}(\underline{s},t)\right]$ defined in [7.56]:

$$\mathrm{d}s\left[\boldsymbol{f}(s,t)\right].\{\hat{\underline{U}}(s)\} \text{ with } \left[\boldsymbol{f}(\underline{s},t)\right] = \left[P, \underline{f}(s,t), \underline{m}(s,t)\right]. \qquad [7.92]$$

7.4.2.4. Virtual rate of work by internal forces

The virtual rate of work by internal forces for the three-dimensional volume $(\mathrm{d}s \times S)$ in this virtual velocity field is given by $-\mathrm{d}s\int_{S(s)}\operatorname{grad}\hat{U}(\underline{x}):\underline{\underline{\sigma}}(\underline{x},t)\mathrm{d}a$. With $\underline{\underline{\sigma}}(\underline{x},t)$ symmetric, $\hat{\underline{\underline{\Omega}}}(s)$ antisymmetric and $\operatorname{grad}\hat{U}(\underline{x})$ given by [7.90], it reduces to

---

3 $\operatorname{grad}\hat{U}(\underline{x}) = \hat{\underline{\underline{\Omega}}}(s) + (\frac{\mathrm{d}\hat{U}(s)}{\mathrm{d}s} - \hat{\underline{\underline{\Omega}}}(s).\underline{t}(s) + \frac{\mathrm{d}\hat{\underline{\underline{\Omega}}}(s)}{\mathrm{d}s}.\underline{PM}) \otimes \dfrac{\underline{t}(s)}{1 - \underline{PM}.\underline{n}(s)/r(s)}.$

$$-ds\int_{S(s)}(\frac{d\hat{\underline{U}}(s)}{ds}-\hat{\underline{\underline{\Omega}}}(s).\underline{t}(s)+\frac{d\hat{\underline{\underline{\Omega}}}(s)}{ds}.\underline{PM})\otimes\underline{t}(s):\underline{\underline{\sigma}}(\underline{x},t)\,da \qquad [7.93]$$

also equal to

$$-ds\int_{S(s)}(\frac{d\hat{\underline{U}}(s)}{ds}-\hat{\underline{\underline{\Omega}}}(s).\underline{t}(s)+\frac{d\hat{\underline{\underline{\Omega}}}(s)}{ds}.\underline{PM}).\underline{\underline{\sigma}}(\underline{x},t).\underline{t}(s)\,da. \qquad [7.94]$$

Recalling that $\left\{\hat{\mathbb{D}}(s)\right\}=\left\{P,\dfrac{d\hat{\underline{U}}(s)}{ds}-\hat{\underline{\underline{\Omega}}}(s).\underline{t}(s),\dfrac{d\hat{\underline{\underline{\Omega}}}(s)}{ds}\right\}$, we see that the first

term of the integrand in [7.94] is just the velocity field generated by the distributor $\left\{\hat{\mathbb{D}}(s)\right\}$ over the cross-section $S(s)$. It follows that the integral is the virtual rate of work by the surface forces $\underline{\underline{\sigma}}(\underline{x},t).\underline{t}(s)\,da$ in the rigid body motion defined by $\left\{\hat{\mathbb{D}}(s)\right\}$. Thence, introducing the resultant wrench of these forces

$$[\Sigma(s,t)]=\left[P,\int_{S(s)}\underline{\underline{\sigma}}(\underline{x},t).\underline{t}(s)\,da,\int_{S(s)}\underline{PM}\wedge\underline{\underline{\sigma}}(\underline{x},t).\underline{t}(s)\,da\right], \qquad [7.95]$$

the virtual rate of work by internal forces for the three-dimensional volume $(ds\times S)$ finally takes the form

$$-[\Sigma(s,t)].\left\{\hat{\mathbb{D}}(s)\right\}ds. \qquad [7.96]$$

7.4.2.5. Matching the models

When writing down the principle of virtual work for both models in associated virtual motions $\hat{\mathbf{U}}$ and $\hat{\underline{U}}$, it appears that the virtual rates of work by quantities of acceleration and external forces are expressed identically for the line element ds and three-dimensional volume element $(ds\times S)$. It follows that the virtual rates of work by internal forces are also identical: $-[\Sigma(s,t)].\left\{\hat{\mathbb{D}}(s)\right\}ds=p_i(\hat{\mathbf{U}})\,dS$.

Recalling that, for the one-dimensional model, the virtual rate of work by internal forces $p_i(\hat{\mathbf{U}})\,ds$ is

$$p_i\left(\hat{\mathbf{U}}\right)ds = -\left[\boldsymbol{X}(s,t)\right].\left\{\hat{\mathbb{D}}(s)\right\}ds, \qquad\qquad [7.97]$$

and taking into account that $\left\{\hat{\mathbb{D}}(s)\right\}$ is arbitrary, we obtain the expressions of the internal force wrench at the field point P of the one-dimensional model in relation to the three-dimensional continuum model:

$$\left[\boldsymbol{X}(s,t)\right]=\left[\boldsymbol{\Sigma}(s,t)\right]=\left[P, \int_{S(s)}\underline{\underline{\sigma}}(\underline{x},t).\underline{t}(s)\,da, \int_{S(s)}\underline{PM}\wedge\underline{\underline{\sigma}}(\underline{x},t).\underline{t}(s)\,da\right]. \qquad [7.98]$$

The *internal force wrench in the one-dimensional model* is identified with the wrench of the contact forces exerted on the transverse cross-section $S(s)$ of the three-dimensional slender solid with outward normal along $\underline{t}(s)$. In terms of reduced elements:

$$\begin{cases} \underline{X}(s,t) = \displaystyle\int_{S(s)}\underline{\underline{\sigma}}(\underline{x},t).\underline{t}(s)\,da \\[4mm] \underline{\Gamma}(s,t) = \displaystyle\int_{S(s)}\underline{PM}\wedge\underline{\underline{\sigma}}(\underline{x},t).\underline{t}(s)\,da. \end{cases} \qquad\qquad [7.99]$$

7.4.3. Terminology and notations

In this mathematical matching process, which yields the local physical meaning of the wrench field $\left[\boldsymbol{X}(s,t)\right]$, the cross-section $S(s)$ plays the major role consistently with the fact that physically it is the very reason for the stiffness of the considered medium (section 7.3.9).

It is therefore natural that the reduced elements of $\left[\boldsymbol{X}(s,t)\right]$ be decomposed into their in-plane and normal components with respect to $S(s)$ in the form (Figure 7.17)

$$\begin{cases} \underline{X}(s,t) = N(s,t)\underline{t}(s)+\underline{V}(s,t) \\[2mm] \underline{\Gamma}(s,t) = T(s,t)\underline{t}(s)+\underline{M}(s,t), \end{cases} \qquad\qquad [7.100]$$

using standard notations, where $\underline{V}(s,t)$ and $\underline{M}(s,t)$ lie in the plane of $S(s)$. The terminology refers to the cross-section:

– $N(s,t)$ is the *normal force*;

– $\underline{V}(s,t)$ is the *shearing force*;

– $T(s,t)$ is the *twisting moment*;

– $\underline{M}(s,t)$ is the *bending moment*.

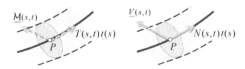

Figure 7.17. *Normal and shearing forces, twisting and bending moments*

Note that the normal force $N(s,t)$ is *positive* when the downstream portion of the system exerts a *traction* on the upstream portion exactly in the same way as in section 7.2.9.

With these notations, we can write the line density of the virtual rate of work by internal forces [7.69] in the explicit form

$$p_i(\hat{\mathbf{U}}) = -N(s,t)\,\hat{D}(s) - \underline{V}(s,t).(\frac{\mathrm{d}\hat{\underline{U}}(s)}{\mathrm{d}s} + \underline{t}(s) \wedge \hat{\underline{\Omega}}(s))$$

$$-T(s,t)\underline{t}(s).\frac{\mathrm{d}\hat{\underline{\Omega}}(s)}{\mathrm{d}s} - \underline{M}(s,t).\frac{\mathrm{d}\hat{\underline{\Omega}}(s)}{\mathrm{d}s},$$

[7.101]

where $\hat{D}(s) = \dfrac{\mathrm{d}\hat{\underline{U}}(s)}{\mathrm{d}s}.\underline{t}(s)$ retains the same meaning as in section 7.2.7 and represents the *virtual rate of stretch* of the director curve.

From [7.101], it appears clearly that if the one-dimensional medium under consideration is assumed to offer resistance neither to angular distortion nor bending or twisting, thus requiring $\underline{V}(s,t)=0$, $T(s,t)=0$, $\underline{M}(s,t)=0$ over $\widehat{AB}$, $p_i(\hat{\mathbf{U}})$ reduces to the expression $p_i(\hat{\mathbf{U}}) = p_i(\hat{\underline{U}}) = -X(s,t)\hat{D}(s)$ obtained in section 7.2.7 for the Statics and Dynamics of wires, with $X(s,t) = N(s,t)$. In plain words, it means that, for the one-dimensional continuum, modeling internal forces by the tension field $N(s,t)$, a scalar field, is a special case of the modeling by the wrench field $[\mathcal{X}(s,t)]$. This is obviously similar to the relationship between the hydrostatic pressure field and the Cauchy tensor field for the three-dimensional continuum (Chapter 6, section 6.7) where assuming that the material offers no resistance to shear stress implies that the stress tensor is necessarily spherical. The comparison is worth being pursued. In both cases, the necessity of a more elaborate model comes from the validation process, showing that the corresponding mathematical compatibility conditions are too constraining and not suited to the expected range of practical applications.

7.5. The Navier–Bernoulli condition

7.5.1. *Virtual rate of angular distortion*

The second term in the expression [7.101] of $p_i(\hat{\mathbf{U}})$ can also be written as

$$-\underline{V}(s,t).(\frac{d\hat{\underline{U}}(s)}{ds} - \hat{\underline{\Omega}}(s) \wedge \underline{t}(s)) = -\underline{V}(s,t).(\frac{d\hat{\underline{U}}(s)}{ds} - \hat{D}(s)\underline{t}(s) - \hat{\underline{\Omega}}(s) \wedge \underline{t}(s)) \quad [7.102]$$

where the cofactor of the shearing force $\underline{V}(s,t)$ can be interpreted as the *virtual rate of angular distortion* as follows.

The virtual motion of the director curve is defined by the virtual velocity field $\hat{\underline{U}}(s)$ independently of $\hat{\underline{\Omega}}(s)$ (see section 7.3.3). At the field point P, the element $d\underline{P} = \underline{t}(s)ds$ along (C) is transported into $(\underline{t}(s)ds + \frac{d\hat{\underline{U}}(s)}{ds}ds)$ and thence $\frac{d\hat{\underline{U}}(s)}{ds}.\underline{t}(s) = \hat{D}(s)$ comes out as the *virtual rate of stretch* of the director curve as previously noted. The virtual rotation rate of the element, which the *virtual rotation rate of* (C) at point P, is obtained from the component of $\frac{d\hat{\underline{U}}(s)}{ds}ds$ orthogonal to $d\underline{P}$ and equal to

$$\frac{d\hat{\underline{U}}(s)}{ds} - \hat{D}(s)\underline{t}(s). \quad\quad\quad [7.103]$$

The virtual vector field $\hat{\underline{\Omega}}(s)$ is the virtual rotation rate of the particle defined as in Figure 7.9 at the field point P: for a unit vector collinear with $\underline{t}(s)$ and considered as part of the particle, the rotation rate is

$$\hat{\underline{\Omega}}(s) \wedge \underline{t}(s). \quad\quad\quad [7.104]$$

Thence, the difference between [7.103] and [7.104]

$$\frac{d\hat{\underline{U}}(s)}{ds} - \hat{D}(s)\underline{t}(s) - \hat{\underline{\Omega}}(s) \wedge \underline{t}(s) \qquad [7.105]$$

actually represents the *virtual angular distortion rate*. This means that the transverse microstructure (particle) being materialized by the plane surface element orthogonal to the director curve and given the virtual motion defined by $\left\{\hat{\underline{U}}(s)\right\}$, [7.105] yields the virtual rate of variation of the angle of that element about the director curve whose virtual motion is defined by $\hat{\underline{U}}(s)$.

7.5.2. The Navier–Bernoulli condition

The *Navier–Bernoulli condition*[4] requires the transverse microstructure to remain orthogonal to the director curve:

$$\frac{d\hat{\underline{U}}(s)}{ds} - \hat{D}(s)\underline{t}(s) - \hat{\underline{\Omega}}(s) \wedge \underline{t}(s) = 0 \quad \text{over } \widehat{AB}. \qquad [7.106]$$

This condition implies

$$\hat{\underline{\Omega}}(s) = \underline{t}(s) \wedge \frac{d\hat{\underline{U}}(s)}{ds} + \hat{\Omega}_t(s)\underline{t}(s) \quad \text{over } \widehat{AB}. \qquad [7.107]$$

The virtual motions complying with [7.106] generate a subspace of the vector space considered for the modeling process, which depends on the vector field $\hat{\underline{U}}(s)$ and the scalar field $\hat{\Omega}_t(s)$. This means that the virtual motion of the microstructure (particle) at the field point P is defined by that of the director curve up to an arbitrary rotation rate about the vector $\underline{t}(s)$ tangent to the director curve.

Note that when the Navier–Bernoulli condition is satisfied, [7.101] reduces to

$$p_i(\hat{\mathbf{U}}) = -N(s,t)\,\hat{D}(s) - T(s,t)\underline{t}(s).\frac{d\hat{\underline{\Omega}}(s)}{ds} - \underline{\mathcal{M}}(s,t).\frac{d\hat{\underline{\Omega}}(s)}{ds}. \qquad [7.108]$$

without any contribution of the shearing force.

---

4 C. Navier (1785–1836). J. Bernoulli (1654–1705).

7.5.3. Discontinuity equations

Equations [7.101]–[7.108] are written in the case of continuous and continuously differentiable fields $\hat{\underline{U}}(s)$ and $\hat{\underline{\Omega}}(s)$. Under the Navier–Bernoulli condition, piecewise continuity and continuous differentiability of these fields are restricted by jump conditions that correspond to [7.106] and [7.107] respectively.

At each point P_j, where $\hat{\underline{U}}(s)$ is discontinuous

$$\left[\!\left[\hat{\underline{U}}(s_j) \right]\!\right] - \left[\!\left[\hat{\underline{U}}(s_j).\underline{t}(s_j) \right]\!\right] \underline{t}(s_j) = 0. \tag{7.109}$$

which means that the jump $\left[\!\left[\hat{\underline{U}}(s_j) \right]\!\right]$ must be *tangent to the director curve* (assumed to be regular).

At each point P_j, where $\dfrac{\mathrm{d}}{\mathrm{d}s}\hat{\underline{U}}(s)$ is discontinuous ($\hat{\underline{U}}(s)$ continuous)

$$\left[\!\left[\hat{\underline{\Omega}}(s_j) \right]\!\right] = \left[\!\left[\hat{\Omega}_t(s_j) \right]\!\right] \underline{t}(s_j) + \underline{t}(s_j) \wedge \left[\!\left[\dfrac{\mathrm{d}\hat{\underline{U}}}{\mathrm{d}s}(s_j) \right]\!\right]. \tag{7.110}$$

In plain words, [7.110] implies that, besides its torsion component $\left[\!\left[\hat{\Omega}_t(s_j) \right]\!\right]$ along $\underline{t}(s_j)$ that may be arbitrary, the jump $\left[\!\left[\hat{\underline{\Omega}}(s_j) \right]\!\right]$ only consists of a bending component that is *determined* by the discontinuity $\left[\!\left[\dfrac{\mathrm{d}\hat{\underline{U}}}{\mathrm{d}s}(s_j) \right]\!\right]$ of the derivative of the virtual velocity field $\hat{\underline{U}}(s)$. This means that P_j is the locus of a *virtual bending hinge* in the considered piecewise continuous and continuously differentiable motion as shown in Figure 7.18. Denoting by $\hat{\theta}(s_j)$ the magnitude of $\left[\!\left[\dfrac{\mathrm{d}\hat{\underline{U}}}{\mathrm{d}s}(s_j) \right]\!\right]$ and $\underline{v}(s_j)$ a unit vector orthogonal to both $\underline{t}(s_j)$ and $\left[\!\left[\dfrac{\mathrm{d}\hat{\underline{U}}}{\mathrm{d}s}(s_j) \right]\!\right]$ in a right-handed triad, we obtain from [7.110]:

$$\left[\!\left[\hat{\underline{\Omega}}(s_j) \right]\!\right] = \left[\!\left[\hat{\Omega}_t(s_j) \right]\!\right] \underline{t}(s_j) + \hat{\theta}(s_j)\underline{v}(s_j). \tag{7.111}$$

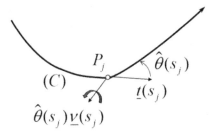

Figure 7.18. *Virtual bending hinge*

7.5.4. *Virtual rate of work by internal forces*

With [7.108]–[7.111], the complete expression for the virtual rate of work by internal forces in Navier–Bernoulli virtual motions is finally written as

$$
\mathcal{P}'_i(\hat{\mathbf{U}}) = -\int_{\overparen{A'B'}} \left(N(s,t)\,\hat{D}(s) + \underline{\Gamma}(s,t).\frac{\mathrm{d}\hat{\underline{\Omega}}(s)}{\mathrm{d}s} \right) \mathrm{d}s
$$
$$
- \sum_{s_{A'}<s_j<s_{B'}} N(s_j,t) \left[\!\left[\, \hat{U}_t(s_j) \,\right]\!\right] + T(s_j,t) \left[\!\left[\, \hat{\Omega}_t(s_j) \,\right]\!\right] \qquad [7.112]
$$
$$
- \sum_{s_{A'}<s_j<s_{B'}} \underline{\mathcal{M}}(s_j,t).\underline{v}(s_j)\hat{\theta}(s_j)
$$

where, as in [7.108], the shearing force does not contribute.

7.5.5. *Comments*

At first glance, the reason for introducing the Navier–Bernoulli condition on virtual motions may seem unclear. Actually, it comes from the fact that many practical problems dealing with beams or arcs that can be modeled as one-dimensional continua are treated with classical constitutive laws (*Thermoelasticity, Viscoelasticity, Elastoplasticity, etc.*) written within the hypothesis that the real motions comply with that condition. This is sometimes referred to as the engineer theory of beams. In the case of the yield design theory [SAL 13] where only virtual motions are considered, the Navier–Bernoulli condition results, through [7.108] and [7.112], from the strength criteria that are adopted when, as a first approximation, they assume the beam to be infinitely resistant to shearing forces.

The comparison with the three-dimensional continuum that was initiated in section 7.4.3 (*in fine*) is worth being reconsidered. In the model presented in Chapter 6 for the classical three-dimensional medium, the particles do not possess any underlying microstructure, they do not exhibit a specific rotation rate different from the rotation rate derived from the virtual velocity field $\hat{\underline{U}}(\underline{x})$. In this sense, *mutatis mutandis*, we may say that the one-dimensional medium with the Navier–Bernoulli condition is analogous to the classical three-dimensional continuum.

7.6. Analysis of systems

7.6.1. *Static determinacy*

We saw in section 7.3.11 that, due to the geometrical modeling being one-dimensional, the field equations of motion and boundary conditions could be integrated explicitly in the form [7.86] or [7.87].

Thus, given the external forces and quantity of acceleration applied to $\overset{\frown}{AB}$, providing these data comply with the fundamental law of dynamics, the field of internal forces is completely and uniquely determined in $\overset{\frown}{AB}$ by [7.86] or [7.87] independently of the mechanical characteristics of the constituent material of the system. Taking the case of equilibrium when the quantities of acceleration are zero, the problem is said to be *statically determinate*. Note that this property is valid in as much as the geometry of the system $\overset{\frown}{AB}$ is, or can be considered as, given and independent of the solution to the problem.

7.6.2. *Systems made of one-dimensional members*

As a matter of fact, a curvilinear one-dimensional element such as $\overset{\frown}{AB}$ is seldom encountered as a system in itself. In practice, three-dimensional complex structures made from assemblages of slender elements such as those presented in Figure 7.1 are common, which can be modeled as two- or three-dimensional systems made of one-dimensional members connected to one another by means of *assembly joints*. The system itself is also connected to the external world by means of *supports* that can be treated in the same way as assembly joints. Without getting into too much detail, we will sketch out a few fundamental results.

The principle of virtual work easily supplies the equations for the analysis of such a system. Considering each member individually as a subsystem, typically

$\widehat{A_i B_i}$, we write the field equation of motion [7.74] and boundary conditions [7.75] at the endpoints A_i and B_i, where the wrenches $\left[\mathcal{R}_{A_i}(t)\right]$ and $\left[\mathcal{R}_{B_i}(t)\right]$ are unknown. Then, considering each node where members $\widehat{A_i B_i}$ are connected, individually as a subsystem, typically denoted by A, (Figure 7.19) we obtain the "equilibrium equation of the node" in the form

$$\sum_i \left[\mathcal{R}_{A_i}(t)\right] = \left[\mathcal{F}(A,t)\right], \qquad\qquad [7.113]$$

with $\left[\mathcal{F}(A,t)\right]$ representing the external force wrench exerted on the system at point A.

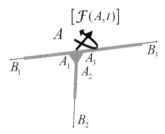

Figure 7.19. *Assembly node*

In the absence of any condition imposed by the mechanical characteristics of the joints at the node A (this corresponds to members connected by a *rigid joint*), this set of equations does not usually make it possible to fully determine the fields of internal forces in all the members of the system. The problem is said to be *statically indeterminate* or *hyperstatic*, with a *finite number of redundant unknowns* in the absence of continuous supports. This number, the *degree of static indeterminacy* of the problem, is the dimension of the vector space of self-equilibrating internal force fields.

Figure 7.20. *Ball and socket assembly joint*

In practice, besides rigid joints, various types of assembly joints are encountered such as the pinned joint, and the ball and socket joint (Figure 7.20) They impose conditions on $\left[\mathcal{R}_{A_i}(t) \right]$, which result in a lower degree of static indeterminacy for the problem that may even lead to static determinacy.

It may also happen that the system becomes *geometrically unstable*, even in the case of static indeterminacy (Figure 7.21), when the affine space of internal force fields that satisfy the equations of the problem with the specified data is empty [SAL 01]. This circumstance can conveniently be analyzed within the framework of the theory of yield design. In such cases, the conditions imposed by the joints at the assembly nodes on the wrenches $\left[\mathcal{R}_{A_i}(t) \right] = -\left[\boldsymbol{X}(A_i, t) \right]$ take the form of *strength criteria* that set to zero the value(s) of some component(s) of its reduced elements. Geometrical instability corresponds to the existence of virtual motions where the virtual rate of work by internal forces is equal to zero while the virtual rate of work by external forces is non-zero, such as in the *panel mechanism* shown in Figure 7.21.

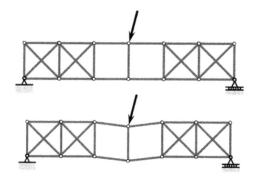

Figure 7.21. *Geometrical instability of a planar truss made of pin-connected members*

Two-dimensional Modeling of Plates and Thin Slabs

8.1. Modeling plates as two-dimensional continua

8.1.1. *Geometrical modeling*

In the same way as for the one-dimensional modeling of wires and beams, the two-dimensional modeling of plates (and thin slabs) proceeds from the specificity of the geometrical characteristics of these structural elements that are *plane* and *slender,* which suggests that a two-dimensional mechanical model built on a *plane director sheet* (D) in the $\mathbb{R}^3$ Euclidean geometric space might provide a relevant description for practical applications (Figure 8.1).

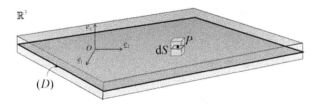

Figure 8.1. *Three-dimensional plate and director sheet*

It is not necessary here to repeat all the arguments related to the relevance assessment of the model, which were presented for the one-dimensional curvilinear continuum in the preceding chapter (section 8.1). Also, the definition of slenderness for the original three-dimensional solid is quite similar. Regarding the director sheet, the terminology "*mean sheet*" will be avoided, taking into account that defining the director sheet geometrically is part of the constitutive equation adopted for the two-dimensional continuum *material* in any practical situation. Thence, the approach adopted in the present construction of the *model* simply refers to the director sheet as being *given*.

A system S is thus described geometrically on a surface S with boundary ∂S in the director sheet (D) and modeled as a set of *particles*, "diluted material points", with surface dS (Figure 8.2). The model stands here as the result of the *smashing* of the three-dimensional plate or slab flat onto the director sheet (D).

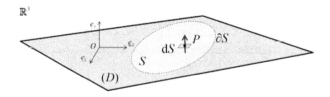

Figure 8.2. *The particle at point P and associated microstructure*

Taking advantage of the experience gained from the modeling of the one-dimensional curvilinear continuum, we can guess that in order to ensure a sufficiently wide domain of relevance, it is necessary that the particles carry, within their geometrical description, additional parameters that reflect the structure of the original three-dimensional solid in a significant way. For this reason, we will directly address the construction of the two-dimensional plane mechanical model *with an oriented microstructure*, which is represented at time t by a segment of a line orthogonal to (D) at the field point P, as pictured out in Figures 8.1 and 8.2. The definition is similar for a subsystem S'.

Let the position vector of the field point P in the director sheet (D) be denoted by $\underline{OP} = \underline{s} = x_i\,\underline{e}_i$, $i = 1,2$ where $(O, \underline{e}_1, \underline{e}_2)$ is the right-handed basis of an orthogonal Cartesian coordinate system in (D) with $\underline{e}_3$ a unit vector orthogonal to (D). The particle at the point P is geometrically characterized by its position vector $\underline{s}$ in (D) and the orientation of the transverse microstructure.

8.1.2. Kinematics: actual and virtual motions

8.1.2.1. Actual motions

Actual motions of S at time t, denoted by $\mathbf{U}(t)$, are defined on S by the evolution of the geometrical parameters of the particles, namely the velocity $\underline{U}(\underline{s}, t)$

of the generic point P and the angular velocity $\underline{\Omega}(\underline{s},t)$ of its attached microstructure. Note that at point P, the rotation rate of the plane tangent to (D) is determined from $\underline{U}(\underline{s},t)$ only and is therefore independent of the rotation rate of the microstructure $\underline{\Omega}(\underline{s},t)$. The fields $\underline{U}(\underline{s},t)$ and $\underline{\Omega}(\underline{s},t)$ are *piecewise continuous and continuously differentiable* on S.

At each point P in S, the velocity distributor $\{\mathbb{U}(\underline{s},t)\} = \{\,P,\,\underline{U}(\underline{s},t),\,\underline{\Omega}(\underline{s},t)\}$ defines the rigid body motion of the particle interpreted as in Figure 8.2, with the general expression for the corresponding velocity field in $\mathbb{R}^3$

$$\forall M \in \mathbb{R}^3,\; \underline{U}(M,t) = \underline{U}(\underline{s},t) + \underline{\Omega}(\underline{s},t) \wedge \underline{PM} = \underline{U}(\underline{s},t) + \underline{\underline{\Omega}}(\underline{s},t).\underline{PM}^{\,1}. \quad [8.1]$$

Comparing this introduction with section 7.3.3 in Chapter 7, it can be anticipated that, by many aspects, this two-dimensional model will be similar to the one-dimensional model studied before, with the important difference that ordinary differential equations along the director curve will be replaced by partial differential equations in the director sheet and imply some mathematical "technicalities". This will require the introduction of tensor quantities and, from the latter point of view, similarities with the classical three-dimensional continuum model will also be encountered.

Due to the specific role played in the model by the direction orthogonal to (D), it is convenient to split the fields $\underline{U}(\underline{s},t)$ and $\underline{\Omega}(\underline{s},t)$ into their in-plane components $\underline{u}(\underline{s},t)$ and $\underline{\omega}(\underline{s},t)$, and out-of-plane components $w(\underline{s},t)\underline{e}_3$ and $\Omega_3(\underline{s},t)\underline{e}_3$ respectively:

$$\left\{ \begin{array}{ll} \underline{U}(\underline{s},t) = \underline{u}(\underline{s},t) + w(\underline{s},t)\underline{e}_3 & \underline{u}(\underline{s},t) \in \mathbb{R}^2 = (D),\; w(\underline{s},t) \in \mathbb{R} \\ \underline{\Omega}(\underline{s},t) = \underline{\omega}(\underline{s},t) + \Omega_3(\underline{s},t)\underline{e}_3 & \underline{\omega}(\underline{s},t) \in \mathbb{R}^2 = (D),\; \Omega_3(\underline{s},t) \in \mathbb{R} \\ \underline{\underline{\Omega}}(\underline{s},t) = \underline{\underline{\omega}}(\underline{s},t) + \Omega_3(\underline{s},t)\underline{\underline{e}}_3 & \underline{\underline{\omega}}(\underline{s},t) = -{}^t\underline{\underline{\omega}}(\underline{s},t) \in \mathbb{R}^2 \times \mathbb{R}^2,\; \underline{\underline{e}}_3 = -\varepsilon_{ij3}\,\underline{e}_i \otimes \underline{e}_j \end{array} \right.$$

$$[8.2]$$

Again, deformation of the system or subsystem is the result of the *variation* of the rigid body motion of the particles on the surface S (respectively S'). The *strain rate* of the two-dimensional continuum at the point P is related to the gradient with respect to $\underline{s}$, of the rigid body motion attached to the particle P, i.e. the gradient of

---

1 $\underline{\underline{\Omega}}(\underline{s},t)$ is the antisymmetric tensor associated with $\underline{\Omega}(\underline{s},t)$ through the vector product.

the velocity distributor $\{\mathbb{U}(\underline{s},t)\}$. Referring to Appendix 3 and taking [8.2] into account, the gradients of $\underline{U}(\underline{s},t)$ and $\underline{\Omega}(\underline{s},t)$ are written as

$$\begin{cases} \underline{\underline{\partial U}}(\underline{s},t) = \underline{\underline{\partial u}}(\underline{s},t) + \underline{e}_3 \otimes \underline{\partial w}(\underline{s},t) & \underline{\underline{\partial u}}(\underline{s},t) \in \mathbb{R}^2 \otimes \mathbb{R}^2 = (D) \otimes (D) \\ \underline{\underline{\partial \Omega}}(\underline{s},t) = \underline{\underline{\partial \omega}}(\underline{s},t) + \underline{e}_3 \otimes \underline{\partial \Omega}_3(\underline{s},t) & \underline{\underline{\partial \omega}}(\underline{s},t) \in \mathbb{R}^2 \otimes \mathbb{R}^2 = (D) \otimes (D), \end{cases} \qquad [8.3]$$

where, in order to avoid confusions later on (section 8.6), the symbol ∂ is chosen to denote the two-dimensional gradients with respect to $\underline{s}$.

The gradient of $\{\mathbb{U}(\underline{s},t)\} = \{\,P,\,\underline{U}(\underline{s},t),\,\underline{\Omega}(\underline{s},t)\}$ is written as a tensorial distributor (Appendix 3, section A3.4.1)

$$\partial\{\mathbb{U}(\underline{s},t)\} = \left\{P,\,\underline{\underline{\partial U}}(\underline{s},t) - \underline{\underline{\Omega}}(\underline{s},t),\,\underline{\underline{\partial \Omega}}(\underline{s},t)\right\} \qquad [8.4]$$

that is

$$\partial\{\mathbb{U}(\underline{s},t)\} = \left\{P,\,\underline{\underline{\partial u}}(\underline{s},t) + \underline{e}_3 \otimes \underline{\partial w}(\underline{s},t) - \underline{\underline{\omega}}(\underline{s},t),\,\underline{\underline{\partial \omega}}(\underline{s},t)\right\}$$
$$+\left\{P,\,-\Omega_3(\underline{s},t)\underline{e}_{3_3},\,\underline{e}_3 \otimes \underline{\partial \Omega}_3(\underline{s},t)\right\}, \qquad [8.5]$$

with $\begin{cases} \underline{\underline{\omega}}(\underline{s},t) = \omega_1(\underline{s},t)\underline{\underline{e}}_1 + \omega_2(\underline{s},t)\underline{\underline{e}}_2 \\ \underline{\underline{e}}_k = -\varepsilon_{ijk}\,\underline{e}_i \otimes \underline{e}_j\,. \end{cases}$

The second line in [8.5] points out the contribution of the out-of-plane component $\Omega_3(\underline{s},t)\underline{e}_3$ of the angular velocity of the microstructure.

This equation means that the rigid body motion rate of the particle $P(\underline{s}+\mathrm{d}\underline{s})$ with respect to the particle $P(\underline{s})$ consists of the *translation rate* with vector

$$\underline{\underline{\partial u}}(\underline{s},t) \cdot \mathrm{d}\underline{s} - \Omega_3(\underline{s},t)\underline{e}_3 \cdot \mathrm{d}\underline{s} + \underline{e}_3(\underline{\partial w}(\underline{s},t) \cdot \mathrm{d}\underline{s}) - \underline{\underline{\omega}}(\underline{s},t) \cdot \mathrm{d}\underline{s} \qquad [8.6]$$

and the rotation rate

$$\underline{\underline{\partial \omega}}(\underline{s},t) \cdot \mathrm{d}\underline{s} + \underline{e}_3 \otimes \underline{\partial \Omega}_3(\underline{s},t) \cdot \mathrm{d}\underline{s}. \qquad [8.7]$$

In [8.6], the term $\partial u(\underline{s},t) \cdot d\underline{s}$ is exactly the two-dimensional counterpart of the term $\mathrm{grad}\,U.\underline{dM}$ in equation [5.4] of Chapter 5 related to the kinematics of the three-dimensional continuum. Thence, the symmetric part of $\partial u(\underline{s},t)$, a two-dimensional tensor that can be denoted by $\underline{d}_2(\underline{s},t)$, comes out as the *strain rate tensor* of a classical *two-dimensional* continuum defined on (D) and the counterpart of $D(s,t)$ in equation [7.51] of Chapter 7. The antisymmetric part is the corresponding *spin tensor* and may be written as $\alpha(\underline{s},t)\underline{e}_{=3}$ so that $\alpha(\underline{s},t)\underline{e}_{=3} \cdot d\underline{s} = \alpha(\underline{s},t)\underline{e}_3 \wedge d\underline{s}$. The vector $\alpha(\underline{s},t)\underline{e}_3$ is the associated spin vector that yields the mean rotation motion of the two-dimensional element at the point $\underline{s}$ and time t in the motion defined by the field $\underline{u}(\underline{s},t)$.

The second term in [8.6] expresses the contribution of the rotation rate of the microstructure about $\underline{e}_3$ to the in-plane motion (Figure 8.3).

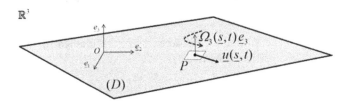

Figure 8.3. *In-plane motion of a particle*

As for the last term, we observe that it is collinear with $\underline{e}_3$ and may also be written as

$$-\underline{\underline{\omega}}(\underline{s},t).d\underline{s} = -\underline{e}_3\left(\underline{e}_3 \wedge \underline{\omega}(\underline{s},t)\right).d\underline{s}, \qquad [8.8]$$

which makes it possible to transform [8.6] into

$$\partial\underline{u}(\underline{s},t).d\underline{s} - \Omega_3(\underline{s},t)\underline{e}_{=3}.d\underline{s} + \underline{e}_3\left(\partial w(\underline{s},t) - \underline{e}_3 \wedge \underline{\omega}(\underline{s},t)\right).d\underline{s}. \qquad [8.9]$$

This equation recalls equation [7.49] in Chapter 7. Its third term is related to the rate of angular distortion of the microstructure with respect to the director sheet (Figure 8.4). The rotation rate of the normal to the director sheet at the point P proceeds from the scalar field $w(\underline{s},t)$ and is equal to $-\underline{e}_3 \wedge \partial w(\underline{s},t)$, while the rotation rate of the microstructure is just $\underline{\omega}(\underline{s},t)$. These two rotation rates are independent from each other as previously noted. Note that, in the particular case, when the condition

$$\partial \underline{w}(\underline{s},t) - \underline{e}_3 \wedge \underline{\omega}(\underline{s},t) = 0, \qquad [8.10]$$

known as the *Kirchhoff–Love condition*, is imposed to the model as an *internal constraint*, it implies that the microstructure remains orthogonal to the director sheet (D) in the evolution of the system[2] (section 8.7).

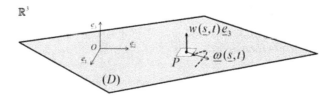

Figure 8.4. *Out-of-plane motion of a particle*

Finally, for the general model considered here, the gradient of the rigid body motion attached to the particle is written as

$$\partial\{\mathbb{U}(\underline{s},t)\} =$$

$$\left\{P, \partial\underline{u}(\underline{s},t) - \underline{\Omega}_3(\underline{s},t)\underline{e}_3 \ + \underline{e}_3 \otimes \left(\partial\underline{w}(\underline{s},t) - \underline{e}_3 \wedge \underline{\omega}(\underline{s},t)\right), \partial\underline{\omega}(\underline{s},t) + \underline{e}_3 \otimes \partial\underline{\Omega}_3(\underline{s},t)\right\}.$$

$$[8.11]$$

8.1.2.2. *Virtual motions*

Virtual motions of the system S or any subsystem S', denoted by $\hat{\mathbb{U}}$, will be defined on S (respectively S') by vector fields $\underline{\hat{U}}(\underline{s})$ and $\underline{\hat{\Omega}}(\underline{s})$ or the velocity distributor $\{\hat{\mathbb{U}}(\underline{s})\}$ in the same way as actual motions. Equivalently, we may say that they are defined by the vector fields $\underline{\hat{u}}(\underline{s})$ and $\underline{\hat{\omega}}(\underline{s})$, and the scalar fields $\hat{w}(\underline{s})$ and $\hat{\Omega}_3(\underline{s})$. They obviously include rigid body motions and encompass actual motions.

---

2 Note the similarity with the Navier–Bernoulli condition, Chapter 7 (section 7.5.2).

As in the preceding examples of the implementation of the virtual work method, these fields are first assumed *continuous and continuously differentiable* (they will be allowed piecewise continuity at a later stage (section 8.5.2)). Their gradient is written as:

$$\partial\{\hat{U}(\underline{s})\} = \left\{P,\ \partial\hat{\underline{u}}(\underline{s}) - \hat{\underline{\Omega}}_3(\underline{s})\underline{e}_3 + \underline{e}_3 \otimes (\partial\hat{w}(\underline{s}) - \underline{e}_3 \wedge \hat{\underline{\omega}}(\underline{s})),\ \partial\hat{\underline{\omega}}(\underline{s}) + \underline{e}_3 \otimes \partial\underline{\Omega}_3(\underline{s})\right\}.$$

[8.12]

They generate a vector space with infinite dimension.

8.2. Virtual rates of work

8.2.1. *Virtual rate of work by quantities of acceleration*

For any subsystem $S' \subseteq S$, the virtual rate of work by quantities of acceleration in the virtual motions defined in section 8.1.2 comes out as the integral on $S' \subseteq S$ of a surface density $[\,\boldsymbol{\rho a}(\underline{s},t))\,].\{\hat{U}(\underline{s})\}$ where $[\,\boldsymbol{\rho a}(\underline{s},t)\,]$ stands as the wrench of quantities of acceleration:

$$\mathcal{A}(\hat{\mathbf{U}}) = \int_S [\,\boldsymbol{\rho a}(\underline{s},t)\,].\{\hat{U}(\underline{s})\}\,dS \qquad\qquad [8.13]$$

$$\mathcal{A}'(\hat{\mathbf{U}}) = \int_{S'} [\,\boldsymbol{\rho a}(\underline{s},t)\,].\{\hat{U}(\underline{s})\}\,dS. \qquad\qquad [8.14]$$

8.2.2. *Virtual rate of work by external forces for the system*

The continuous linear form $\mathcal{P}_e(\hat{\mathbf{U}})$ expressing the virtual rate of work by external forces for the system consists of two main contributions: one is related to the surface S, and the other to the boundary ∂S.

8.2.2.1. *Surface forces*

Surface forces are modeled by a surface density $\underline{f}(\underline{s},t)$ corresponding to the distributed external forces exerted on S (Figure 8.5). Again, taking into account the specific role of the direction normal to (D), this surface density will be split into its normal (or out-of-plane) and in-plane components:

$$\begin{cases} \underline{f}(\underline{s},t) = \underline{f}_D(\underline{s},t) + f_3(\underline{s},t)\underline{e}_3 \\ \underline{f}_D(\underline{s},t) = f_1(\underline{s},t)\underline{e}_1 + f_2(\underline{s},t)\underline{e}_2. \end{cases}$$

[8.15]

Occasionally, an external moment surface density $\underline{h}_D(\underline{s},t)$, parallel to (D), will be considered and a component $c(\underline{s},t)$ along $\underline{e}_3$ may be introduced *pro memoria*.

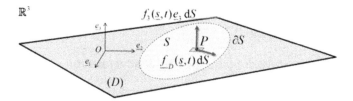

Figure 8.5. *Distributed forces acting on the system*

In the same way as in Chapter 7 (section 7.3.5), we introduce the wrench field $\left[\boldsymbol{f}(\underline{s},t)\right]$ defined from $\underline{f}(\underline{s},t)$ and $\underline{h}_D(\underline{s},t) + c(\underline{s},t)\underline{e}_3$ by

$$\forall P \in S, \left[\boldsymbol{f}(\underline{s},t)\right] = \left[P, \underline{f}(\underline{s},t), \underline{h}_D(\underline{s},t) + c(\underline{s},t)\underline{e}_3\right]$$

[8.16]

and the contribution of the distributed external forces acting on S to the virtual rate of work is written as:

$$\mathcal{P}_e(\hat{\mathbf{U}}) = \int_S \left[\boldsymbol{f}(\underline{s},t)\right].\left\{\hat{\mathbb{U}}(\underline{s})\right\} dS = \int_S \left(\underline{f}(\underline{s},t).\hat{\underline{U}}(\underline{s}) + \underline{h}_D(\underline{s},t).\hat{\underline{\Omega}}(\underline{s}) + c(\underline{s},t)\,\Omega_3\right) dS.$$

[8.17]

Line densities of in-plane or out-of-plane external forces, denoted by $\underline{\phi}_D(\underline{s},t)$ and $\phi_3(\underline{s},t)\underline{e}_3$, or in-plane external moments $\underline{\gamma}_D(\underline{s},t)$, will also be considered[3] in section 8.5.1. They define the line density wrench

$$\left[\boldsymbol{\mathcal{F}}(\underline{s},t)\right] = \left[P, \underline{\phi}_D(\underline{s},t) + \phi_3(\underline{s},t)\underline{e}_3, \underline{\gamma}_D(\underline{s},t)\right].$$

[8.18]

---

3 Such forces may result from the connections of the system with other structural elements such as stiffeners.

8.2.2.2. *Boundary forces*

As shown in Figure 8.6, boundary forces are modeled by a *line density of force* on ∂S, which will be split into its in-plane and out-of-plane components so that the distributed force acting on the element $\mathrm{d}\ell$ with outward normal $\underline{n}(\underline{s})$ is written as

$$\underline{T}(\underline{s},t)\,\mathrm{d}\ell + R(\underline{s},t)\underline{e}_3\,\mathrm{d}\ell, \tag{8.19}$$

and a *line density of moment* on ∂S, with in-plane component $\underline{H}(\underline{s},t)$ and out-of-plane component $C(\underline{s},t)$ *pro memoria*, defining the distributed moment on the element $\mathrm{d}\ell$

$$\underline{H}(\underline{s},t)\,\mathrm{d}\ell + C(\underline{s},t)\underline{e}_3\,\mathrm{d}\ell. \tag{8.20}$$

These reduced elements define the wrench $\left[\boldsymbol{\mathcal{R}}(\underline{s},t)\right]$:

$$\begin{cases} \left[\boldsymbol{\mathcal{R}}(\underline{s},t)\right] = \left[P, \underline{T}(\underline{s},t) + R(\underline{s},t)\underline{e}_3,\ \underline{H}(\underline{s},t) + C(\underline{s},t)\underline{e}_3\right] \\ \underline{T}(\underline{s},t) \in (D) = \mathbb{R}^2,\ \underline{H}(\underline{s},t) \in (D) = \mathbb{R}^2. \end{cases} \tag{8.21}$$

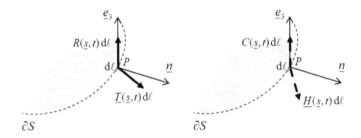

Figure 8.6. *Distributed boundary forces*

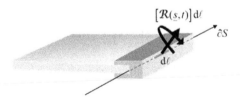

Figure 8.7. *Plate element supported by a beam*

From the physical viewpoint, these forces may result from the connection of the plate with a support along ∂S (Figure 8.7). In the case of a free boundary, boundary forces are equal to zero.

Finally, the virtual rate of work by external forces for the system in any virtual motion $\hat{\mathbf{U}}$ defined by the vector fields $\hat{\underline{u}}(\underline{s})$ and $\hat{\underline{\omega}}(\underline{s})$, and the scalar fields $\hat{w}(\underline{s})$ and $\hat{\Omega}_3(\underline{s})$ is written as

$$
\begin{aligned}
\mathcal{P}_e(\hat{\underline{u}}, \hat{w}, \hat{\underline{\omega}}, \hat{\Omega}_3) = & \int_S \underline{f}_D(\underline{s},t).\hat{\underline{u}}(\underline{s})\,dS + \int_L \underline{\phi}_D(\underline{s},t).\hat{\underline{u}}(\underline{s})\,dL + \int_{\partial S} \underline{T}(\underline{s},t).\hat{\underline{u}}(\underline{s})\,d\ell \\
& + \int_S f_3(\underline{s},t)\,\hat{w}(\underline{s})\,dS + \int_L \phi_3(\underline{s},t)\,\hat{w}(\underline{s})\,dL + \int_{\partial S} R(\underline{s},t)\,\hat{w}(\underline{s})\,d\ell \\
& + \int_S \underline{h}_D(\underline{s},t).\hat{\underline{\omega}}(\underline{s})\,dS + \int_L \underline{\gamma}_D(\underline{s},t).\hat{\underline{\omega}}(\underline{s})\,dL + \int_{\partial S} \underline{H}(\underline{s},t).\hat{\underline{\omega}}(\underline{s})\,d\ell \\
& + \int_S c(\underline{s},t)\,\hat{\Omega}_3(\underline{s})\,dS + \int_{\partial S} C(\underline{s},t)\,\hat{\Omega}_3(\underline{s})\,d\ell
\end{aligned}
$$

[8.22]

or, equivalently,

$$
\begin{aligned}
\mathcal{P}_e(\hat{\mathbf{U}}) = & \int_S [\boldsymbol{f}(\underline{s},t)].\{\hat{\mathbb{U}}(\underline{s})\}\,dS + \int_L [\boldsymbol{F}(\underline{s},t)].\{\hat{\mathbb{U}}(\underline{s})\}\,dL \\
& + \int_{\partial S} [\boldsymbol{R}(\underline{s},t)].\{\hat{\mathbb{U}}(\underline{s})\}\,d\ell.
\end{aligned}
$$

[8.23]

8.2.3. Virtual rate of work by external forces for subsystems

For $S' \subset S$, a subsystem of S *stricto sensu* defined on $S' \subset S$, the expression for $\mathcal{P}'_e(\hat{\mathbf{U}})$ is similar to [8.23]:

$$
\begin{aligned}
\mathcal{P}'_e(\hat{\mathbf{U}}) = & \int_{S'} [\boldsymbol{f}^{S'}(\underline{s},t)].\{\hat{\mathbb{U}}(\underline{s})\}\,dS + \int_{L \cap S'} [\boldsymbol{F}^{S'}(\underline{s},t)].\{\hat{\mathbb{U}}(\underline{s})\}\,dL \\
& + \int_{\partial S'} [\boldsymbol{R}^{S'}(\underline{s},t)].\{\hat{\mathbb{U}}(\underline{s})\}\,d\ell.
\end{aligned}
$$

[8.24]

Consistently with the physical origin of this modeling process (section 8.1.1), it is assumed that the particles of S *do not exert any action at a distance* on each other. Thence, the force wrench surface density $\left[\boldsymbol{f}^{S'}(\underline{s},t) \right]$ and line density $\left[\boldsymbol{F}^{S'}(\underline{s},t) \right]$ do not depend on S'

$$\forall S' \subset S, \begin{cases} \left[\boldsymbol{f}^{S'}(s,t) \right] = \left[\boldsymbol{f}(s,t) \right] \\ \left[\boldsymbol{F}^{S'}(\underline{s},t) \right] = \left[\boldsymbol{F}(\underline{s},t) \right] \end{cases} \tag{8.25}$$

and $\mathcal{P}'_e(\hat{\mathbf{U}})$ is finally written in the form

$$\mathcal{P}'_e(\hat{\mathbf{U}}) = \int_{S'} \left[\boldsymbol{f}(\underline{s},t) \right] . \left\{ \hat{\mathbf{U}}(\underline{s}) \right\} dS + \int_{L \cap S'} \left[\boldsymbol{F}(\underline{s},t) \right] . \left\{ \hat{\mathbf{U}}(\underline{s}) \right\} dL$$
$$+ \int_{\partial S'} \left[\boldsymbol{R}^{S'}(\underline{s},t) \right] . \left\{ \hat{\mathbf{U}}(\underline{s}) \right\} d\ell \tag{8.26}$$

with

$$\begin{cases} \left[\boldsymbol{R}^{S'}(\underline{s},t) \right] = \left[P, \underline{T}^{S'}(\underline{s},t) + V^{S'}(\underline{s},t)\underline{e}_3, \underline{H}^{S'}(\underline{s},t) + C^{S'}(\underline{s},t)\underline{e}_3 \right] \\ \underline{T}^{S'}(\underline{s},t) \in (D) = \mathbb{R}^2, \ \underline{H}^{S'}(\underline{s},t) \in (D) = \mathbb{R}^2. \end{cases} \tag{8.27}$$

The force wrench field $\left[\boldsymbol{R}^{S'}(\underline{s},t) \right]$ defined on $\partial S'$ models the external forces exerted on the boundary $\partial S'$ by the world external to S'. Upon the same arguments as before, they are only exerted by the *constituent particles of S external* to S'.

8.2.4. *Virtual rate of work by internal forces*

As in the preceding modeling processes for the three-dimensional and one-dimensional continua, we postulate that, for any subsystem $S' \subseteq S$, the continuous linear form $\mathcal{P}'_i(\hat{\mathbf{U}})$ is the integral over S' of a surface density $p_i(\hat{\mathbf{U}})$ independent of S', which is a linear form of the local values of the virtual fields $\hat{\underline{U}}(\underline{s})$ and $\hat{\underline{\Omega}}(\underline{s})$, and their gradients:

$$\forall \hat{\mathbf{U}} \text{ m. v.}, \ \forall S' \subseteq S, \ \mathcal{P}'_i(\hat{\mathbf{U}}) = \int_{S'} p_i(\hat{\mathbf{U}}) dS \tag{8.28}$$

Equivalently, $p_i(\hat{\mathbf{U}})$ at the field point P is a linear form in $\{\hat{U}(\underline{s})\}$ and $\partial\{\hat{U}(\underline{s})\}$, which may be written as

$$p_i(\hat{\mathbf{U}}) = -[A(\underline{s},t)].\{\hat{U}(\underline{s})\} - \left\langle [\boldsymbol{X}(\underline{s},t)] \,\middle|\, \partial\{\hat{U}(\underline{s})\} \right\rangle, \qquad [8.29]$$

with $[A(\underline{s},t)]$ a wrench of force, $[\boldsymbol{X}(\underline{s},t)] = \left[P, \underline{\underline{X}}(\underline{s},t), \underline{\underline{\Gamma}}(\underline{s},t) \right]$ a tensorial wrench and $\left\langle [\boldsymbol{X}(\underline{s},t)] \,\middle|\, \partial\{\hat{U}(\underline{s})\} \right\rangle$ its contracted product with the tensorial distributor $\partial\{\hat{U}(\underline{s})\}$. From the definition in Appendix 3 (equation [A3.34]), we have the general expression

$$\left\langle [\boldsymbol{X}(\underline{s},t)] \,\middle|\, \partial\{\hat{U}(\underline{s})\} \right\rangle = {}^t\underline{\underline{X}}(\underline{s},t):(\partial\hat{U}(\underline{s}) - \hat{\underline{\underline{\Omega}}}(\underline{s})) + {}^t\underline{\underline{\Gamma}}(\underline{s},t):\partial\hat{\underline{\underline{\Omega}}}(\underline{s}). \qquad [8.30]$$

8.2.5. *Tensorial wrench field of internal forces*

Implementing the principle of virtual work in order to specify the expression [8.28] of the virtual rate of work by internal forces, we refer to the dual statement of the law of mutual actions:

$$\begin{cases} \forall S' \subseteq S, \ \forall \hat{U} \text{ r.b.v.m. of } S', \\ P_i'(\hat{\mathbf{U}}) = -\int_{S'} \left([A(\underline{s},t)].\{\hat{U}(\underline{s})\} + \left\langle [\boldsymbol{X}(\underline{s},t)] \,\middle|\, \partial\{\hat{U}(\underline{s})\} \right\rangle \right) dS = 0. \end{cases} \qquad [8.31]$$

As stated in section 8.1.2, any rigid body motion of the system $S' \subseteq S$ is characterized by the rigid body motion $\{\hat{U}(\underline{s})\}$ of the current particle $P(\underline{s})$ being constant on S'. It then follows from [8.31] that $[A(\underline{s},t)]$ is zero on S and $p_i(\hat{\mathbf{U}})$ reduces to

$$p_i(\hat{\mathbf{U}}) = -\left\langle [\boldsymbol{X}(\underline{s},t)] \,\middle|\, \partial\{\hat{U}(\underline{s})\} \right\rangle, \qquad [8.32]$$

where $[\boldsymbol{X}(\underline{s},t)] = \left[P, \underline{\underline{X}}(\underline{s},t), \underline{\underline{\Gamma}}(\underline{s},t) \right]$ is the *tensorial wrench field* that models *internal forces*.

8.3. Equations of motion

8.3.1. *Statement of the principle of virtual work*

We now write down the second statement of the principle of virtual work under the assumption that no line densities of in-plane or out-of-plane external forces are exerted. From [8.13], [8.23], [8.26] and [8.32], we obtain for S

$$\left\{ \begin{array}{l} \text{in a Galilean frame,} \\[4pt] \forall \hat{\boldsymbol{U}} \text{ v.m.,} \\[4pt] \int_S \left[\boldsymbol{f}(\underline{s},t)\right].\left\{\hat{U}(\underline{s})\right\}\mathrm{d}S + \int_{\partial S}\left[\boldsymbol{R}(\underline{s},t)\right].\left\{\hat{U}(\underline{s})\right\}\mathrm{d}\ell \\[8pt] -\int_S \left\langle \left[\boldsymbol{X}(\underline{s},t)\right] \middle| \partial\left\{\hat{U}(\underline{s})\right\}\right\rangle \mathrm{d}S = \int_S \left[\boldsymbol{\rho a}(\underline{s},t)\right].\left\{\hat{U}(\underline{s})\right\}\mathrm{d}S \end{array}\right. \qquad [8.33]$$

and for any $S' \subset S$

$$\left\{ \begin{array}{l} \text{in a Galilean frame,} \\[4pt] \forall S' \subset S,\ \forall \hat{\boldsymbol{U}} \text{ v.m.,} \\[4pt] \int_{S'} \left[\boldsymbol{f}(\underline{s},t)\right].\left\{\hat{U}(\underline{s})\right\}\mathrm{d}S + \int_{\partial S'}\left[\boldsymbol{R}^{S'}(\underline{s},t)\right].\left\{\hat{U}(\underline{s})\right\}\mathrm{d}\ell \\[8pt] -\int_{S'} \left\langle \left[\boldsymbol{X}(\underline{s},t)\right] \middle| \partial\left\{\hat{U}(\underline{s})\right\}\right\rangle \mathrm{d}S = \int_{S'} \left[\boldsymbol{\rho a}(\underline{s},t)\right].\left\{\hat{U}(\underline{s})\right\}\mathrm{d}S. \end{array}\right. \qquad [8.34]$$

Assuming the wrench field of internal forces to be *continuously differentiable with respect to* $\underline{s}$ and using the identity (Appendix 3, equation [A3.55])

$$\mathrm{div}\left(\left[\boldsymbol{X}(\underline{s},t)\right].\left\{\hat{U}(\underline{s})\right\}\right) = \left\langle \left[\boldsymbol{X}(\underline{s},t)\right] \middle| \partial\left\{\hat{U}(\underline{s})\right\}\right\rangle$$
$$+ {}^t\underline{\underline{X}}(\underline{s},t):\hat{\underline{\underline{\Omega}}}(\underline{s}) + \hat{\underline{U}}(\underline{s}).\mathrm{div}\,\underline{\underline{X}}(\underline{s},t) + \hat{\underline{\underline{\Omega}}}(\underline{s}).\mathrm{div}\,\underline{\underline{\Gamma}}(\underline{s},t),$$
$$[8.35]$$

while keeping in mind the difference of status between $\left[\boldsymbol{R}(\underline{s},t)\right]$ and $\left[\boldsymbol{R}^{S'}(\underline{s},t)\right]$, we may write [8.33] and [8.34] in the form

in a Galilean frame,

$\forall S' \subseteq S, \forall \hat{\mathbf{U}}$ v.m.,

$$\int_{S'} \left[\boldsymbol{f}(\underline{s},t) \right] . \left\{ \hat{\mathrm{U}}(\underline{s}) \right\} \mathrm{d}S + \int_{\partial S'} \left[\boldsymbol{\mathcal{R}}^{S'}(\underline{s},t) \right] . \left\{ \hat{\mathrm{U}}(\underline{s}) \right\} \mathrm{d}\ell$$

$$- \int_{S'} \mathrm{div} \left(\left[\boldsymbol{X}(\underline{s},t) \right] . \left\{ \hat{\mathrm{U}}(\underline{s}) \right\} \right) \mathrm{d}S \qquad\qquad [8.36]$$

$$+ \int_{S'} \left({}^{\mathrm{t}}\underline{\underline{X}}(\underline{s},t) : \hat{\underline{\underline{\Omega}}}(\underline{s}) + \hat{\underline{U}}(\underline{s}) . \mathrm{div}\, \underline{\underline{X}}(\underline{s},t) + \hat{\underline{\underline{\Omega}}}(\underline{s}) . \mathrm{div}\, \underline{\underline{\Gamma}}(\underline{s},t) \right) \mathrm{d}S$$

$$= \int_{S'} \left[\rho \boldsymbol{a}(\underline{s},t) \right] . \left\{ \hat{\mathrm{U}}(\underline{s}) \right\} \mathrm{d}S.$$

Finally, through the *divergence theorem* applied to the vector field $\left[\boldsymbol{X}(\underline{s},t) \right] . \left\{ \hat{\mathrm{U}}(\underline{s}) \right\}$, the principle is conveniently expressed for the dualization procedure:

in a Galilean frame,

$\forall S' \subseteq S, \forall \hat{\mathbf{U}}$ v.m.,

$$\int_{S'} \left[\boldsymbol{f}(\underline{s},t) \right] . \left\{ \hat{\mathrm{U}}(\underline{s}) \right\} \mathrm{d}S + \int_{\partial S'} \left[\boldsymbol{\mathcal{R}}^{S'}(\underline{s},t) \right] . \left\{ \hat{\mathrm{U}}(\underline{s}) \right\} \mathrm{d}\ell$$

$$- \int_{\partial S'} \left(\left[\boldsymbol{X}(\underline{s},t) \right] . \left\{ \hat{\mathrm{U}}(\underline{s}) \right\} \right) . \underline{n}(\underline{s}) \, \mathrm{d}\ell \qquad\qquad [8.37]$$

$$+ \int_{S'} \left({}^{\mathrm{t}}\underline{\underline{X}}(\underline{s},t) : \hat{\underline{\underline{\Omega}}}(\underline{s}) + \hat{\underline{U}}(\underline{s}) . \mathrm{div}\, \underline{\underline{X}}(\underline{s},t) + \hat{\underline{\underline{\Omega}}}(\underline{s}) . \mathrm{div}\, \underline{\underline{\Gamma}}(\underline{s},t) \right) \mathrm{d}S$$

$$= \int_{S'} \left[\rho \boldsymbol{a}(\underline{s},t) \right] . \left\{ \hat{\mathrm{U}}(\underline{s}) \right\} \mathrm{d}S.$$

8.3.2. *Boundary equations*

Recalling that[4]

$$\int_{\partial S'} \left(\left[\boldsymbol{X}(\underline{s},t) \right] . \left\{ \hat{\mathrm{U}}(\underline{s}) \right\} \right) . \underline{n}(\underline{s}) \, \mathrm{d}\ell = \int_{\partial S'} \left(\left[\boldsymbol{X}(\underline{s},t) . \underline{n}(\underline{s}) \right] \right) . \left\{ \hat{\mathrm{U}}(\underline{s}) \right\} \mathrm{d}\ell, \qquad [8.38]$$

---

4 Appendix 3, equation [A3.30].

with the same arguments as in Chapters 6 and 7 applied to [8.37], we derive the boundary condition on ∂S

$$\forall M \in \partial S, \ [\boldsymbol{X}(\underline{s},t)].\underline{n}(\underline{s}) = [\boldsymbol{R}(\underline{s},t)] \tag{8.39}$$

and we determine the line density $\left[\boldsymbol{R}^{S'}(\underline{s},t)\right]$ on $\partial S'$

$$\forall \partial S', \ \forall M \in \partial S', \left[\boldsymbol{R}^{S'}(\underline{s},t)\right] = [\boldsymbol{X}(\underline{s},t)].\underline{n}(\underline{s}). \tag{8.40}$$

8.3.3. *Specifying the tensorial wrench of internal forces*

In these equations, the unit vector $\underline{n}(\underline{s})$ lies in (D) while $[\boldsymbol{R}(\underline{s},t)]$ and $\left[\boldsymbol{R}^{S'}(\underline{s},t)\right]$ are written as

$$\begin{cases} [\boldsymbol{R}(\underline{s},t)] = \left[P, \underline{T}(\underline{s},t) + R(\underline{s},t)\underline{e}_3, \ \underline{H}(\underline{s},t) + C(\underline{s},t)\underline{e}_3\right] \\ \left[\boldsymbol{R}^{S'}(\underline{s},t)\right] = \left[P, \underline{T}^{S'}(\underline{s},t) + V^{S'}(\underline{s},t)\underline{e}_3, \ \underline{H}^{S'}(\underline{s},t) + C^{S'}(\underline{s},t)\underline{e}_3\right] \end{cases} \tag{8.41}$$

as specified in [8.21] and [8.27]. It follows that the reduced elements of the tensorial wrench $[\boldsymbol{X}(\underline{s},t)] = \left[P, \underline{\underline{X}}(\underline{s},t), \underline{\underline{\Gamma}}(\underline{s},t)\right]$ must take the form

$$\begin{cases} \underline{\underline{X}}(\underline{s},t) = \underline{\underline{N}}(\underline{s},t) + \underline{e}_3 \otimes \underline{V}(\underline{s},t) \\ \underline{\underline{\Gamma}}(\underline{s},t) = \underline{\underline{H}}(\underline{s},t) + \underline{e}_3 \otimes \underline{G}(\underline{s},t) \end{cases} \tag{8.42}$$

with

$$\begin{cases} \underline{\underline{N}}(\underline{s},t) \in (D) \times (D) = \mathbb{R}^2 \times \mathbb{R}^2, \ \underline{V}(\underline{s},t) \in (D) = \mathbb{R}^2 \\ \underline{\underline{H}}(\underline{s},t) \in (D) \times (D) = \mathbb{R}^2 \times \mathbb{R}^2, \ \underline{G}(\underline{s},t) \in (D) = \mathbb{R}^2. \end{cases} \tag{8.43}$$

Incidentally, it is worth noting that, putting together [8.42], [8.43] and [8.44], the surface density of virtual rate of work by internal forces becomes:

$$p_i(\hat{\mathbf{U}}) = -{}^t\underline{\underline{N}}(\underline{s},t):(\partial\underline{\hat{u}}(\underline{s}) - \hat{\Omega}_3(\underline{s})\underline{e}_3) - \underline{V}(\underline{s},t).(\partial\hat{w}(\underline{s}) - \underline{e}_3.\underline{\hat{\omega}}(\underline{s}))$$
$$- {}^t\underline{\underline{H}}(\underline{s},t):\partial\underline{\hat{\omega}}(\underline{s}) - \underline{G}(\underline{s},t).\partial\hat{\Omega}_3(\underline{s}). \tag{8.44}$$

With [8.42] and [8.43], the *boundary condition* on ∂S and the expression of the *line density* $\left[\mathcal{R}^{S'}(\underline{s},t) \right]$ on $\partial S'$ take the form

$$
\begin{cases}
\underline{\underline{N}}(\underline{s},t).\underline{n}(\underline{s}) = \underline{T}(\underline{s},t), \;\; \underline{V}(\underline{s},t).\underline{n}(\underline{s}) = R(\underline{s},t) \\[2mm]
\underline{\underline{H}}(\underline{s},t).\underline{n}(\underline{s}) = \underline{H}(\underline{s},t), \;\; \underline{G}(\underline{s},t).\underline{n}(\underline{s}) = C(\underline{s},t)
\end{cases}
\tag{8.45}
$$

$$
\begin{cases}
\underline{T}^{S'}(\underline{s},t) = \underline{\underline{X}}(\underline{s},t).\underline{n}(\underline{s}), \;\; V^{S'}(\underline{s},t) = \underline{V}(\underline{s},t).\underline{n}(\underline{s}) \\[2mm]
\underline{H}^{S'}(\underline{s},t) = \underline{\underline{H}}(\underline{s},t).\underline{n}(\underline{s}), \;\; C^{S'}(\underline{s},t) = \underline{G}(\underline{s},t).\underline{n}(\underline{s}).
\end{cases}
\tag{8.46}
$$

8.3.4. Field equations of motion

The field equations of motion for the system are obtained from the surface integrals in [8.37] by means of the identity (Appendix 3, section A3.4.4)

$$
{}^t\underline{\underline{X}}(\underline{s},t):\underline{\underline{\hat{\Omega}}}(\underline{s}) = \underline{\underline{\hat{\Omega}}}(\underline{s}).({}^t\underline{\underline{X}}(\underline{s},t):\underline{\underline{e}}_i\,)\underline{e}_i),
\tag{8.47}
$$

which yields:

$$
\begin{cases}
\text{in a Galilean frame,} \\[1mm]
\forall \hat{\mathbb{U}} \text{ v.m.,} \\[2mm]
\displaystyle \int_S \left[f(\underline{s},t) \right].\left\{ \hat{\mathbb{U}}(\underline{s}) \right\} dS - \int_S \left[\boldsymbol{\rho a}(\underline{s},t) \right].\left\{ \hat{\mathbb{U}}(\underline{s}) \right\} dS + \\[3mm]
\displaystyle + \int_S \hat{U}(\underline{s}).\operatorname{div} \underline{\underline{X}}(\underline{s},t)\, dS + \int_S \underline{\underline{\hat{\Omega}}}(\underline{s}).\left(\operatorname{div}\underline{\underline{\Gamma}}(\underline{s},t) + ({}^t\underline{\underline{X}}(\underline{s},t):\underline{\underline{e}}_i\,)\underline{e}_i \right) dS = 0.
\end{cases}
\tag{8.48}
$$

This equation can also be written in the compact form

$$
\begin{cases}
\text{in a Galilean frame,} \\[1mm]
\forall \hat{\mathbb{U}} \text{ v.m.,} \\[2mm]
\displaystyle \int_S \left(\left[f(\underline{s},t) \right] - \left[\boldsymbol{\rho a}(\underline{s},t) \right] \right).\left\{ \hat{\mathbb{U}}(\underline{s}) \right\} dS + \int_S \left\{ \hat{\mathbb{U}}(\underline{s}) \right\}.\operatorname{div}\left[\mathbf{X}(\underline{s},t) \right] dS = 0.
\end{cases}
\tag{8.49}
$$

with the definition of $\text{div}\left[\boldsymbol{X}(\underline{s},t)\right]$ given in Appendix 3 (section A3.4.4)

$$\text{div}\left[\boldsymbol{X}(\underline{s},t)\right]=\left[M,\,\text{div}\,\underline{\underline{X}}(\underline{s},t),\,\text{div}\,\underline{\underline{\Gamma}}(\underline{s},t)+\left(^{t}\underline{\underline{X}}(\underline{s},t):\underline{e}_{i}\,\right)\underline{e}_{i}\,\right)\right] \qquad [8.50]$$

and the field equation of motion is obtained as a *conservation law in terms of wrenches:*

$$\begin{cases} \text{in a Galilean frame, } \forall M \in S, \\ \\ \text{div}\left[\boldsymbol{X}(\underline{s},t)\right]+\left[\boldsymbol{f}(\underline{s},t)\right]-\left[\boldsymbol{\rho a}(\underline{s},t)\right]=0. \end{cases} \qquad [8.51]$$

As it is assumed that the particles of S do not exert any action at a distance on each other, which implies $\left[\boldsymbol{f}^{S'}(s,t)\right]=\left[\boldsymbol{f}(s,t)\right]$, the field equation of motion for any subsystem $S' \subset S$ is obviously identical to [8.51].

Comments

The field equation of motion [8.51] and boundary conditions [8.39] and [8.40] are similar to those established in Chapter 7 (section 7.3.8) for the curvilinear one-dimensional continuum. Again, consistently with the initial assumptions about external forces acting on a subsystem, it comes out that the wrench of external forces $\left[\boldsymbol{R}^{S'}(\underline{s},t)\right]$ acting at a point M of the boundary $\partial S'$ of a subsystem, proceeds from the knowledge of the internal force wrench field $\left[\boldsymbol{X}(\underline{s},t)\right]$ determined in the system through [8.51] and [8.39], and therefore only depends on the subsystem through the normal $\underline{n}(\underline{s})$ to $\partial S'$ at that point. Thence, from now on, the notation $\left[\text{R}^{S'}(\underline{s},t)\right]=\left[P,\,\underline{T}^{S'}(\underline{s},t)+V^{S'}(\underline{s},t)\underline{e}_{3},\,\underline{H}^{S'}(\underline{s},t)+C^{S'}(\underline{s},t)\underline{e}_{3}\,\right]$ can be substituted by

$$\left[\boldsymbol{R}(\underline{s},\underline{n},t)\right]=\left[P,\,\underline{T}(\underline{s},\underline{n},t)+V(\underline{s},\underline{n},t)\underline{e}_{3},\,\underline{H}(\underline{s},\underline{n},t)+C(\underline{s},\underline{n},t)\underline{e}_{3}\,\right] \qquad [8.52]$$

This result is obviously consistent with the intuitive idea of the two-dimensional model being the result of the smashing onto (D) of a plane and slender three-dimensional system.

8.3.5. *Field equations of motion in terms of reduced elements*

As explained later on in section 8.6.3, the wrench of quantities of acceleration at point P usually reduces to $\left[\boldsymbol{\rho a}(\underline{s},t)\right]=\left[P,\,\rho(\underline{s},t)\underline{a}(\underline{s},t),\,0\right]$ where $\underline{a}(s,t)$, the *acceleration* of the particle at the point P, may be split into its in-plane and out-of-plane components, $\underline{a}(s,t) = \underline{a}_D(s,t) + a_3(\underline{s},t)\underline{e}_3$, while $\boldsymbol{\rho}(\underline{s},t)$ is the mass surface density.

Then, with [8.42] and [8.43] expressing the reduced elements of $[\boldsymbol{X}(\underline{s},t)]$, the equation [8.51] yields four vector equations for $\underline{\underline{N}}(\underline{s},t)$, $\underline{V}(\underline{s},t)$, $\underline{\underline{H}}(\underline{s},t)$ and $\underline{G}(\underline{s},t)$:

$$\left\{ \begin{array}{l} \text{in a Galilean frame,} \\ \text{div}\,\underline{\underline{N}}(\underline{s},t) + \underline{f}_D(\underline{s},t) - \boldsymbol{\rho}(\underline{s},t)\,\underline{a}_D(\underline{s},t) = 0 \end{array} \right. \qquad [8.53]$$

$$\left\{ \begin{array}{l} \text{in a Galilean frame,} \\ \text{div}\,\underline{V}(\underline{s},t) + f_3(\underline{s},t) - \boldsymbol{\rho}(\underline{s},t)\,a_3(\underline{s},t) = 0 \end{array} \right. \qquad [8.54]$$

$$\text{div}\,\underline{\underline{H}}(\underline{s},t) + \underline{V}(\underline{s},t).\underline{e}_3 + \underline{h}_D(\underline{s},t) = 0 \;^5 \qquad [8.55]$$

$$\text{div}\,\underline{G}(\underline{s},t) + {}^t\underline{\underline{N}}(\underline{s},t){:}\underline{e}_3 + c(\underline{s},t) = 0 \;^6. \qquad [8.56]$$

Again, in comparison with the curvilinear one-dimensional continuum, we identify [8.53] and [8.54] as the counterparts of the first line in equation [7.77] of Chapter 7 (section 7.3.8) and, [8.55] and [8.56] as the counterparts of the second line.

8.3.6. Entailment of the equations of motion

8.3.6.1. In-plane equations

In section 8.2.2, the components $c(\underline{s},t)$ and $C(\underline{s},t)$ along $\underline{e}_3$ of the surface density and line density of external moments on S and ∂S were introduced *pro memoria*, meaning that, as a general rule, the model is considered with the assumption that they are equal to zero and $C^{S'}(\underline{s},t)$ in section 8.2.3 as well. It then follows from [8.45] and [8.46] that the field $\underline{G}(\underline{s},t)$ is identically zero.

---

5 $((\underline{V} \otimes \underline{e}_3){:}\underline{e}_k)\underline{e}_k = -V_1\underline{e}_2 + V_2\underline{e}_1 = \underline{V}.\underline{e}_3 = -\underline{e}_3 \wedge \underline{V}.$

6 Since $\underline{\underline{N}} = N_{ij}\underline{e}_i \otimes \underline{e}_j$, $i,j = 1,2$, we have $({}^t\underline{\underline{N}}{:}\underline{e}_k)\underline{e}_k = ({}^t\underline{\underline{N}}{:}\underline{e}_3)\underline{e}_3 = (N_{12} - N_{21})\underline{e}_3.$

Within this set of hypotheses, it comes out from [8.56] that the two-dimensional tensor field $\underline{\underline{N}}(\underline{s},t)$ is *symmetric* and only governed by the in-plane field equation [8.53] and the boundary condition [8.45]. These equations

$$
\begin{cases}
\underline{\underline{N}}(\underline{s},t) = {}^t\underline{\underline{N}}(\underline{s},t) \\[4pt]
\text{in a Galilean frame,} \\[4pt]
\operatorname{div}\underline{\underline{N}}(\underline{s},t) + \underline{f}_D(\underline{s},t) - \rho(\underline{s},t)\underline{a}_D(\underline{s},t) = 0 \text{ in } S \\[4pt]
\underline{\underline{N}}(\underline{s},t).\underline{n}(\underline{s}) = \underline{T}(\underline{s},t) \text{ on } \partial S
\end{cases}
\qquad [8.57]
$$

define the mechanical modeling of a *classical two-dimensional continuum*, a *plane membrane*, whose density of virtual rate of work by internal forces [8.44] reduces to

$$
p_i(\hat{\mathbf{U}}) = p_i(\hat{\underline{u}}) = -\underline{\underline{N}}(\underline{s},t):\partial\hat{\underline{\underline{u}}}(\underline{s}). \qquad [8.58]
$$

It is worth noting that from the *dimensional analysis* viewpoint, the components of $\underline{\underline{N}}(\underline{s},t)$ behave as *force linear densities* (see also section 8.6.6).

8.3.6.2. *Out-of-plane equations*

The field equations [8.54] and [8.55] together with the boundary conditions [8.45] constitute the set of out-of-plane equations of motion.

***In* S**

$$
\begin{cases}
\text{in a Galilean frame,} \\[4pt]
\operatorname{div}\underline{V}(\underline{s},t) + f_3(\underline{s},t) - \rho(\underline{s},t)\,a_3(\underline{s},t) = 0
\end{cases}
\qquad [8.59]
$$

$$
\operatorname{div}\underline{\underline{H}}(\underline{s},t) + \underline{V}(\underline{s},t).\underline{e}_3 + \underline{h}_D(\underline{s},t) = 0 \qquad [8.60]
$$

***On* ∂S**

$$
\begin{cases}
\underline{V}(\underline{s},t).\underline{n}(\underline{s}) = R(\underline{s},t) \\[4pt]
\underline{\underline{H}}(\underline{s},t).\underline{n}(\underline{s}) = \underline{H}(\underline{s},t).
\end{cases}
\qquad [8.61]
$$

On $\partial S'$

$$\begin{cases} V(\underline{s},\underline{n},t) = \underline{V}(\underline{s},t).\underline{n}(\underline{s}) \\ \underline{H}(\underline{s},\underline{n},t) = \underline{\underline{H}}(\underline{s},t).\underline{n}(\underline{s}). \end{cases}$$
[8.62]

It is common practice to introduce new variables $\underline{\underline{M}}(\underline{s},t)$ and $\underline{m}_D(\underline{s},t)$ defined by

$$\underline{\underline{M}}(\underline{s},t) = \underline{e}_3 . \underline{\underline{H}}(\underline{s},t) \text{ and } \underline{m}_D(\underline{s},t) = \underline{e}_3 . \underline{h}_D(\underline{s},t)$$
[8.63]

with the result that [8.60] is equivalent to[7]

$$\text{div } \underline{\underline{M}}(\underline{s},t) + \underline{V}(\underline{s},t) + \underline{m}_D(\underline{s},t) = 0$$
[8.64]

while [8.61] and [8.62] become

$$\text{on } \partial S \begin{cases} \underline{V}(\underline{s},t).\underline{n}(\underline{s}) = R(\underline{s},t) \\ \underline{\underline{M}}(\underline{s},t).\underline{n}(\underline{s}) = \underline{e}_3 . \underline{H}(\underline{s},t) \end{cases}$$
[8.65]

and

$$\text{on } \partial S' \begin{cases} V(\underline{s},\underline{n},t) = \underline{V}(\underline{s},t).\underline{n}(\underline{s}) \\ \underline{H}(\underline{s},\underline{n},t) = -\underline{e}_3 . \underline{\underline{M}}(\underline{s},t).\underline{n}(\underline{s}). \end{cases}$$
[8.66]

The second rank tensor $\underline{\underline{M}}(\underline{s},t)$ is named the *tensor of internal moments*.

Despite the terminology that might be somewhat misleading the components of $\underline{\underline{M}}(\underline{s},t)$ behave as *forces* from the *dimensional viewpoint*, while the components of $\underline{V}(\underline{s},t)$ appear as *force linear densities* (see also section 8.6.6).

---

7 More justification for the introduction of tensor $\underline{\underline{M}}$ will appear in section 8.6.6, when the two-dimensional model is derived from the original three-dimensional slender solid within the three-dimensional continuum mechanics framework.

With $\underline{\underline{G}}(\underline{s},t) = 0$ and $\underline{\underline{N}}(\underline{s},t) = {}^t\underline{\underline{N}}(\underline{s},t)$, the surface density of virtual rate of work by internal forces [8.44] is now written as

$$p_i(\hat{\mathbf{U}}) = -\underline{\underline{N}}(\underline{s},t):\partial\hat{\underline{u}}(\underline{s}) - \underline{V}(\underline{s},t).\left(\partial\hat{w}(\underline{s}) - \underline{e}_3.\hat{\underline{\omega}}(\underline{s})\right) - {}^t\underline{\underline{M}}(\underline{s},t):\left(\underline{e}_3.\partial\hat{\underline{\omega}}(\underline{s})\right). \quad [8.67]$$

8.4. Physical interpretation and classical presentation

8.4.1. *Internal forces*

Equations [8.62] and [8.66] make it clear that the basic physical concept at the origin of this modeling process lies in the assumptions made in section 8.2.3 about the external forces exerted on a subsystem (Figure 8.8).

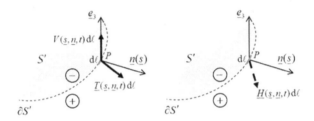

Figure 8.8. *Boundary external forces on a subsystem*

With external forces exerted on the system modeled as indicated in section 8.2.1, the classical presentation of the two-dimensional modeling of plates starts from the assumption that the particles of the system exert no action upon one another at a distance, which implies that they only interact through contact actions.

At a point P, adjacent particles interact along an arc of a line $d\ell$ with normal $\underline{n}(\underline{s})$ which defines the (+) and (−) sides as in Chapter 6 (section 6.5.2). The *contact forces* exerted by the particle on the (+) side on the particle on the (−) side, depending on $\underline{n}(\underline{s})$, are assumed to be *proportional to* $d\ell$ and modeled as follows, with $c(\underline{s},t)$ and $C(\underline{s},t)$ equal to zero.

– A force with components $\underline{T}(\underline{s},\underline{n},t)d\ell \in (D)$ and $V(\underline{s},\underline{n},t)d\ell$ along $\underline{e}_3$

$$\underline{T}(\underline{s},\underline{n},t)d\ell + V(\underline{s},\underline{n},t)\underline{e}_3\, d\ell \quad [8.68]$$

where $V(\underline{s},\underline{n},t)d\ell$ is named the *shear force* acting on the element with normal $\underline{n}(\underline{s})$.

– A moment with no component along $\underline{e}_3$

$$\underline{H}(\underline{s},\underline{n},t)\,d\ell \in (D).$$ [8.69]

8.4.2. *Field equations of motion*

Establishing the equations of motion within this framework refers to the same arguments as in the case of the three-dimensional continuum with

– the infinitely small tetrahedron in $\mathbb{R}^3$ being substituted by an infinitely small triangle in (D) in order to prove the linear dependence on $\underline{T}(\underline{s},\underline{n},t)$, $V(\underline{s},\underline{n},t)$ and $\underline{H}(\underline{s},\underline{n},t)$ with respect to $\underline{n}(\underline{s})$;

– the infinitely small parallelepiped being substituted by an infinitely small parallelogram in order to establish the partial differential equations for the internal force fields.

8.4.2.1. *In-plane equations of motion*

Exactly as for the three-dimensional continuum, writing down the *fundamental law of dynamics*, to the first significant order of magnitude, for an infinitely small triangle, and considering the in-plane components of the overall resultant force and quantity of acceleration, proves that $\underline{T}(\underline{s},\underline{n},t)$ depends linearly on $\underline{n}(\underline{s})$ through a tensor $\underline{\underline{N}}(\underline{s},t) \in (D)\times(D)$:

$$\underline{T}(\underline{s},\underline{n},t) = \underline{\underline{N}}(\underline{s},t).\underline{n}(\underline{s}).$$ [8.70]

Then, with $c(\underline{s},t) = 0$, expressing the equilibrium of moments about $\underline{e}_3$ for an infinitely small parallelogram proves the symmetry of $\underline{\underline{N}}(\underline{s},t) \in (D)\times(D)$. The equilibrium of the in-plane components of the overall resultant force and quantity of acceleration for this element yields the in-plane differential equation of motion. Finally:

$$\begin{cases} \underline{\underline{N}}(\underline{s},t) = N_{ij}(\underline{s},t)\underline{e}_i \otimes \underline{e}_j \ \ i,j=1,2 \\ N_{ij}(\underline{s},t) = N_{ji}(\underline{s},t) \\ \text{in a Galilean frame,} \\ \operatorname{div}\underline{\underline{N}}(\underline{s},t) + \underline{f}_D(\underline{s},t) - \rho(\underline{s},t)\underline{a}_D(\underline{s},t) = 0 \ \text{in } S. \end{cases}$$ [8.71]

8.4.2.2. *Out-of-plane equations of motion*

8.4.2.2.1. Shear forces

The small triangle argument applied to the out-of-plane components also proves the linear dependence on $V(\underline{s},\underline{n},t)$ with respect to $\underline{n}(\underline{s})$ through a linear operator, a vector, $\underline{V}(\underline{s},t) \in (D)$:

$$\begin{cases} V(\underline{s},\underline{n},t) = \underline{V}(\underline{s},t) \cdot \underline{n}(\underline{s}) \\ \underline{V}(\underline{s},t) = V_i(\underline{s},t)\underline{e}_i, \ i = 1,2. \end{cases} \qquad [8.72]$$

The component $V_1(\underline{s},t)$ (respectively $V_2(\underline{s},t)$) is the magnitude of the line density of shear force $V_1(\underline{s},t)\underline{e}_3$ on the element with normal $\underline{e}_1$ (respectively $\underline{e}_2$) at point P (Figure 8.9). The vector $\underline{V}(\underline{s},t) \in (D)$ is usually named the *shear force vector*, although it is clear that it represents *a linear operator and not a force*.

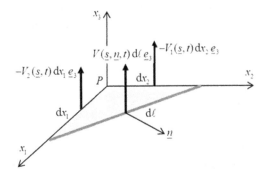

Figure 8.9. *The infinitely small triangle argument applied to shear forces*

Then, writing down the equilibrium of the out-of-plane components of the overall resultant force and quantity of acceleration for the infinitely small parallelogram element (Figure 8.10), we retrieve the differential equation of motion along $\underline{e}_3$, which concerns the shear force field, as already written in [8.59]:

$$\operatorname{div}\underline{V}(\underline{s},t) + f_3(\underline{s},t) - \rho(\underline{s},t)a_3(\underline{s},t) = 0 \qquad [8.73]$$

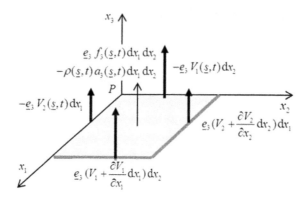

Figure 8.10. *Equilibrium of the resultant forces along $\underline{e}_3$*

8.4.2.2.2. Internal moments

In the same way as before, the global equilibrium of the moments applied to an infinitely small triangle proves that $\underline{H}(\underline{s},\underline{n},t)$ is a linear function of $\underline{n}(\underline{s})$ through the second rank tensor $\underline{\underline{H}}(\underline{s},t) \in (D) \otimes (D)$ (Figure 8.11):

$$\begin{cases} \underline{H}(\underline{s},\underline{n},t) = \underline{\underline{H}}(\underline{s},t).\underline{n}(\underline{s}) \\ \underline{\underline{H}}(\underline{s},t) = H_{ji}(\underline{s},t)\underline{e}_j \otimes \underline{e}_i \quad i,j = 1,2. \end{cases} \qquad [8.74]$$

The components $H_{11}(\underline{s},t)$ and $H_{21}(\underline{s},t)$ (respectively $H_{12}(\underline{s},t)$ and $H_{22}(\underline{s},t)$) of $\underline{\underline{H}}(\underline{s},t)$ are the components along $\underline{e}_1$ and $\underline{e}_2$ of $\underline{H}_1(\underline{s},t)$, the line density of internal moment on the line element dx_2 with normal $\underline{e}_1$ (respectively $\underline{H}_2(\underline{s},t)$, dx_1, $\underline{e}_2$) as shown in Figure 8.11.

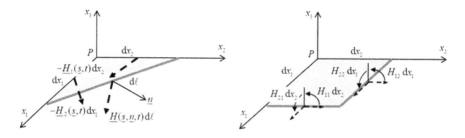

Figure 8.11. *The infinitely small triangle argument applied to internal moments. Components of $\underline{\underline{H}}(\underline{s},t)$*

Then, considering the global equilibrium of the moments applied to an infinitely small parallelogram, we obtain (Figure 8.12)

$$\frac{\partial \underline{H}_2(\underline{s},t)}{\partial x_2} + \frac{\partial \underline{H}_1(\underline{s},t)}{\partial x_1} + V_2(\underline{s},t)\underline{e}_1 - V_1(\underline{s},t)\underline{e}_2 + \underline{h}_D(\underline{s},t) = 0 \qquad [8.75]$$

or explicitly

$$\begin{cases} \dfrac{\partial H_{11}(\underline{s},t)}{\partial x_1} + \dfrac{\partial H_{12}(\underline{s},t)}{\partial x_2} + V_2(\underline{s},t) + h_1(\underline{s},t) = 0 \\[2mm] \dfrac{\partial H_{21}(\underline{s},t)}{\partial x_1} + \dfrac{\partial H_{22}(\underline{s},t)}{\partial x_2} - V_1(\underline{s},t) + h_2(\underline{s},t) = 0. \end{cases} \qquad [8.76]$$

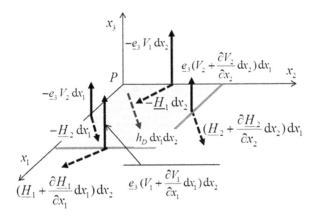

Figure 8.12. *Equilibrium of moments applied to an infinitely small parallelogram*

This equation can be written in the compact form

$$\mathrm{div}\,\underline{\underline{H}}(\underline{s},t) + \underline{e}_3 \wedge \underline{V}(\underline{s},t) + \underline{h}_D(\underline{s},t) = 0, \qquad [8.77]$$

which is [8.60]. With $\underline{\underline{M}}(\underline{s},t) = \underline{e}_3 \cdot \underline{\underline{H}}(\underline{s},t)$, the *tensor of internal moments*, whose components are shown in Figure 8.13, and $\underline{m}_D(\underline{s},t) = \underline{e}_3 \wedge \underline{h}_D(\underline{s},t) = \underline{e}_3 \cdot \underline{h}_D(\underline{s},t)$, we retrieve [8.64]:

$$\mathrm{div}\,\underline{\underline{M}}(\underline{s},t) + \underline{V}(\underline{s},t) + \underline{m}_D(\underline{s},t) = 0. \qquad [8.78]$$

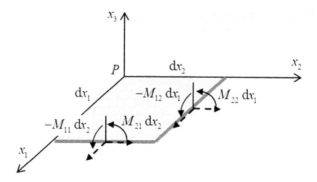

Figure 8.13. *Components of* $\underline{\underline{M}}(\underline{s},t)$

8.4.3. *Boundary equations*

The boundary conditions on ∂S follow from the fundamental law of dynamics applied to an infinitely small parallelogram element parallel to ∂S at point P, with infinitely small thickness along $\underline{n}(\underline{s})$ compared with $d\ell$ (Figure 8.14). From [8.70], [8.72] and [8.74], we derive:

$$
\begin{cases}
\forall P \in \partial S, \\[4pt]
\underline{T}(\underline{s},t) + R(\underline{s},t)\underline{e}_3 - \underline{\underline{N}}(\underline{s},t).\underline{n}(\underline{s}) - \underline{V}(\underline{s},t).\underline{n}(\underline{s})\underline{e}_3 = 0 \\[4pt]
\underline{H}(\underline{s},t) - \underline{\underline{H}}(\underline{s},t).\underline{n}(\underline{s}) = 0,
\end{cases}
\qquad [8.79]
$$

which is identical to [8.45].

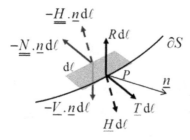

Figure 8.14. *Boundary conditions*

Comments

As has already been pointed out, the presentation in the section above is quite similar to the stress vector approach in Chapter 6 (section 6.5). It is unescapable in any textbook due to its pedagogical importance. Indeed, the mathematical technicalities that had to be introduced and overcome in section 8.3 make it difficult not to miss the guiding physical thread of the reasoning. Thence, thanks to the intuitive and somewhat heuristic arguments together with the illustrative figures provided in this section, it is easier to visualize the modeling process.

A first remark is that, under the assumptions made about the external forces applied to the system (sections 8.2.2, 8.2.3 and 8.3.7), the in-plane and out-of-plane problems are *decoupled*. The in-plane problem describes a classical plane two-dimensional continuum, with a symmetric tensor field modeling the internal forces, the membrane internal force tensor field.

Another important remark concerns *a result that has not been established yet!* Looking at Figures 8.12 and 8.13 and comparing them to Chapter 6 (section 6.5.2), we would have expected the internal moment tensor to be proven symmetric but, as a matter of fact, *no symmetry result* can be established at this stage. For this reason, the explicit forms of [8.78] in Cartesian and polar coordinates are respectively:

$$\left\{ \begin{array}{l} \dfrac{\partial M_{xx}}{\partial x} + \dfrac{\partial M_{xy}}{\partial y} + V_x + (m_D)_x = 0 \\[4mm] \dfrac{\partial M_{yx}}{\partial x} + \dfrac{\partial M_{yy}}{\partial y} + V_y + (m_D)_y = 0 \end{array} \right.$$

[8.80]

$$\left\{ \begin{array}{l} \dfrac{\partial M_{rr}}{\partial r} + \dfrac{1}{r}\dfrac{\partial M_{r\theta}}{\partial \theta} + \dfrac{M_{rr} - M_{\theta\theta}}{r} + V_r + (m_D)_r = 0 \\[4mm] \dfrac{\partial M_{\theta r}}{\partial r} + \dfrac{1}{r}\dfrac{\partial M_{\theta\theta}}{\partial \theta} + \dfrac{M_{r\theta} + M_{\theta r}}{r} + V_\theta + (m_D)_\theta = 0. \end{array} \right.$$

[8.81]

8.5. Piecewise continuous fields

8.5.1. *Piecewise continuous field of internal forces*

We now assume that the wrench field of internal forces $[\mathcal{X}(\underline{s},t)]$ is piecewise continuous and continuously differentiable with discontinuities $[\![[\mathcal{X}(\underline{s},t)]]\!]$ when crossing a discontinuity line L in S following the outward normal $\underline{n}(\underline{s})$:

$$\llbracket [\boldsymbol{X}(\underline{s},t)] \rrbracket = \left[P, \llbracket \underline{\underline{X}}(\underline{s},t) \rrbracket, \llbracket \underline{\underline{\Gamma}}(\underline{s},t) \rrbracket \right] \tag{8.82}$$

A reasoning similar to the one detailed in Chapter 6 (section 6.4.2) may now be applied to the field differential equations [8.57], [8.59] and [8.60] or [8.64]. Assuming the actual fields $\underline{U}(\underline{s},t)$ and $\underline{\Omega}(\underline{s},t)$ to be *continuous and continuously differentiable* on S, we obtain the jump conditions when crossing L following $\underline{n}(\underline{s})$:

$$\llbracket \underline{\underline{N}}(\underline{s},t) \rrbracket \cdot \underline{n}(\underline{s}) + \underline{\phi}_D(\underline{s},t) = 0, \tag{8.83}$$

$$\llbracket \underline{V}(\underline{s},t) \rrbracket \cdot \underline{n}(\underline{s}) + \phi_3(\underline{s},t) = 0, \tag{8.84}$$

$$\llbracket \underline{\underline{H}}(\underline{s},t) \rrbracket \cdot \underline{n}(\underline{s}) + \underline{\gamma}_D(\underline{s},t) = 0$$

or   [8.85]

$$\llbracket \underline{\underline{M}}(\underline{s},t) \rrbracket \cdot \underline{n}(\underline{s}) + \underline{m}_D(\underline{s},t) = 0.$$

These jump conditions may be written in a compact form in terms of wrenches

$$\llbracket [\boldsymbol{X}(\underline{s},t)] \rrbracket \cdot \underline{n}(\underline{s}) + [\boldsymbol{F}(\underline{s},t)] = 0. \tag{8.86}$$

consistently with [8.51].

In plain words, a *line density of external forces* along a line L generates a discontinuity of the internal force field when crossing that line, which is governed by [8.86]. In the absence of such line densities of external forces, the internal force field may nevertheless exhibit a discontinuity when crossing a line L following its normal $\underline{n}(\underline{s})$, but [8.86] implies that the internal force wrench on that line is continuous.

8.5.2. *Piecewise continuous virtual motions*

As in preceding occasions (Chapters 6 and 7), having completed the modeling process using continuous and continuously differentiable virtual motions as mathematical test functions, we now examine the expression to be adopted for the principle of virtual work on the vector space of *piecewise continuous and continuously differentiable virtual motions*. From sections 8.2.1–8.2.4, it clearly comes out that only the expressions of $\mathcal{P}_i(\hat{\mathbf{U}})$ and $\mathcal{P}'_i(\hat{\mathbf{U}})$ are concerned and should be completed. The rationale is similar to the proof given in Chapter 6 and will only be sketched out here.

Let $\hat{\mathbf{U}}$ defined by the velocity distributor field $\{\hat{\mathbb{U}}(\underline{s})\} = \{P, \underline{\hat{U}}(\underline{s}), \underline{\hat{\Omega}}(\underline{s})\}$ denote such a virtual motion and $L_{\hat{\mathbf{u}}}$ a discontinuity line in S for the fields $\underline{\hat{U}}$ and $\underline{\hat{\Omega}}$ so that, at point P when crossing $L_{\hat{\mathbf{u}}}$ following $\underline{n}(\underline{s})$,

$$\llbracket\{\hat{\mathbb{U}}(\underline{s})\}\rrbracket = \{P, \llbracket\underline{\hat{U}}(\underline{s})\rrbracket, \llbracket\underline{\hat{\Omega}}(\underline{s})\rrbracket\}. \tag{8.87}$$

It is assumed that $L_{\hat{\mathbf{u}}}$ bears no line density of external forces.

$L_{\hat{\mathbf{u}}}$ may be used to delimitate two complementary subsystems S_1' and S_2' of the system S as indicated in Figure 8.15 and the principle of virtual work may be written classically on S_1' and S_2' with $p_i(\hat{\mathbf{U}}) = -\langle[\boldsymbol{X}(\underline{s},t)] \,|\, \partial\{\hat{\mathbb{U}}(\underline{s})\}\rangle$ as the expression of the surface density of virtual rate of work. We obtain respectively

$$\begin{cases} \mathcal{P}_i'(\hat{\mathbf{U}}) = -\int_{S_1'} \langle[\boldsymbol{X}(\underline{s},t)] \,|\, \partial\{\hat{\mathbb{U}}(\underline{s})\}\rangle \, dS \text{ for } S_1' \\ \mathcal{P}_i'(\hat{\mathbf{U}}) = -\int_{S_2'} \langle[\boldsymbol{X}(\underline{s},t)] \,|\, \partial\{\hat{\mathbb{U}}(\underline{s})\}\rangle \, dS \text{ for } S_2'. \end{cases} \tag{8.88}$$

Regarding the expression of the virtual rate of work by external forces for S_1', we have

$$\mathcal{P}_e'(\hat{\mathbf{U}}) = \int_{S_1'} [\boldsymbol{f}(\underline{s},t)].\{\hat{\mathbb{U}}(\underline{s})\} \, dS + \int_{\partial S_1} [\boldsymbol{R}(\underline{s},t)].\{\hat{\mathbb{U}}(\underline{s})\} \, d\ell$$

$$+ \int_{L_{\hat{\mathbf{u}}}} \{\hat{\mathbb{U}}_1(\underline{s})\}.[\boldsymbol{X}(\underline{s},t)].\underline{n}(\underline{s}) \, dL_{\hat{\mathbf{u}}} \tag{8.89}$$

and, taking the continuity of $[\boldsymbol{X}(\underline{s},t)].\underline{n}(\underline{s})$ into account, a similar expression for S_2' but for $+\int_{L_{\hat{\mathbf{u}}}} \{\hat{\mathbb{U}}_1(\underline{s})\}.[\boldsymbol{X}(\underline{s},t)].\underline{n}(\underline{s}) \, dL_{\hat{\mathbf{u}}}$ being replaced by $-\int_{L_{\hat{\mathbf{u}}}} \{\hat{\mathbb{U}}_2(\underline{s})\}.[\boldsymbol{X}(\underline{s},t)].\underline{n}(\underline{s}) \, dL_{\hat{\mathbf{u}}}$. It follows, as in Chapter 6, that the expression of the virtual rate of work by internal forces for the system must now be completed in the form

$$\mathcal{P}_i(\hat{\mathbf{U}}) = -\int_S \left\langle [\boldsymbol{X}(\underline{s},t)] \middle| \partial\{\hat{\mathbb{U}}(\underline{s})\} \right\rangle dS - \int_{L_{\hat{\mathbb{U}}}} [\![\{\hat{\mathbb{U}}(\underline{s})\}]\!].[\boldsymbol{X}(\underline{s},t)].\underline{n}(\underline{s}) dL_{\hat{\mathbb{U}}} \qquad [8.90]$$

or explicitly

$$\mathcal{P}_i(\hat{\mathbf{U}}) = -\int_S ({}^t\underline{\underline{X}}(\underline{s},t):(\partial\hat{\underline{U}}(\underline{s}) - \hat{\underline{\Omega}}(\underline{s})) + {}^t\underline{\underline{\Gamma}}(\underline{s},t):\partial\hat{\underline{\Omega}}(\underline{s})) dS$$
$$- \int_{L_{\hat{\mathbb{U}}}} \{P, [\![\hat{\underline{U}}(\underline{s})]\!], [\![\hat{\underline{\Omega}}(\underline{s})]\!]\}.\left[P, \underline{\underline{X}}(\underline{s},t).\underline{n}(\underline{s}), \underline{\underline{\Gamma}}(\underline{s},t).\underline{n}(\underline{s})\right] dL_{\hat{\mathbb{U}}}. \qquad [8.91]$$

that is, from [8.67],

$$\mathcal{P}_i(\hat{\mathbf{U}}) = -\int_S (\underline{\underline{N}}(\underline{s},t):\partial\hat{\underline{u}}(\underline{s}) + \underline{V}(\underline{s},t).(\partial\hat{w}(\underline{s}) - \underline{e}_3.\hat{\underline{\omega}}(\underline{s})) + {}^t\underline{\underline{M}}(\underline{s},t):\underline{e}_3.\partial\hat{\underline{\omega}}(\underline{s})) dS$$
$$- \int_{L_{\hat{\mathbb{U}}}} ([\![\hat{\underline{u}}(\underline{s})]\!].\underline{\underline{N}}(\underline{s},t).\underline{n}(\underline{s}) + [\![\hat{w}(\underline{s})]\!]\underline{V}(\underline{s},t).\underline{n}(\underline{s}) + (\underline{e}_3.[\![\hat{\underline{\omega}}(\underline{s})]\!]).\underline{\underline{M}}(\underline{s},t).\underline{n}(\underline{s})) dL_{\hat{\mathbb{U}}}.$$
$$[8.92]$$

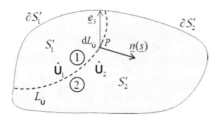

Figure 8.15. *Discontinuous virtual motion*

8.6. Matching the model with the three-dimensional continuum

8.6.1. *The matching procedure*

Trying to establish a connection between the classical three-dimensional continuum where internal forces are modeled by the Cauchy tensor stress field and the present two-dimensional modeling of plates and thin slabs is a natural follow-up to the initial guiding idea of this modeling process being derived from the thought experiment where a three-dimensional plane and slender solid is smashed flat onto a director sheet (D). This may be done, as in [SAL 13]:

– either through a direct "micro-macro" process, where the field and boundary equations for the Cauchy stress field in the three-dimensional model are integrated following the $\underline{e}_3$ direction throughout the solid to obtain the equations of motions for the two-dimensional model together with the expressions of the reduced elements of the internal force wrench field;

– or following the virtual work approach in the same way as in Chapter 7 for the curvilinear one-dimensional model.

Choosing the latter option, we will, for any virtual motion $\hat{\mathbf{U}}$, define a virtual motion for the three-dimensional solid through a virtual velocity field generated by the velocity distributor field $\left\{\hat{\mathbb{U}}(\underline{s})\right\}=\left\{P,\ \underline{\hat{U}}(\underline{s}),\ \underline{\hat{\Omega}}(\underline{s})\right\}$ which defines $\hat{\mathbf{U}}$. Then, we will identify the virtual rates of work by quantities of acceleration, external forces and internal forces in this virtual motion for the cylindrical volume, denoted by $(\mathrm{d}S \times A'A)$, delimited by the lower and upper end surfaces of the three-dimensional solid (Figure 8.16) with the corresponding quantities for the particle $\mathrm{d}S$ in $\hat{\mathbf{U}}$.

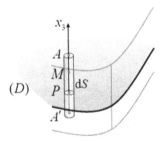

Figure 8.16. *Cylindrical volume* $(\mathrm{d}S \times A'A)$ *in the three-dimensional solid*

8.6.2. *Three-dimensional virtual velocity field*

Given the continuous and continuously differentiable virtual velocity distributor field $\left\{\hat{\mathbb{U}}(\underline{s})\right\}=\left\{P,\ \underline{\hat{u}}(\underline{s})+\hat{w}(\underline{s})\underline{e}_3,\ \underline{\hat{\Omega}}(\underline{s})\right\}$, we consider the virtual velocity field defined as follows *within the three-dimensional solid.*

Let M be a field point in the three-dimensional solid located on the $\underline{e}_3$ axis passing through the field point P in (D), so that its position vector $\underline{x}$ is

$\underline{x} = \underline{s} + x_3\, \underline{e}_3$. The virtual velocity at point M is generated by the velocity distributor $\left\{\hat{\mathbb{U}}(\underline{s})\right\}$

$$\underline{\hat{U}}(\underline{x}) = \underline{\hat{U}}(\underline{s}, x_3) = \underline{\hat{U}}(\underline{s}) + \underline{\hat{\underline{\Omega}}}(\underline{s}) \wedge \underline{PM} = \underline{\hat{U}}(\underline{s}) + \underline{\hat{\underline{\Omega}}}(\underline{s}).\underline{PM}, \qquad [8.93]$$

which means that the normal $A'A$ to (D) at point P in the three-dimensional solid is given the rigid body virtual motion defined by $\left\{\hat{\mathbb{U}}(\underline{s})\right\}$. More explicitly

$$\underline{\hat{U}}(\underline{s}, x_3) = \underline{\hat{u}}(\underline{s}) + \hat{w}(\underline{s})\underline{e}_3 + \underline{\hat{\underline{\Omega}}}(\underline{s}).x_3\,\underline{e}_3 = \underline{\hat{u}}(\underline{s}) + \hat{w}(\underline{s})\underline{e}_3 + \underline{\hat{\underline{\omega}}}(\underline{s}).x_3\,\underline{e}_3 \qquad [8.94]$$

with
$$\begin{cases} \underline{\hat{U}}(\underline{s}) = \underline{\hat{u}}(\underline{s}) + \hat{w}(\underline{s})\underline{e}_3 & \underline{\hat{u}}(\underline{s}) \in (D) = \mathbb{R}^2,\; \hat{w}(\underline{s}) \in \mathbb{R} \\[6pt] \underline{\hat{\underline{\Omega}}}(\underline{s}) = \underline{\hat{\omega}}(\underline{s}) + \hat{\Omega}_3(\underline{s})\underline{e}_3 & \underline{\hat{\omega}}(\underline{s}) \in (D) = \mathbb{R}^2,\; \hat{\Omega}_3(\underline{s}) \in \mathbb{R} \\[6pt] \underline{\hat{\underline{\Omega}}}(\underline{s}) = \underline{\hat{\underline{\omega}}}(\underline{s}) + \hat{\Omega}_3(\underline{s})\underline{\underline{e}}_3 & \underline{\hat{\underline{\omega}}}(\underline{s}) = -^t\underline{\hat{\underline{\omega}}}(\underline{s}) \in \mathbb{R}^2 \times \mathbb{R}^2. \end{cases} \qquad [8.95]$$

The gradient of this virtual velocity field at point M comes from

$$\underline{\underline{\operatorname{grad}}}\,\underline{\hat{U}}(\underline{x}).\mathrm{d}\underline{x} = \underline{\underline{\partial\hat{u}}}(\underline{s}).\mathrm{d}\underline{s} + \underline{e}_3\,\partial\hat{w}(\underline{s}).\mathrm{d}\underline{s} + \underline{\hat{\underline{\omega}}}(\underline{s}).\underline{e}_3\,\mathrm{d}x_3 + (\partial\underline{\hat{\underline{\omega}}}(\underline{s}).\mathrm{d}\underline{s}) \wedge \underline{e}_3\,x_3 \qquad [8.96]$$

and may be written as

$$\underline{\underline{\operatorname{grad}}}\,\underline{\hat{U}}(\underline{x}) = \underline{\underline{\partial\hat{u}}}(\underline{s}) + \underline{e}_3 \otimes \underline{\partial\hat{w}}(\underline{s}) - \underline{\underline{e}}_3\,.\underline{\hat{\underline{\omega}}}(\underline{s}) \otimes \underline{e}_3 - x_3\,\underline{\underline{e}}_3\,.\partial\underline{\hat{\underline{\omega}}}(\underline{s}), \qquad [8.97]$$

with the explicit meaning

$$\underline{\underline{\operatorname{grad}}}\,\underline{\hat{U}}(\underline{x}) = \frac{\partial\hat{u}_i}{\partial x_j}\underline{e}_i \otimes \underline{e}_j + \frac{\partial\hat{w}}{\partial x_j}\underline{e}_3 \otimes \underline{e}_j - \hat{\omega}_j\,\underline{\underline{e}}_3\,.\underline{e}_j \otimes \underline{e}_3 - x_3\,\frac{\partial\hat{\omega}_i}{\partial x_j}\underline{\underline{e}}_3\,.\underline{e}_i \otimes \underline{e}_j,\; i, j = 1, 2.$$
$$[8.98]$$

8.6.3. *Virtual rate of work by quantities of acceleration*

At the field point M in the cylindrical volume, the quantity of acceleration of the volume element $\mathrm{d}\Omega = \mathrm{d}S\,\mathrm{d}x_3$ is $\rho(\underline{x}, t)\underline{a}(\underline{x}, t)\,\mathrm{d}S\,\mathrm{d}x_3$. Thence, in the virtual velocity field [8.93], the virtual rate of work by quantities of acceleration for the cylindrical volume $(dS \times A'A)$ is equal to:

$$dS\left[P, \int_{A'A} \rho(\underline{x},t)\underline{a}(\underline{x},t)\,dx_3, \int_{A'A} x_3\,\underline{e}_3 \cdot \rho(\underline{x},t)\underline{a}(\underline{x},t)\,dx_3\right] \cdot \left\{\hat{\mathbb{U}}(\underline{s})\right\} \text{ with } \underline{x} = \underline{s} + x_3\,\underline{e}_3.$$

[8.99]

In consideration of the slenderness of the three-dimensional solid, the resultant moment term in the wrench is usually considered negligible and the resultant force term is reduced to $\rho(\underline{s},t)\underline{a}(\underline{s},t)$ with $\rho(\underline{s},t) = \int_{A'A} \rho(\underline{x},t)\,dx_3$ the mass surface density.

8.6.4. Virtual rate of work by external forces

The virtual rate of work by external forces acting in the cylindrical volume $(dS \times A'A)$ and on its lower and upper end surfaces is also expressed by a duality product similar to [8.99] where the wrench $\left[\boldsymbol{f}(\underline{s},t)\right]$ has been already defined in [8.16]:

$$\begin{cases} dS\left[\boldsymbol{f}(\underline{s},t)\right] \cdot \left\{\hat{\mathbb{U}}(\underline{s})\right\} \\ \left[\boldsymbol{f}(\underline{s},t)\right] = \left[P,\ \underline{f}(\underline{s},t),\ \underline{h}_D(\underline{s},t) + c(\underline{s},t)\underline{e}_3\right]. \end{cases}$$

[8.100]

8.6.5. Virtual rate of work by internal forces

The virtual rate of work by internal forces in the cylindrical volume $(dS \times A'A)$ is equal to

$$dS\int_{A'A} -\underline{\underline{\sigma}}(\underline{x},t):\text{grad}\,\hat{U}(\underline{x})\,dx_3 \text{ with } \underline{x} = \underline{s} + x_3\,\underline{e}_3.$$

[8.101]

Let $\underline{\underline{\sigma}}_D(\underline{x},t) \in \mathbb{R}^2 \otimes \mathbb{R}^2$ denote the second rank symmetric tensor derived

from $\underline{\sigma}(\underline{x},t)$ through $\underline{\underline{\sigma}}_D(\underline{x},t)=\sigma_{ij}(\underline{x},t)\underline{e}_i\otimes\underline{e}_j$, $i,j=1,2$. Recalling   that
$\underline{e}_k\cdot\underline{e}_j=-\varepsilon_{ijk}\underline{e}_i$, [8.101] comes out as the sum of the following terms ($j=1,2$):

$$-\mathrm{d}S\int_{A'A}\underline{\underline{\sigma}}_D(\underline{x},t):\partial\hat{u}(\underline{s})\,\mathrm{d}x_3-\mathrm{d}S\int_{A'A}\sigma_{j3}(\underline{x},t)\underline{e}_j\cdot\partial\hat{w}\,\mathrm{d}x_3$$

$$\text{[8.102]}$$

$$+\mathrm{d}S\int_{A'A}\sigma_{j3}(\underline{x},t)\underline{e}_j\cdot(\underline{e}_3\cdot\hat{\underline{\omega}}(\underline{s}))\,\mathrm{d}x_3+\mathrm{d}S\int_{A'A}x_3\,\underline{\underline{\sigma}}_D(\underline{x},t):(\underline{e}_3\cdot\partial\hat{\underline{\omega}}(\underline{s}))\,\mathrm{d}x_3\,.$$

8.6.6. *Identifying the reduced elements of the internal force wrench field*

Writing down the principle of virtual work for both models in associated virtual motions $\hat{\mathbf{U}}$ and $\hat{\underline{U}}$, it appears that the virtual rates of work by quantities of acceleration and external forces exhibit identical expressions for the surface element $\mathrm{d}S$ on the one side and cylindrical volume $(\mathrm{d}S\times A'A)$ on the other. It follows that the virtual rates of work by internal forces [8.102] and $p_i(\hat{\mathbf{U}})\,\mathrm{d}S$ derived from [8.67] are also identical:

$$\begin{cases}\forall\hat{u},\,\forall\hat{w},\,\forall\hat{\underline{\omega}},\,j=1,2,\\[4pt]-\mathrm{d}S\int_{A'A}\underline{\underline{\sigma}}_D(\underline{x},t):\partial\hat{u}(\underline{s})\,\mathrm{d}x_3-\mathrm{d}S\int_{A'A}\sigma_{j3}(\underline{x},t)\underline{e}_j\cdot\partial\hat{w}\,\mathrm{d}x_3\\[4pt]+\mathrm{d}S\int_{A'A}\sigma_{j3}(\underline{x},t)\underline{e}_j\cdot(\underline{e}_3\cdot\hat{\underline{\omega}}(\underline{s}))\,\mathrm{d}x_3+\mathrm{d}S\int_{A'A}x_3\,\underline{\underline{\sigma}}_D(\underline{x},t):(\underline{e}_3\cdot\partial\hat{\underline{\omega}}(\underline{s}))\,\mathrm{d}x_3\\[4pt]=-\underline{\underline{N}}(\underline{s},t):\partial\hat{u}(\underline{s})-\underline{V}(\underline{s},t)\cdot(\partial\hat{w}(\underline{s})-\underline{e}_3\cdot\hat{\underline{\omega}}(\underline{s}))-{}^t\underline{\underline{M}}(\underline{s},t):(\underline{e}_3\cdot\partial\hat{\underline{\omega}}(\underline{s})).\end{cases}\qquad\text{[8.103]}$$

We thus obtain the expressions of the reduced elements of the internal force tensorial wrench at the field point P of the two-dimensional model ($j=1,2$):

$$\begin{cases}\underline{\underline{N}}(\underline{s},t)=\int_{A'A}\underline{\underline{\sigma}}_D(\underline{x},t)\,\mathrm{d}x_3\\[6pt]\underline{V}(\underline{s},t)=\int_{A'A}\sigma_{j3}(\underline{x},t)\underline{e}_j\,\mathrm{d}x_3\\[6pt]\underline{\underline{M}}(\underline{s},t)=-\int_{A'A}x_3\,{}^t\underline{\underline{\sigma}}_D(\underline{x},t)\,\mathrm{d}x_3=-\int_{A'A}x_3\,\underline{\underline{\sigma}}_D(\underline{x},t)\,\mathrm{d}x_3\,.\end{cases}\qquad\text{[8.104]}$$

These expressions are obviously in line with the dimensional analysis viewpoint referred to in section 8.3.7. They confirm that, for *any line element* $d\ell$ *with normal* $\underline{n}(\underline{s})$ at the field point P, *the wrench of contact forces for the two-dimensional model is equal to the wrench of the contact forces acting on* $(d\ell \times A'A)$ *in the three-dimensional medium.*

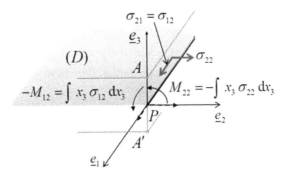

Figure 8.17. *Components of* $\underline{\underline{M}}(\underline{s},t)$ *and* $\underline{\underline{\sigma}}(\underline{x},t)$

Note that the expression of $\underline{\underline{M}}(\underline{s},t)$ in [8.104] explains the introduction of that tensor after $\underline{\underline{H}}(\underline{s},t)$, as shown in Figure 8.17.

Comments

This matching procedure first shows that there is no inconsistency between the three-dimensional continuum model and the modeling of plates as two-dimensional continua.

Incidentally, it should be observed that deriving the two-dimensional model from the three-dimensional one inevitably leads to $\underline{\underline{M}}(\underline{s},t)$ being a *symmetric tensor* while, as pointed out in section 8.4.4, this restriction *does not come out* from the direct virtual work approach.

However, it is essential to retain that the general two-dimensional model is mechanically consistent and valid independently of the possibility of performing the matching process. For this reason, it may cover a wider range of practical applications than could be anticipated from its derivation through the matching process.

8.7. The Kirchhoff–Love condition

8.7.1. *Virtual rate of angular distortion: Kirchhoff–Love condition*

The Kirchhoff–Love condition was already introduced in section 8.1.2 when it was noted that the term $\partial \hat{w}(\underline{s}) - \underline{e}_3 \wedge \hat{\underline{\omega}}(\underline{s}) = \partial \hat{w}(\underline{s}) - \underline{e}_{\underline{3}} \cdot \hat{\underline{\omega}}(\underline{s})$ represents the *virtual rate of angular distortion* of the microstructure with respect to the director sheet in the virtual motion defined by the independent vector fields $\hat{\underline{u}}(\underline{s}), \hat{\underline{\omega}}(\underline{s})$, and scalar fields $\hat{w}(\underline{s}), \hat{\Omega}_3(\underline{s})$. The virtual rotation rate of the normal to the director sheet at the point P is determined uniquely from the scalar field $\hat{w}(\underline{s})$ and equal to $-\underline{e}_3 \wedge \partial \hat{w}(\underline{s})$ while the rotation rate of the microstructure is just $\hat{\underline{\omega}}(\underline{s})$.

Similarly to the Navier–Bernoulli condition, the Kirchhoff–Love condition[8] requires the transverse microstructure to remain orthogonal to the director sheet, i.e.

$$\partial \hat{w}(\underline{s}) - \underline{e}_3 \wedge \hat{\underline{\omega}}(\underline{s}) = \partial \hat{w}(\underline{s}) - \underline{e}_{\underline{3}} \cdot \hat{\underline{\omega}}(\underline{s}) = 0 \text{ over } S, \qquad [8.105]$$

a condition relating the vector field $\hat{\underline{\omega}}(\underline{s})$ and the scalar field $\hat{w}(\underline{s})$ which is equivalent to

$$\hat{\underline{\omega}}(\underline{s}) = -\underline{e}_{\underline{3}} \cdot \partial \hat{w}(\underline{s}). \qquad [8.106]$$

With the Kirchhoff–Love condition, the expression [8.67] of $p_i\,(\hat{\mathbf{U}})$ reduces to

$$p_i\,(\hat{\mathbf{U}}) = -\underline{\underline{N}}(\underline{s},t) : \partial \hat{\underline{u}}(\underline{s}) - {}^t\underline{\underline{M}}(\underline{s},t) : \underline{e}_{\underline{3}} \cdot \partial \hat{\underline{\omega}}(\underline{s}) \qquad [8.107]$$

without any contribution of the shearing force $\underline{V}(\underline{s},t)$ in the same way as in Chapter 7 (section 7.5.2) for the curvilinear one-dimensional continuum with the Navier–Bernoulli condition. Also, the Kirchhoff–Love condition implies that

$$\underline{e}_{\underline{3}} \cdot \partial \hat{\underline{\omega}}(\underline{s}) = \partial^2 \hat{w}(\underline{s}) = \hat{\underline{\underline{\chi}}}(\underline{s}), \qquad [8.108]$$

where $\hat{\underline{\underline{\chi}}}(\underline{s})$, a symmetric second rank tensor, is the *virtual rate of curvature tensor* of the director sheet in the considered virtual motion. It follows that, in [8.107],

---

8 A.E.H. Love (1863–1940) and G. Kirchhoff (1824–1887).

only the *symmetric part* of tensor $\underline{\underline{M}}(\underline{s},t)$ contributes to the virtual work by internal forces:

$$\begin{cases} p_i(\hat{\mathbf{U}}) = -\underline{\underline{N}}(\underline{s},t):\partial\hat{\underline{u}}(\underline{s}) - \underline{\underline{M}}(\underline{s},t):\partial^2\hat{w}(\underline{s}) \\ p_i(\hat{\mathbf{U}}) = -\underline{\underline{N}}(\underline{s},t):\partial\hat{\underline{u}}(\underline{s}) - \underline{\underline{M}}(\underline{s},t):\hat{\underline{\underline{\chi}}}(\underline{s}). \end{cases} \qquad [8.109]$$

8.7.2. Discontinuity equations

Referring to Figure 8.15, at a point P of $L_{\hat{u}}$ where the virtual motion is discontinuous, the jump conditions associated with [8.105] regarding the scalar field $\hat{w}(\underline{s})$ are first

$$[\![\hat{w}(\underline{s})]\!] = 0 \qquad [8.110]$$

and then

$$[\![\hat{\underline{\omega}}(\underline{s})]\!] = -\underline{e}_3.[\![\partial\hat{w}(\underline{s})]\!]. \qquad [8.111]$$

It follows from [8.110] that $\hat{w}(\underline{s})$ is continuous when crossing $L_{\hat{u}}$. Thence, its derivative $\partial\hat{w}(\underline{s}).\underline{t}(\underline{s})$ along $L_{\hat{u}}$ is continuous[9]. With $(\underline{t}(\underline{s}), \underline{n}(\underline{s}), \underline{e}_3)$, a right-handed triad of unit vectors, the continuity of this tangent derivative implies that $[\![\partial\hat{w}(\underline{s})]\!]$ is written as

$$[\![\partial\hat{w}(\underline{s})]\!] = \hat{\theta}(\underline{s})\underline{n}(\underline{s}) \qquad [8.112]$$

and then

$$[\![\hat{\underline{\omega}}(\underline{s})]\!] = -\underline{e}_3.\hat{\theta}(\underline{s})\underline{n}(\underline{s}) = \hat{\theta}(\underline{s})\underline{t}(\underline{s}). \qquad [8.113]$$

---

9 This result is often referred to as Hadamard's compatibility condition (J. Hadamard, 1865–1963).

As shown in Figure 8.18, the discontinuity line $L_{\hat{\mathrm{u}}}$ comes out as a virtual *hinge line* where the virtual rotation rate about the tangent $\underline{t}(\underline{s})$ is equal to the jump of the normal derivative, $[\![\dfrac{\partial \hat{w}}{\partial n}(\underline{s})]\!] = \hat{\theta}(\underline{s})$.

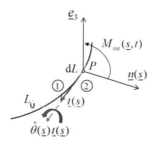

Figure 8.18. *Hinge line in a discontinuous virtual motion*

8.7.3. *Virtual rate of work by internal forces*

From the preceding results, we can now derive the expression to be adopted for the virtual rate of work by internal forces in the case of piecewise continuous and continuously differentiable Kirchhoff–Love virtual motions. Equation [8.92] becomes

$$\mathcal{P}_i(\hat{\mathbf{U}}) = -\int_S (\underline{\underline{N}}(\underline{s},t):\partial\hat{\underline{u}}(\underline{s}) + \underline{\underline{M}}(\underline{s},t):\partial^2\hat{w}(\underline{s}))\,\mathrm{d}S$$
$$-\int_{L_{\hat{\mathrm{u}}}} ([\![\hat{\underline{u}}(\underline{s})]\!].\underline{\underline{N}}(\underline{s},t).\underline{n}(\underline{s}) + [\![\partial\hat{w}(\underline{s})]\!].\underline{\underline{M}}(\underline{s},t).\underline{n}(\underline{s}))\,\mathrm{d}L_{\hat{\mathrm{u}}}\,, \qquad [8.114]$$

that is, with [8.112],

$$\mathcal{P}_i(\hat{\mathbf{U}}) = -\int_S (\underline{\underline{N}}(\underline{s},t):\partial\hat{\underline{u}}(\underline{s}) + \underline{\underline{M}}(\underline{s},t):\partial^2\hat{w}(\underline{s}))\,\mathrm{d}S$$
$$-\int_{L_{\hat{\mathrm{u}}}} ([\![\hat{\underline{u}}(\underline{s})]\!].\underline{\underline{N}}(\underline{s},t).\underline{n}(\underline{s}) + \hat{\theta}(\underline{s})\,M_{nn}(\underline{s},t))\,\mathrm{d}L_{\hat{\mathrm{u}}}\,. \qquad [8.115]$$

Comments

Comments similar to those in Chapter 7 (section 7.5.5) about the Navier–Bernoulli condition can be made here. As a general rule, the Kirchhoff–Love condition is part of all classical constitutive equations adopted for the two-dimensional plane continuum such as Elasticity and Elastoplasticity. As already pointed out, it implies that only the symmetric part of the tensor of internal moments contributes to the virtual rate of work by internal forces and, as a result, this tensor will be considered symmetric. As a matter of fact, the latter assumption is no loss of generality since the constitutive laws mentioned above are most often derived from the their three-dimensional counterpart through the matching process described in section 8.6. This is the case, for instance, when studying metallic plates.

When dealing with thin reinforced concrete slabs modeled as two-dimensional plane continuums, the symmetry of the internal moment tensor is taken for granted. Load carrying capacity analyses are performed within the framework of the theory of yield design with strength criteria involving the internal moment tensor only. As explained in [SAL 13], it follows that the virtual motions to be considered in the corresponding upper bound approaches *must comply with the Kirchhoff–Love condition*. Among them, piecewise continuous and continuously differentiable motions with hinge lines are the most popular, especially through Johansen's yield line theory [JOH 31, JOH 52] based upon Johansen's strength criteria for isotropic or anisotropic[10] plates. Other strength criteria are used, derived from their three-dimensional counterparts, for load carrying capacity analyses of metallic plates (Tresca's plates and von Mises' plates).

8.8. An illustrative example: circular plate under a distributed load

8.8.1. *Load carrying capacity*

As an illustrative example, we consider a circular plate S with center O and radius R that is built-in in a fixed rigid support along its boundary ∂S (Figure 8.19). It is subjected to a uniformly distributed vertical load with surface density $f_3(\underline{s},t)\underline{e}_3 = -p\,\underline{e}_3$, $p > 0$. This plate is the two-dimensional modeling of either a *homogeneous* metallic plate, or a thin *homogeneous* reinforced concrete slab, whose resistance of the constituent material element is defined through a strength criterion imposed on the reduced elements of the internal force wrench.

It is anticipated that, as in Chapter 6 (sections 6.10.8 and 6.10.9), the implementation of the principle of virtual work would make it possible to derive some assessment of the load carrying capacity of the plate within the two-dimensional modeling framework based upon the knowledge of the resistance of its constituent material assumed to be *isotropic*.

---

10 See [SAV 98].

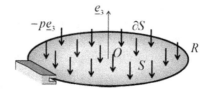

Figure 8.19. *Circular plate under a uniformly distributed load*

8.8.1.1. *Virtual motions*

As a first simple approach, using polar coordinates (r,θ) with center O, we consider the pure Kirchhoff–Love virtual motions defined in [8.116], where $\hat{\delta}$ stands for the virtual deflection rate in the center of the plate and which depends on two geometrical parameters a and b (Figure 8.20):

$$\begin{cases} 0 < a < b < R, \ \hat{\delta} > 0 \\ \hat{\underline{u}}(\underline{s}) = 0 \\ \hat{w}(\underline{s}) = -\hat{\delta} \ \text{ for } 0 \le r \le a \\ \hat{w}(\underline{s}) = -\hat{\delta}\dfrac{b-r}{b-a} \ \text{ for } a \le r \le b \\ \hat{w}(\underline{s}) = 0 \ \text{ for } b \le r \le R. \end{cases} \qquad [8.116]$$

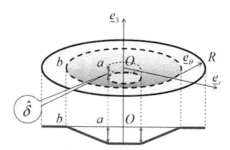

Figure 8.20. *Virtual motions*

The corresponding virtual velocity distributor field is continuous and piecewise continuously differentiable, exhibiting positive and negative rotation rate jumps,

$\hat{\theta}_a = \dfrac{\hat{\delta}}{b-a}$ and $\hat{\theta}_b = -\dfrac{\hat{\delta}}{b-a}$ respectively, when crossing the circles $(r = a)$ and

$(r = b)$, which are the hinge lines in the virtual motion as a consequence of $[8.112]^{11}$

$$\begin{cases} r = a: \; [\![\hat{\underline{\omega}}]\!] = -\underline{e}_3 \cdot [\![\partial \hat{w}]\!] = \dfrac{\hat{\delta}}{b-a}(-\underline{e}_\theta) \\[4mm] r = b: \; [\![\hat{\underline{\omega}}]\!] = -\underline{e}_3 \cdot [\![\partial \hat{w}]\!] = -\dfrac{\hat{\delta}}{b-a}(-\underline{e}_\theta). \end{cases}$$

[8.117]

From the general expression of $\hat{\underline{\underline{\chi}}}(\underline{s})$ in polar coordinates

$$\begin{aligned} \hat{\underline{\underline{\chi}}} &= \frac{\partial^2 \hat{w}}{\partial r^2} \underline{e}_r \otimes \underline{e}_r + (\frac{1}{r}\frac{\partial^2 \hat{w}}{\partial r \partial \theta} - \frac{1}{r^2}\frac{\partial \hat{w}}{\partial \theta})(\underline{e}_r \otimes \underline{e}_\theta + \underline{e}_\theta \otimes \underline{e}_r) \\ &+ (\frac{1}{r}\frac{\partial \hat{w}}{\partial r} + \frac{1}{r^2}\frac{\partial^2 \hat{w}}{\partial \theta^2})\underline{e}_\theta \otimes \underline{e}_\theta \end{aligned}$$

[8.118]

we obtain here

$$\begin{cases} \hat{\underline{\underline{\chi}}}(\underline{s}) = 0 \;\; \text{for} \;\; 0 < r < a \text{ and } b < r < R \\[4mm] \hat{\underline{\underline{\chi}}}(\underline{s}) = \dfrac{\hat{\delta}}{r(b-a)}\underline{e}_\theta \otimes \underline{e}_\theta \;\; \text{for} \;\; a < r < b. \end{cases}$$

[8.119]

8.8.1.2. Virtual rate of work by external forces

The virtual rate of work by external forces (the uniformly distributed load p) amounts to:

$$\mathcal{P}_e(\hat{\mathbf{U}}) = \mathcal{P}_e(\hat{w}) = \frac{\pi}{3}p\hat{\delta}(a^2 + ab + b^2).$$

[8.120]

8.8.1.3. Virtual rate of work by internal forces

With the expression [8.116] for the virtual velocity field, the virtual rate of work by internal forces [8.115] reduces to

---

11 It is recalled that, according to Figure 8.18, the triad $(\underline{t},\underline{n},\underline{e}_3)$ is right-handed which implies here that $\underline{t}$ is opposite to $\underline{e}_\theta$ if $\underline{n}$ lies along $\underline{e}_r$.

$$\mathcal{P}_i(\hat{\mathbf{U}}) = \mathcal{P}_i(\hat{w}) = -\int_S \underline{\underline{M}}(\underline{s},t):\underline{\underline{\partial^2 \hat{w}}}(\underline{s})\,dS - \int_{L_{\hat{U}}} \hat{\theta}(\underline{s})\,M_{nn}(\underline{s},t)\,dL_{\hat{U}} \qquad [8.121]$$

explicitly here:

$$
\begin{aligned}
\mathcal{P}_i(\hat{w}) = -&\iint\limits_{\substack{0<\theta<2\pi \\ a<r<b}} M_{\theta\theta}(r,\theta,t)\frac{\hat{\delta}}{b-a}\,dr\,d\theta \\
-&\int\limits_{0<\theta<2\pi} M_{rr}(a,\theta,t)\frac{\hat{\delta}}{b-a}a\,d\theta + \int\limits_{0<\theta<2\pi} M_{rr}(b,\theta,t)\frac{\hat{\delta}}{b-a}b\,d\theta.
\end{aligned}
\qquad [8.122]
$$

8.8.1.4. Virtual rate of work by quantities of acceleration

Looking for an assessment of load carrying capacity of the plate amounts to investigating the compatibility of the system *equilibrium equations* under the applied load with the constraints imposed by the strength criterion of the constituent material. The virtual rate of work by quantities of acceleration is thus set to zero.

8.8.1.5. Implementation of the principle of virtual work

The principle of virtual work for any internal wrench field in equilibrium with the applied uniformly distributed load $-p\underline{e}_3$, $p > 0$ is then written as:

$$
\begin{aligned}
\mathcal{P}_e(\hat{w}) + \mathcal{P}_i(\hat{w}) = &\frac{\pi}{3}p\hat{\delta}(a^2+ab+b^2) - \iint\limits_{\substack{0<\theta<2\pi \\ a<r<b}} M_{\theta\theta}(r,\theta)\frac{\hat{\delta}}{b-a}\,dr\,d\theta \\
-&\int\limits_{0<\theta<2\pi} M_{rr}(a,\theta)\frac{\hat{\delta}}{b-a}a\,d\theta + \int\limits_{0<\theta<2\pi} M_{rr}(b,\theta)\frac{\hat{\delta}}{b-a}b\,d\theta = 0.
\end{aligned}
$$

$$[8.123]$$

Whatever the strength criterion of the *isotropic* constituent element of the *homogeneous* plate, it defines a maximum positive value, namely m_0^+, and a minimum negative value $-m_0^-$, for $M_{nn}, \forall\underline{n}$. It follows from [8.123] that, to be sustained by the plate, the load *must comply* with the constraint

$$p \le \frac{6(m_0^+ + m_0^-)b}{b^3 - a^3}. \qquad [8.124]$$

Minimizing this upper bound with respect to the parameters a and b, we obtain, as a minimum value

$$p \le \frac{6(m_0^+ + m_0^-)}{R^2},$$

[8.125]

which corresponds to $b \to R$ and $a \to 0$. In other words, the minimizing virtual motion in that case tends to be "*conical*" with the "*negative*" hinge line along the boundary ∂S and the "*positive*" hinge line at the apex O (Figure 8.21).

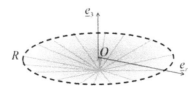

Figure 8.21. *Conical Kirchhoff–Love virtual motion*

It is worth noting that, in this limit virtual motion, the contribution of the apex to [8.122] reduces to zero and thence the upper bound [8.125] for the load carrying capacity of the plate only results from the distributed contribution where the virtual velocity distributor field is continuous and piecewise continuously differentiable, and the contribution of the negative hinge line along ∂S. As a result, if the boundary condition along ∂S were that of a simple support allowing for free rotation of the plate, the upper bound would just be reduced to

$$p \le \frac{6 m_0^+}{R^2}.$$

[8.126]

Also, it comes out from [8.125] that the upper bound on the total load $p\pi R^2$ applied to the plate *does not depend on the size of the plate*, a remarkable result that could be anticipated from dimensional analysis.

8.8.2. *Resistance of the two-dimensional plate element*

In pure Kirchhoff–Love virtual motions such as [8.116] (pure Kirchhoff–Love bending), the expression [8.115] of the virtual rate of work by internal forces only involves the tensor of internal moments $\underline{M}(\underline{s},t)$. It follows that, looking for an upper bound for the load carrying capacity of a plate with this class of virtual motions, only the strength criterion related to $\underline{\underline{M}}(\underline{s},t)$ shall be considered. Setting aside the membrane internal force tensor field, this simple classical approach discards the influence of $\underline{V}(\underline{s},t)$ which amounts to assuming that the shear force vector suffers no limitation. Usually, this point is addressed in a second run.

Conversely, as explained in [SAL 13], as long as the strength criterion retained for the plate element only involves the tensor of internal moments, the only virtual motions to be considered in order to obtain a non-trivial upper bound are pure Kirchhoff–Love bending motions.

The strength criterion defining the admissible tensors of internal moments $\underline{\underline{M}}(\underline{s},t)$ may be derived, through the matching process described in section 8.6.6 when it is feasible, from the three-dimensional strength criterion of the constituent material. This is the case, classically, for transversally homogeneous metallic plates with a von Mises[12] or Tresca criterion. For instance, for a plate with thickness h made of a Tresca constituent material (Chapter 6, section 6.10.6), the strength criterion obtained through [8.104] for the two-dimensional analysis is written as

$$\mathrm{Sup}\left\{\left|M_1\right|,\left|M_2\right|,\left|M_1-M_2\right|\right\}\leq m_0(\underline{s}),\tag{8.127}$$

where M_1 and M_2 are the principal values of the symmetric tensor $\underline{\underline{M}}(\underline{s},t)$, and $m_0(\underline{s})$ is derived from the resistance $\sigma_0(\underline{s})$ of the three-dimensional constituent material under simple tension: $m_0(\underline{s})=\sigma_0(\underline{s})\dfrac{h^2}{4}$.

This process is not feasible in the case of thin reinforced concrete slabs for instance and the strength criterion proceeds directly both from experiments and (possibly heuristic) theoretical analyses at the macroscopic level.

The most famous criterion is known as *Johansen's criterion* for isotropic plates written in the form

$$\begin{cases} -m_0^-(\underline{s})\leq M_1\leq m_0^+(\underline{s}) \\ -m_0^-(\underline{s})\leq M_2\leq m_0^+(\underline{s}), \end{cases}\tag{8.128}$$

where m_0^+ and m_0^- may be different, depending on the reinforcement in the upper and lower layers. A more sophisticated expression of Johansen's criterion deals with

---

12 R. von Mises (1883–1953).

orthotropic plates, an extensive analysis of which is given in [SAV 98] and briefly outlined in [SAL 13].

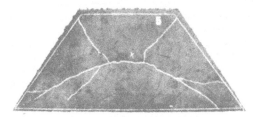

Figure 8.22. *Bending of a reinforced plaster slab (M. Milicevic, Mons, Belgium)*

An abundant literature has been devoted to the analysis of plates by means of Johansen's yield line theory of which the preceding sections present a simplified example. It is worth noting that it was essentially driven by the design of reinforced concrete slabs for which observations of actual collapse mechanisms quite often suggested the existence of hinge lines as in Figure 8.22, showing the results of a reduced scale experiment. It is not to be forgotten anyhow that this approach can only provide upper bounds for the load carrying capacity as recalled by equations [8.125] and [8.126].

8.8.3. *An equilibrated internal force wrench distribution*

This example presents the interesting specificity that an internal force wrench distribution can be built up, which is in equilibrium with the maximum value [8.125] of the prescribed load. Using polar coordinates, it reads

$$\begin{cases} \underline{\underline{M}}(r,\theta) = m_0^+ (\underline{e}_r \otimes \underline{e}_r + \underline{e}_\theta \otimes \underline{e}_\theta) - (m_0^+ - m_0^-)\frac{r^2}{R^2}\underline{e}_r \otimes \underline{e}_r \\ \underline{V}(r,\theta) = -3(m_0^+ - m_0^-)\frac{r}{R^2}\underline{e}_r , \end{cases}$$ [8.129]

which satisfies the field equation of equilibrium [8.64] with the value $p = \dfrac{6(m_0^+ + m_0^-)}{R^2}$ for the load and, at the same time, complies with the strength criterion [8.127] or [8.128].

Referring to the terminology introduced by [BIS 53] and adopted in [SAL 13] for yield design problems, bringing together [8.116] and [8.129] provides a *complete solution* to the yield design problem of the Tresca or Johansen circular plate under a uniformly distributed load. *Nevertheless, this result should not be considered as proof that the plate will actually sustain the load* $pR^2 = 6(m_0^+ + m_0^-)$.

Appendix 1

Introduction to Tensor Calculus

A1.1. Vector space. Euclidean space

A1.1.1. *Vector space*

Let E be a vector space of finite dimension n over the field $\mathbb{R}$ with the following operations:

– Addition of vectors

$$\underline{u} \in E, \underline{v} \in E \mapsto \underline{w} = \underline{u} + \underline{v} \in E. \tag{A1.1}$$

– Scalar multiplication

$$\underline{u} \in E, \lambda \in \mathbb{R} \mapsto \underline{w} = \lambda \underline{u} \in E. \tag{A1.2}$$

The dimension n of the vector space E is the highest integer for which it is possible to find n vectors in E, namely $\underline{e}_1 \neq 0$, $\underline{e}_2 \neq 0$ ..., $\underline{e}_n \neq 0$ such that

$$\alpha_1 \underline{e}_1 + \alpha_2 \underline{e}_2 ... + \alpha_n \underline{e}_n = 0 \Rightarrow \alpha_1 = 0, \alpha_2 = 0 ..., \alpha_n = 0. \tag{A1.3}$$

Such a set of vectors is a basis $\left\{ \underline{e}_i \right\}$ in E and any vector $\underline{u}$ can be expressed in only one way as

$$\forall \underline{u} \in E, \underline{u} = u_1 \underline{e}_1 + u_2 \underline{e}_2 ... + u_n \underline{e}_n = u_i \underline{e}_i \tag{A1.4}$$

where the scalars u_i are the components of $\underline{u}$ in the basis $\left\{ \underline{e}_i \right\}$.

In this equation, the *Cartesian summation convention* or "Dummy index convention" is used, which means that in any product of terms a repeated suffix is held to be summed over $1, 2..., n$.

A1.1.2. *Scalar product of two vectors*

Choosing a basis $\{\underline{e}_i\}$, let E be endowed with the standard scalar product (also "inner" or "dot") product (to be read "$\underline{u}$ dot $\underline{v}$"):

$$\begin{cases} E \times E \to \mathbb{R} \\ \underline{u} \in E, \underline{v} \in E \mapsto \underline{u}.\underline{v} \in \mathbb{R}. \end{cases} \qquad [A1.5]$$

The basis $\{\underline{e}_i\}$ is said to be *orthonormal*, if

$$\underline{e}_i \cdot \underline{e}_j = \delta_{ij}, \qquad [A1.6]$$

where δ_{ij} is the *Kronecker delta* (or Kronecker symbol)

$$\begin{cases} \delta_{ij} = 1 \text{ if } i = j \\ \delta_{ij} = 0 \text{ if } i \neq j. \end{cases} \qquad [A1.7]$$

Then, the scalar product $\underline{u}.\underline{v}$ is written as

$$\begin{cases} E \times E \to \mathbb{R} \\ \underline{u} \in E, \underline{v} \in E \mapsto \underline{u}.\underline{v} = u_i v_i \in \mathbb{R}. \end{cases} \qquad [A1.8]$$

Note that the components of any vector $\underline{u}$ in the orthonormal basis $\{\underline{e}_i\}$ are obtained through the scalar products with the vectors of the orthonormal basis

$$\underline{u} = u_i \underline{e}_i \iff \forall i = 1..., n, \, u_i = \underline{u}.\underline{e}_i. \qquad [A1.9]$$

A1.1.3. *Change of orthonormal basis*

Defining the new orthonormal basis $\{\underline{e}'_i\}$ in terms of the basis $\{\underline{e}_k\}$ through the components of the vectors $\underline{e}'_i$ in the basis $\{\underline{e}_k\}$, namely $\alpha_{ik} = \underline{e}'_i \cdot \underline{e}_k$ according to [A1.9], we write that the new basis is orthonormal: $\underline{e}'_i \cdot \underline{e}'_j = \delta_{ij}$. It follows that

$$\alpha_{ik}\alpha_{jk} = \delta_{ij}.$$ [A1.10]

Writing that

$$\underline{e}'_i = \alpha_{ik}\underline{e}_k \text{ and } (\underline{e}_k = \alpha_{ik}\underline{e}'_i)$$ [A1.11]

we obtain the components of a vector $\underline{u}$ in the basis $\{\underline{e}'_i\}$ in terms of the components in the basis $\{\underline{e}_k\}$:

$$u'_i = \alpha_{ik}u_k.$$ [A1.12]

As a result

$$u'_i v'_i = \alpha_{ik}u_k\alpha_{i\ell}u_\ell = \delta_{k\ell}u_k u_\ell = u_k u_k,$$ [A1.13]

which proves that the scalar product is invariant in any change of orthonormal basis: the definition given initially is intrinsic.

Also, writing that the basis $\{\underline{e}_k\}$ as expressed in [A1.11] is orthonormal in terms of the new basis, $\underline{e}_k \cdot \underline{e}_\ell = \delta_{k\ell}$, we obtain the counterpart of [A1.10]:

$$\alpha_{jk}\alpha_{j\ell} = \delta_{k\ell}.$$ [A1.14]

A1.1.4. *Vector product or cross product, triple scalar product*

This analysis is *only* devoted to the three-dimensional case ($n = 3$). Given an orthonormal basis ($\underline{e}_1, \underline{e}_2, \underline{e}_3$), the vector product or *cross product* of two vectors $\underline{u}$ and $\underline{v}$ is denoted by $\underline{u} \wedge \underline{v}$ or $\underline{u} \times \underline{v}$ (to be read as "$\underline{u}$ cross $\underline{v}$") and defined by

$$\underline{u} \wedge \underline{v} = \varepsilon_{ijk}u_i v_j\underline{e}_k$$ [A1.15]

with ε_{ijk} the *permutation symbol*

$$\varepsilon_{ijk} = \begin{cases} 0 & \text{if any two of } i,j,k \text{ are the same} \\ 1 & \text{if } ijk \text{ is an even permutation of } 1,2,3 \\ -1 & \text{if } ijk \text{ is an odd permutation of } 1,2,3. \end{cases} \qquad \text{[A1.16]}$$

In other words,

$$\underline{u} \wedge \underline{v} = (u_2 v_3 - u_3 v_2)\underline{e}_1 + (u_3 v_1 - u_1 v_3)\underline{e}_2 + (u_1 v_2 - u_2 v_1)\underline{e}_3. \qquad \text{[A1.17]}$$

The *triple scalar product* of three vectors $\underline{u}, \underline{v}, \underline{w}$, sometimes denoted by $[\underline{u}, \underline{v}, \underline{w}]$, is the result of the scalar product of vector $\underline{u}$ with the cross product of vectors $\underline{v}$ and $\underline{w}$

$$[\underline{u}, \underline{v}, \underline{w}] = \underline{u}.(\underline{v} \wedge \underline{w}) = \varepsilon_{ijk} u_i v_j w_k. \qquad \text{[A1.18]}$$

It is worth noting that $[\underline{u}, \underline{v}, \underline{w}] = \underline{u}.(\underline{v} \wedge \underline{w}) = (\underline{u} \wedge \underline{v}).\underline{w}$ and, more generally, an even permutation applied to vectors $\underline{u}, \underline{v}, \underline{w}$ does not change the triple scalar product, whereas an odd permutation results in changing the sign.

From the geometrical viewpoint:

Denoting by θ the angle between the directions $\underline{u}$ and $\underline{v}$, we have the following expressions:

– *Scalar product*

$$\underline{u}.\underline{v} = |\underline{u}||\underline{v}| \cos \theta \qquad \text{[A1.19]}$$

– *Vector product*

$$\underline{u} \wedge \underline{v} = |\underline{u}||\underline{v}| \sin \theta \, \underline{n} \qquad \text{[A1.20]}$$

with $\underline{n}$ a unit vector orthogonal to $\underline{u}$ and $\underline{v}$, such that the triad $(\underline{u}, \underline{v}, \underline{n})$ be right handed, and θ counted positive about $\underline{n}$. Thence, the magnitude of $\underline{u} \wedge \underline{v}$ along $\underline{n}$ is the measure of the area of the parallelogram built upon $\underline{u}$ and $\underline{v}$ counted positive as the triad $(\underline{u}, \underline{v}, \underline{n})$ is right handed.

– Triple scalar product

As a consequence of [A1.20], the triple scalar product $[\underline{u},\underline{v},\underline{w}]=(\underline{u}\wedge\underline{v}).\underline{w}$ is equal to $\left|\underline{u}\right|\left|\underline{v}\right|\sin\theta\,\underline{n}.\underline{w}$ that is the algebraic measure of the volume of the parallelepiped built upon $\underline{u},\underline{v},\underline{w}$ counted positive if this triad is right handed.

A1.1.5. *Euclidean space*

A Euclidean space is defined as a vector space endowed with a scalar product as above.

A1.2. First rank Euclidean tensors

A first rank Euclidean tensor u is a *linear form* on the Euclidean space E:

$$\begin{cases} E\xrightarrow{\ u\ }\mathbb{R} \\ \underline{v}\in E\mapsto u(\underline{v})\in\mathbb{R}. \end{cases} \tag{A1.21}$$

u is an element of E^*, the dual space of E. Taking advantage of the Euclidean character of E, we can identify the linear form u with the vector $\underline{u}$ in E defined as follows through the scalar product:

$$\forall \underline{v}\in E,\ \underline{u}.\underline{v}=u(\underline{v}). \tag{A1.22}$$

This is the *canonical identification* of the Euclidean space E with its dual space E^*. A first rank Euclidean tensor will thus be equivalently represented by a vector $\underline{u}$ or a linear form u.

A1.3. Second rank Euclidean tensors

A1.3.1. *Definition*

A second rank Euclidean tensor $\underline{\underline{T}}$ is a *bilinear form* on $E\times E$

$$\begin{cases} E\times E\xrightarrow{\ \underline{\underline{T}}\ }\mathbb{R} \\ (\underline{u},\underline{v})\in E\times E\mapsto \underline{\underline{T}}(\underline{u},\underline{v})\in\mathbb{R}. \end{cases} \tag{A1.23}$$

In a basis $\left\{ \underline{e}_i \right\}$:

$$\begin{cases} \underline{\underline{T}}(\underline{u},\underline{v}) = u_i v_j \, \underline{\underline{T}}(\underline{e}_i,\underline{e}_j) \\ \underline{\underline{T}}(\underline{u},\underline{v}) = T_{ij} u_i v_j \ \text{ with } T_{ij} = \underline{\underline{T}}(\underline{e}_i,\underline{e}_j). \end{cases}$$

[A1.24]

A1.3.2. *Tensor product of two vectors*

Given two first rank Euclidean tensors $\underline{a} \in E$, $\underline{b} \in E$, their tensor product, denoted by $\underline{a} \otimes \underline{b}$, is the second rank Euclidean tensor defined by

$$\begin{cases} E \times E \xrightarrow{\ \underline{a} \otimes \underline{b}\ } \mathbb{R} \\ (\underline{u},\underline{v}) \in E \times E \mapsto (\underline{a} \otimes \underline{b})(\underline{u},\underline{v}) = (\underline{a} \cdot \underline{u})(\underline{b} \cdot \underline{v}) \in \mathbb{R}. \end{cases}$$

[A1.25]

Note that the tensor product is distributive with respect to the addition of vectors and *non-commutative*.

Such tensors as $\underline{\underline{T}} = \underline{a} \otimes \underline{b}$ defined here above, which can be written in the form of the tensor product of two vectors, are called *product tensors*.

As a particular case, the tensor product of two vectors $\underline{e}_i$ and $\underline{e}_j$ of the orthonormal basis is the second rank Euclidean tensor $\underline{e}_i \otimes \underline{e}_j$ such that

$$\begin{cases} E \times E \xrightarrow{\ \underline{e}_i \otimes \underline{e}_j\ } \mathbb{R} \\ (\underline{u},\underline{v}) \in E \times E \mapsto (\underline{e}_i \otimes \underline{e}_j)(\underline{u},\underline{v}) = (\underline{e}_i \cdot \underline{u})(\underline{e}_j \cdot \underline{v}) = u_i v_j. \end{cases}$$

[A1.26]

Thence, from [A1.24], we can write

$$\underline{\underline{T}}(\underline{u},\underline{v}) = T_{ij}\, u_i v_j = T_{ij}\, (\underline{e}_i \otimes \underline{e}_j)(\underline{u},\underline{v})$$

[A1.27]

and consequently

$$\underline{\underline{T}} = T_{ij}\, \underline{e}_i \otimes \underline{e}_j.$$

[A1.28]

A1.3.3. *Components of a second rank Euclidean tensor*

From [A1.28], we see that the second rank Euclidean tensors as defined by [A1.23] generate a vector space of dimension n^2 and the tensor products $\underline{e}_i \otimes \underline{e}_j$ constitute a basis of this vector space in the form of n^2 product tensors. T_{ij} are the components of $\underline{\underline{T}}$.

Consider the particular case when $\underline{\underline{T}}$ is the tensor product of two vectors. The components T_{ij} derived from [A1.24] and [A1.25] are written as

$$\begin{cases} T_{ij} = \underline{\underline{T}}(\underline{e}_i, \underline{e}_j) = (\underline{a} \otimes \underline{b})(\underline{e}_i, \underline{e}_j) \\ \qquad = (\underline{a}.\underline{e}_i)(\underline{b}.\underline{e}_j) = a_i\, b_j\,. \end{cases} \qquad \text{[A1.29]}$$

Note that the scalar product of two vectors is a bilinear form on $E \times E$ represented by the second rank Euclidean tensor $\underline{\underline{1}}$ called the metric tensor

$$\underline{\underline{1}}(\underline{u}, \underline{v}) = \underline{u}.\underline{v} = u_i v_i\,. \qquad \text{[A1.30]}$$

Its components given by [A1.29] are the Kronecker delta δ_{ij} :

$$\underline{\underline{1}} = \underline{e}_i \otimes \underline{e}_i = \delta_{ij}\, \underline{e}_i \otimes \underline{e}_j\,. \qquad \text{[A1.31]}$$

A1.3.4. *Transpose tensor*

Given a second rank Euclidean tensor $\underline{\underline{T}}$, the transpose tensor denoted by ${}^t\underline{\underline{T}}$ is defined from $\underline{\underline{T}}$ by permuting the roles of $\underline{u}$ and $\underline{v}$ in the bilinear form [A1.23]:

$$\forall(\underline{u}, \underline{v}) \in E \times E \mapsto {}^t\underline{\underline{T}}(\underline{u}, \underline{v}) = \underline{\underline{T}}(\underline{v}, \underline{u}). \qquad \text{[A1.32]}$$

In the particular case, when $\underline{\underline{T}}$ is the tensor product of two vectors, we obtain

$${}^t(\underline{a} \otimes \underline{b})(\underline{u}, \underline{v}) = (\underline{a} \otimes \underline{b})(\underline{v}, \underline{u}) = (\underline{a}.\underline{v})(\underline{b}.\underline{u}) \qquad \text{[A1.33]}$$

hence

$$^t(\underline{a} \otimes \underline{b}) = \underline{b} \otimes \underline{a}, \tag{A1.34}$$

which amounts to permuting $\underline{a}$ and $\underline{b}$.

It follows that in the general case, with $\underline{\underline{T}} = T_{ij}\,\underline{e}_i \otimes \underline{e}_j$ in an orthonormal basis, we obtain

$$\begin{cases} {}^t\underline{\underline{T}} = T_{ij}\,\underline{e}_j \otimes \underline{e}_i \\[2mm] {}^t\underline{\underline{T}} = T_{ji}\,\underline{e}_i \otimes \underline{e}_j. \end{cases} \tag{A1.35}$$

A1.3.5. Symmetric and antisymmetric second rank Euclidean tensors

– A *symmetric* second rank Euclidean tensor is defined from the characteristic property of a symmetric bilinear form

$$\forall(\underline{u},\underline{v}) \in E \times E \mapsto \underline{\underline{T}}(\underline{u},\underline{v}) = \underline{\underline{T}}(\underline{v},\underline{u}) \tag{A1.36}$$

in other words,

$$\underline{\underline{T}} = {}^t\underline{\underline{T}}, \; T_{ij} = T_{ji}. \tag{A1.37}$$

– An *antisymmetric* (or skew-symmetric) second rank Euclidean tensor is defined as an antisymmetric bilinear form

$$\forall(\underline{u},\underline{v}) \in E \times E \mapsto \underline{\underline{T}}(\underline{u},\underline{v}) = -\underline{\underline{T}}(\underline{v},\underline{u}) \tag{A1.38}$$

thence

$$\underline{\underline{T}} = -{}^t\underline{\underline{T}}, \; T_{ij} = -T_{ji}. \tag{A1.39}$$

– Any second rank Euclidean tensor $\underline{\underline{T}}$ can be uniquely expressed as the sum of a symmetric tensor $\underline{\underline{T}}_s$ and an antisymmetric tensor $\underline{\underline{T}}_a$

$$\begin{cases} \underline{\underline{T}} = \underline{\underline{T}}_s + \underline{\underline{T}}_a \\[2mm] \underline{\underline{T}}_s = \frac{1}{2}(\underline{\underline{T}} + {}^t\underline{\underline{T}}), \; \underline{\underline{T}}_a = \frac{1}{2}(\underline{\underline{T}} - {}^t\underline{\underline{T}}). \end{cases} \tag{A1.40}$$

In terms of components:

$$(T_s)_{ij} = \frac{1}{2}(T_{ij} + T_{ji}), \quad (T_a)_{ij} = \frac{1}{2}(T_{ij} - T_{ji}). \hspace{2cm} [\text{A1.41}]$$

A1.4. Linear mappings

A1.4.1. *Linear mapping associated with a second rank Euclidean tensor*

Let $\underline{\varphi}$ be a linear mapping from E into E with components φ_{ij}

$$\begin{cases} E \xrightarrow{\ \varphi\ } E \\ \underline{v} \mapsto \underline{\varphi}(\underline{v}) \\ v_j \, \underline{e}_j \mapsto \varphi_{ij} v_j \, \underline{e}_i. \end{cases} \hspace{2cm} [\text{A1.42}]$$

Through the canonical identification of E with its dual space E^*, we can associate $\underline{\varphi}$ with the second rank Euclidean tensor $\underline{\underline{T}}$ defined by duality in the following way:

$$\begin{cases} \forall(\underline{u},\underline{v}) \in E \times E, \ \underline{\underline{T}}(\underline{u},\underline{v}) = \underline{u}.\underline{\varphi}(\underline{v}) \\ \hspace{1.2cm} T_{ij}\, u_i v_j = u_i \varphi_{ij} v_j, \end{cases} \hspace{2cm} [\text{A1.43}]$$

which yields the components of $\underline{\underline{T}}$

$$T_{ij} = \varphi_{ij}. \hspace{2cm} [\text{A1.44}]$$

Conversely, any second rank Euclidean tensor can be associated through [A1.43] with a linear mapping $\underline{\varphi}$ from E into E.

From [A1.31], it comes out that the metric tensor is associated with the *Identity* linear mapping

$$\begin{cases} E \xrightarrow{\ Id\ } E \\ \underline{v} \mapsto \underline{Id}(\underline{v}) = \underline{v}. \end{cases} \hspace{2cm} [\text{A1.45}]$$

A1.4.2. *Inverse linear mapping. Inverse tensor*

Given the linear mapping φ associated with the tensor $\underline{\underline{T}}$, the inverse linear mapping φ^{-1} is such that

$$
\begin{cases}
(\varphi^{-1} \circ \varphi)(\underline{v}) = Id(\underline{v}) = \underline{v} \\
(\varphi^{-1})_{ki}\,\varphi_{ij}\,v_j\,\underline{e}_k = v_k\,\underline{e}_k .
\end{cases}
\tag{A1.46}
$$

The second rank tensor associated with φ^{-1} is the inverse tensor of tensor $\underline{\underline{T}}$ denoted by $\underline{\underline{T}}^{-1}$ and the relationship between the components of $\underline{\underline{T}}$ and $\underline{\underline{T}}^{-1}$ comes out straightforwardly from [A1.46]:

$$
(T^{-1})_{ki}\,T_{ij} = \delta_{kj}.
\tag{A1.47}
$$

A1.4.3. *Linear mapping associated with a symmetric second rank Euclidean tensor*

– The linear mapping associated with a symmetric second rank Euclidean tensor $\underline{\underline{T}}$ has n real *eigenvalues* $\lambda_1, \lambda_2 ..., \lambda_n$.

– The *eigenvectors* are real.

– The eigenvectors corresponding to two distinct eigenvalues are orthogonal.

– Thence, if the n real eigenvalues $\lambda_1, \lambda_2 ..., \lambda_n$ are distinct, an orthonormal basis of n eigenvectors $(\underline{u}_1, \underline{u}_2 ..., \underline{u}_n)$ corresponding to these eigenvalues can be constructed

$$
\begin{cases}
\varphi(\underline{u}_1) = \lambda_1 \underline{u}_1 \\
\varphi(\underline{u}_2) = \lambda_2 \underline{u}_2 \\
... \\
\varphi(\underline{u}_n) = \lambda_n \underline{u}_n .
\end{cases}
\tag{A1.48}
$$

– The result holds also in the case of multiple eigenvalues.

– As a result, in this orthonormal basis $(\underline{u}_1, \underline{u}_2 ..., \underline{u}_n)$, the tensor is written as

$$\begin{cases} \underline{\underline{T}} = \underline{\underline{T}}(\underline{u}_i, \underline{u}_j)\underline{u}_i \otimes \underline{u}_j = \underline{u}_i \cdot \underline{\varphi}(\underline{u}_j)\underline{u}_i \otimes \underline{u}_j \\[2mm] \underline{\underline{T}} = \lambda_j(\underline{u}_j \cdot \underline{u}_i)\underline{u}_i \otimes \underline{u}_j = \lambda_j \delta_{ij}\, \underline{u}_i \otimes \underline{u}_j \\[2mm] \underline{\underline{T}} = \lambda_1 \underline{u}_1 \otimes \underline{u}_1 + \lambda_2 \underline{u}_2 \otimes \underline{u}_2 \ldots + \lambda_n \underline{u}_n \otimes \underline{u}_n. \end{cases} \qquad [\text{A}1.49]$$

– The directions corresponding to the eigenvectors $(\underline{u}_1, \underline{u}_2 \ldots, \underline{u}_n)$ are called the *principal axes* of the tensor $\underline{\underline{T}}$ and the eigenvalues are the *principal values* of $\underline{\underline{T}}$.

A1.4.4. *Linear mapping associated with an antisymmetric second rank Euclidean tensor*

Only the three dimensional case is considered here $(n = 3)$.

The linear mapping φ associated with an antisymmetric second rank Euclidean tensor $\underline{\underline{\Omega}}$, defined according to [A1.43] by

$$\forall (\underline{u}, \underline{v}) \in E \times E, \ \underline{\underline{\Omega}}(\underline{u}, \underline{v}) = \underline{u} \cdot \underline{\varphi}(\underline{v}) \qquad [\text{A}1.50]$$

can be expressed in the form of a vector product

$$\forall \underline{v} \in E, \ \underline{\varphi}(\underline{v}) = \underline{\Omega} \wedge \underline{v}. \qquad [\text{A}1.51]$$

In order to prove this result, we write and combine [A1.50] and [A1.51] in terms of components in a right-handed basis

$$\Omega_{ik} u_i v_k = \varepsilon_{ijk} u_i \Omega_j v_k \qquad [\text{A}1.52]$$

hence

$$\Omega_{ik} = \varepsilon_{ijk} \Omega_j, \qquad [\text{A}1.53]$$

which yields six equations to determine the three components of $\underline{\Omega}$ from the components of $\underline{\underline{\Omega}}$, such as $\Omega_3 = -\Omega_{12}$ and $\Omega_3 = \Omega_{21}$. These equations are compatible since $\underline{\underline{\Omega}}$ is antisymmetric and yield

$$\underline{\Omega} = -(\Omega_{23}\,\underline{e}_1 + \Omega_{31}\,\underline{e}_2 + \Omega_{12}\,\underline{e}_3) = -\frac{1}{2}\varepsilon_{ijk}\Omega_{ij}\,\underline{e}_k. \qquad [A1.54]$$

Conversely, [A1.53] may also be written as[1],

$$\begin{cases} \underline{\underline{\Omega}} = -\varepsilon_{ijk}\Omega_k\,\underline{e}_i \otimes \underline{e}_j = \Omega_1\,\underline{e}_{=1} + \Omega_2\,\underline{e}_{=2} + \Omega_3\,\underline{e}_{=3} \\ \text{with } \underline{e}_{=k} = -\varepsilon_{ijk}\,\underline{e}_i \otimes \underline{e}_j. \end{cases} \qquad [A1.55]$$

A1.5. Change of orthonormal basis. Invariants

A1.5.1. *Change of orthonormal basis: components of a tensor in the new basis*

The change of orthonormal basis was already defined in section A1.1.3 through the *direction cosines* $\alpha_{ik} = \underline{e}'_i \cdot \underline{e}_k$ between the new basis $\{\underline{e}'_i\}$ and the original one $\{\underline{e}_k\}$:

$$\begin{cases} \underline{e}'_i = \alpha_{ik}\underline{e}_k\,, & \alpha_{ik}\alpha_{jk} = \delta_{ij} \\ \underline{e}_k = \alpha_{ik}\underline{e}'_i\,, & \alpha_{ij}\alpha_{ik} = \delta_{jk}. \end{cases} \qquad [A1.56]$$

The components of a tensor $\underline{\underline{T}}$ in the new basis are obtained easily from their counterparts in the initial one by expanding $\underline{\underline{T}}$ in terms of the tensor products $\underline{e}_k \otimes \underline{e}_h$ on the one side and $\underline{e}'_i \otimes \underline{e}'_j$ on the other side and taking the distributivity of the tensor product into account:

$$\begin{cases} \underline{\underline{T}} = T_{kh}\,\underline{e}_k \otimes \underline{e}_h = T'_{ij}\,\underline{e}'_i \otimes \underline{e}'_j \\ T_{kh}\,\alpha_{ik}\underline{e}'_i \otimes \alpha_{jh}\underline{e}'_j = T'_{ij}\,\underline{e}'_i \otimes \underline{e}'_j \end{cases} \qquad [A1.57]$$

hence

$$T'_{ij} = \alpha_{ik}\alpha_{jh}T_{kh} \qquad [A1.58]$$

and conversely

---

1 $^t\underline{e}_{=k} = -\underline{e}_{=k} = \underline{e}_{=k}^{-1}; \ \underline{e}_{=k} \cdot \underline{e}_j = -\varepsilon_{ijk}\,\underline{e}_i.$

$$T_{ij} = \alpha_{ki}\alpha_{hj}T'_{kh}. \qquad\qquad\qquad\text{[A1.59]}$$

Note that when only orthonormal bases are used, Euclidean tensors are called Cartesian tensors.

A1.5.2. *Invariants*

The definition of the eigenvalues and eigenvectors (section A1.4.3) of the linear mapping associated with the tensor $\underline{\underline{T}}$ is obviously intrinsic, meaning that it is independent of the choice of the orthonormal basis $\{\underline{e}_i\}$. The eigenvalues $\lambda_1, \lambda_2..., \lambda_n$ are obtained through the solution of the equation

$$\det[T_{ij} - \lambda\,\delta_{ij}] = 0, \qquad\qquad\qquad\text{[A1.60]}$$

which states that the *characteristic polynomial* in λ, obtained as the determinant of the linear mapping

$$\underline{v} \in E \mapsto \underline{\varphi}(\underline{v}) - \lambda\underline{v} \in E, \qquad\qquad\qquad\text{[A1.61]}$$

is equal to zero. It follows that the n coefficients of this polynomial which are functions of the components are independent of the choice of the orthonormal basis. They form a basis for the n polynomial invariants of degrees 1 to n in T_{ij}. Among these:

$$\det[T_{ij}] = \det\underline{\underline{T}} \;\; \text{is the invariant of degree } n \qquad\qquad\text{[A1.62]}$$

$$T_{ii} = \operatorname{tr}\underline{\underline{T}}\,{}^2 \;\; (\text{read ``\textit{trace}''}) \text{ is the invariant of degree 1} \qquad\text{[A1.63]}$$

that can also be written through [A1.24]

$$\operatorname{tr}\underline{\underline{T}} = \underline{\underline{T}}(\underline{e}_i,\underline{e}_i)\,{}^2. \qquad\qquad\qquad\text{[A1.64]}$$

The invariance of [A1.62] and [A1.63] can also be proven directly through [A1.56] and [A1.57], e.g. for $\operatorname{tr}\underline{\underline{T}}$:

$$T'_{ii} = \alpha_{ik}\alpha_{ih}T_{kh} = \delta_{kh}T_{kh} = T_{kk}. \qquad\qquad\text{[A1.65]}$$

---

2 Dummy index convention.

The *trace* operation of [A1.63] and [A1.64] is called the *contraction* of the second rank Euclidean tensor $\underline{\underline{T}}$ (see sections A1.7.1 and A1.7.6).

Note that *in the three-dimensional case* ($n = 3$), $\det[T_{ij}] = \det \underline{\underline{T}}$ admits the following expression:

$$\det[T_{ij}] = \det \underline{\underline{T}} = \varepsilon_{ijk} T_{1i} T_{2j} T_{3k} = \varepsilon_{ijk} T_{i1} T_{j2} T_{k3} \qquad [A1.66]$$

where ε_{ijk} is the permutation symbol defined by [A1.16]. Hence, with the triple scalar product defined in [A1.18],

$$\det \underline{\underline{T}} = [T_{1i} \underline{e}_i, T_{2j} \underline{e}_j, T_{3k} \underline{e}_k] = [T_{i1} \underline{e}_i, T_{j2} \underline{e}_j, T_{k3} \underline{e}_k]. \qquad [A1.67]$$

A1.6. Rank p Euclidean tensors

A1.6.1. *Definition*

A rank p Euclidean tensor is a p-linear form on $E \times E \ldots \times E = E^p$. For example, for $p = 3$, we have for the third rank Euclidean tensor $\underline{\underline{\underline{T}}}$

$$\begin{cases} E \times E \times E \xrightarrow{\ \underline{\underline{\underline{T}}}\ } \mathbb{R} \\ (\underline{u}, \underline{v}, \underline{w}) \in E \times E \times E \mapsto \underline{\underline{\underline{T}}}(\underline{u}, \underline{v}, \underline{w}) \in \mathbb{R} \end{cases} \qquad [A1.68]$$

and

$$\underline{\underline{\underline{T}}}(\underline{u}, \underline{v}, \underline{w}) = \underline{\underline{\underline{T}}}(\underline{e}_i, \underline{e}_j, \underline{e}_k) u_i v_j w_k = T_{ijk} u_i v_j w_k. \qquad [A1.69]$$

A Euclidean tensor of rank p will be denoted by a letter (most often upper case) underlined p times. Obviously, this notation would soon become awkward if p were large but it proves quite convenient for the applications to Mechanics, where the rank of the tensors to be introduced is seldom higher than two, as it makes it easy to distinguish, at a glance, the nature of the mathematical objects arising in the formulas.

A1.6.2. *Tensor product of two tensors*

As an example, let $\underline{\underline{T}}$ and $\underline{\underline{\underline{T}}}'$ be two Euclidean tensors of rank 2 and 3 respectively. The tensor product $\underline{\underline{T}} \otimes \underline{\underline{\underline{T}}}' = \mathcal{T}$ is the five-linear form defined in the same way as [A1.26] by

$$\mathcal{T}(\underline{u}^1, \underline{u}^2 ..., \underline{u}^5) = (\underline{\underline{T}} \otimes \underline{\underline{\underline{T}}}')(\underline{u}^1, \underline{u}^2 ..., \underline{u}^5) = \underline{\underline{T}}(\underline{u}^1, \underline{u}^2) \underline{\underline{\underline{T}}}'(\underline{u}^3, \underline{u}^4, \underline{u}^5) \qquad \text{[A1.70]}$$

$$\mathcal{T}_{ijk\ell m} u_i^1 u_j^2 u_k^3 u_\ell^4 u_m^5 = T_{ij} u_i^1 u_j^2 T'_{k\ell m} u_k^3 u_\ell^4 u_m^5 \quad i = 1, 2 ..., n ..., m = 1, 2 ..., n. \qquad \text{[A1.71]}$$

Note that the tensor product is distributive with respect to addition, but not commutative, and can obviously be iterated.

Considering now the tensor products $\underline{e}_i \otimes \underline{e}_j \otimes \underline{e}_k \otimes \underline{e}_\ell \otimes \underline{e}_m$, we have

$$\begin{cases} (\underline{e}_i \otimes \underline{e}_j ... \otimes \underline{e}_m)(\underline{u}^1, \underline{u}^2 ..., \underline{u}^5) = u_i^1 \, u_j^2 \, u_k^3 \, u_\ell^4 \, u_m^5 \\[2mm] i = 1, 2 ..., n, j = 1, 2 ..., n, ..., \ m = 1, 2 ..., n \end{cases} \qquad \text{[A1.72]}$$

and, for any Euclidean tensor $\mathcal{T}$ of rank 5 defined as in section A1.6.1, we may write

$$\mathcal{T} = \mathcal{T}_{ijk\ell m} \underline{e}_i \otimes \underline{e}_j ... \otimes \underline{e}_m. \qquad \text{[A1.73]}$$

These tensor products constitute a basis of the vector space $E \otimes E ... \otimes E$ of dimension n^5 generated by the five-linear forms on E^5.

A1.7. Contracted products

The tensor product of two tensors $\underline{\underline{T}}$ and $\underline{\underline{\underline{T}}}'$ of rank p and p' respectively, introduced in section A1.6.2, results in a tensor of rank $p + p'$. The contracted product that will be defined now will result in a tensor of rank $p + p' - 2$.

A1.7.1. *Contracted product of two first rank Euclidean tensors*

Consider two first rank Euclidean tensors $\underline{u}$ and $\underline{v}$. The contracted product of these two tensors results in a scalar (Euclidean tensor of rank $1+1-2=0$). It is the dot product $\underline{u}.\underline{v}$ defined in section A1.2. In an orthonormal basis $\{\underline{e}_i\}$:

$$\underline{u}.\underline{v} = u_i v_i. \tag{A1.74}$$

Note that this contracted product also appears as the result of the contraction (section A1.5.2) of the product tensor $\underline{u} \otimes \underline{v}$ in the form

$$\underline{u}.\underline{v} = u_i v_i = \operatorname{tr}(\underline{u} \otimes \underline{v}). \tag{A1.75}$$

A1.7.2. *Contracted product of a second rank and a first rank Euclidean tensor*

Consider $\underline{\underline{T}}$ and $\underline{v}$, second and first rank Euclidean tensors respectively. Their contracted product results in a first rank Euclidean tensor, a linear form on E. It is denoted by $\underline{\underline{T}}.\underline{v}$ and defined as follows:

$$\underline{u} \mapsto (\underline{\underline{T}}.\underline{v}).\underline{u} = \underline{\underline{T}}(\underline{u}, \underline{v}) \tag{A1.76}$$

where the second dot stands for the scalar product of the linear forms $\underline{\underline{T}}.\underline{v}$ and $\underline{u}$.

In an orthonormal basis $\{\underline{e}_i\}$

$$(\underline{\underline{T}}.\underline{v}).\underline{u} = T_{ij} u_i v_j, \tag{A1.77}$$

hence

$$\underline{\underline{T}}.\underline{v} = T_{ij} v_j \underline{e}_i. \tag{A1.78}$$

We observe from [A1.78] that $\underline{\underline{T}}.\underline{v}$ is just the image of $\underline{v}$ through the linear mapping associated with $\underline{\underline{T}}$ (section A1.4.1):

$$\forall \underline{v} \in E, \underline{\underline{T}}.\underline{v} = \underline{\varphi}(\underline{v}). \tag{A1.79}$$

From the definition, it obviously comes out that the contracted product is not commutative. Thence, equation [A1.78] is said to represent the *right-hand side contracted product* of a second rank and a first rank Euclidean tensor. The *left-hand side contracted product* $\underline{u}.\underline{\underline{T}}$ is defined in a similar way:

$$\underline{v} \mapsto \underline{v}.(\underline{u}.\underline{\underline{T}}) = \underline{\underline{T}}(\underline{u},\underline{v}). \qquad\qquad [A1.80]$$

Comparing this definition with [A1.32], we obtain

$$\begin{cases} \underline{v}.(\underline{u}.\underline{\underline{T}}) = \underline{\underline{T}}(\underline{u},\underline{v}) = {}^{t}\underline{\underline{T}}(\underline{v},\underline{u}) \\ \Rightarrow \quad \underline{u}.\underline{\underline{T}} = {}^{t}\underline{\underline{T}}.\underline{u}, \end{cases} \qquad\qquad [A1.81]$$

which shows that $\underline{u}.\underline{\underline{T}}$ is the image of $\underline{u}$ through the linear mapping associated with ${}^{t}\underline{\underline{T}}$.

From [A1.76] and [A1.80], we derive

$$\underline{\underline{T}}(\underline{u},\underline{v}) = \underline{u}.(\underline{\underline{T}}.\underline{v}) = (\underline{u}.\underline{\underline{T}}).\underline{v} \qquad\qquad [A1.82]$$

that can be written unambiguously as

$$\underline{\underline{T}}(\underline{u},\underline{v}) = \underline{u}.\underline{\underline{T}}.\underline{v}. \qquad\qquad [A1.83]$$

It is worth considering the case when $\underline{\underline{T}}$ is a product tensor $\underline{\underline{T}} = \underline{a} \otimes \underline{b}$. From definition [A1.76] and equation [A1.83], we get

$$\begin{cases} ((\underline{a} \otimes \underline{b}).\underline{v}).\underline{u} = (\underline{a} \otimes \underline{b})(\underline{u},\underline{v}) \\ ((\underline{a} \otimes \underline{b}).\underline{v}).\underline{u} = (\underline{u}.\underline{a})(\underline{b}.\underline{v}) \\ \Rightarrow (\underline{a} \otimes \underline{b}).\underline{v} = \underline{a}(\underline{b}.\underline{v}), \end{cases} \qquad\qquad [A1.84]$$

which shows that the linear form $(\underline{a} \otimes \underline{b}).\underline{v}$ is collinear with $\underline{a}$, with a magnitude equal to the scalar product of $\underline{b}$ and $\underline{v}$.

In the same way,

$$\underline{u}.(\underline{a} \otimes \underline{b}) = (\underline{u}.\underline{a})\underline{b} \qquad\qquad [A1.85]$$

is the linear form collinear with $\underline{b}$ whose magnitude is the scalar product of $\underline{a}$ and $\underline{u}$.

For instance, in an orthonormal basis, with $\underline{\underline{T}} = T_{ij}\,\underline{e}_i \otimes \underline{e}_j$, we derive

$$\begin{cases} \underline{\underline{T}}.\underline{v} = T_{ij}\,\underline{e}_i \otimes \underline{e}_j.\underline{v} = T_{ij}\,\underline{e}_i(\underline{e}_j.\underline{v}) = T_{ij}\,\underline{e}_i(\underline{e}_j.\underline{e}_k)v_k = T_{ij}\,\underline{e}_i v_j \\ \underline{u}.\underline{\underline{T}} = \underline{u}.T_{ij}\,\underline{e}_i \otimes \underline{e}_j = T_{ij}(\underline{u}.\underline{e}_i)\underline{e}_j = T_{ij}u_k(\underline{e}_k.\underline{e}_i)\underline{e}_j = T_{ij}\,\underline{e}_j u_i\,. \end{cases}$$

[A1.86]

A1.7.3. Contracted product of two second rank tensors

Let $\underline{\underline{T}}$ and $\underline{\underline{T}}'$ be two second rank Euclidean tensors associated with the linear mappings φ and φ' respectively. The contracted product of $\underline{\underline{T}}$ and $\underline{\underline{T}}'$ (in that order), denoted by $\underline{\underline{T}}.\underline{\underline{T}}'$, is the second rank Euclidean tensor associated with the linear mapping $\varphi \circ \varphi'$:

$$\forall \underline{v} \in E,\ (\underline{\underline{T}}.\underline{\underline{T}}').\underline{v} = \varphi \circ \varphi'(\underline{v}).$$

[A1.87]

Explicitly, in an orthonormal basis, with $\underline{\underline{T}} = T_{ik}\,\underline{e}_i \otimes \underline{e}_k$ and $\underline{\underline{T}}' = T'_{\ell j}\,\underline{e}_\ell \otimes \underline{e}_j$,

$$\begin{cases} \forall \underline{v} \in E,\ (\underline{\underline{T}}.\underline{\underline{T}}').\underline{v} = \varphi \circ \varphi'(\underline{v}) \\ \qquad\qquad = \varphi(T'_{\ell j}v_j\,\underline{e}_\ell) \\ \qquad\qquad = T_{i\ell}T'_{\ell j}v_j\,\underline{e}_i \end{cases}$$

[A1.88]

and

$$\underline{\underline{T}}.\underline{\underline{T}}' = T_{i\ell}T'_{\ell j}\,\underline{e}_i \otimes \underline{e}_j.$$

[A1.89]

It is worth noting that this result is formally and conveniently obtained through

$$T_{ik}T'_{\ell j}(\underline{e}_i \otimes \underline{e}_k).(\underline{e}_\ell \otimes \underline{e}_j) = T_{ik}T'_{\ell j}\,\underline{e}_i(\underline{e}_k.\underline{e}_\ell) \otimes \underline{e}_j$$

[A1.90]

where $(\underline{e}_k.\underline{e}_\ell)$ is the scalar product of the adjacent vectors of $\underline{\underline{T}}$ and $\underline{\underline{T}}'$ in the tensor product $\underline{\underline{T}} \otimes \underline{\underline{T}}' = T_{ik}T'_{\ell j}(\underline{e}_i \otimes \underline{e}_k) \otimes (\underline{e}_\ell \otimes \underline{e}_j)$.

From   [A1.89],   we   also   derive   the   following   expression   for
$(\underline{\underline{T}}.\underline{\underline{T'}})(\underline{u},\underline{v}) = \underline{u}.(\underline{\underline{T}}.\underline{\underline{T'}}).\underline{v}$:

$$\underline{u}.(\underline{\underline{T}}.\underline{\underline{T'}}).\underline{v} = (\underline{u}.\underline{\underline{T}}).(\underline{\underline{T'}}.\underline{v}) = ({}^{t}\underline{\underline{T}}.\underline{u}).(\underline{\underline{T'}}.\underline{v}). \qquad [A1.91]$$

– *Inverse of the contracted product of two second rank Euclidean tensors*

From [A1.87], it follows that the inverse $(\underline{\underline{T}}.\underline{\underline{T'}})^{-1}$ of the contracted product
$\underline{\underline{T}}.\underline{\underline{T'}}$ is the second rank Euclidean tensor associated with the inverse of the linear
mapping $\underline{\varphi} \circ \underline{\varphi'}$ that is $(\underline{\varphi} \circ \underline{\varphi'})^{-1} = \underline{\varphi'}^{-1} \circ \underline{\varphi}^{-1}$; hence

$$(\underline{\underline{T}}.\underline{\underline{T'}})^{-1} = \underline{\underline{T'}}^{-1}.\underline{\underline{T}}^{-1}. \qquad [A1.92]$$

– *Transpose of the contracted product of two second rank Euclidean tensors*

The transpose tensors of $\underline{\underline{T}}$ and $\underline{\underline{T'}}$ (section A1.3.4) are written as, in an
orthonormal basis,

$$\begin{cases} {}^{t}\underline{\underline{T}} = T_{i\ell}\,\underline{e}_{\ell} \otimes \underline{e}_{i} \\ {}^{t}\underline{\underline{T'}} = T'_{kj}\,\underline{e}_{j} \otimes \underline{e}_{k} \end{cases} \qquad [A1.93]$$

and the transpose of the contracted product $\underline{\underline{T}}.\underline{\underline{T'}} = T_{ik}T'_{kj}\underline{e}_{i} \otimes \underline{e}_{j}$ is

$$ {}^{t}(\underline{\underline{T}}.\underline{\underline{T'}}) = T_{ik}T'_{kj}\underline{e}_{j} \otimes \underline{e}_{i}. \qquad [A1.94]$$

We observe that the right-hand member of this equation may be written as the
result of the contracted product of ${}^{t}\underline{\underline{T'}}$ and ${}^{t}\underline{\underline{T}}$

$$T_{ik}T'_{kj}\underline{e}_{j} \otimes \underline{e}_{i} = (T'_{kj}\,\underline{e}_{j} \otimes \underline{e}_{k}).(T_{i\ell}\,\underline{e}_{\ell} \otimes \underline{e}_{i}) \qquad [A1.95]$$

hence

$$ {}^{t}(\underline{\underline{T}}.\underline{\underline{T'}}) = {}^{t}\underline{\underline{T'}}.{}^{t}\underline{\underline{T}}. \qquad [A1.96]$$

A1.7.4. *Contracted product of two Euclidean tensors*

Considering two Euclidean tensors T and T' of rank p and p' respectively, their contracted product is defined in the same way as [A1.90] for second rank Euclidean tensors. As an example, with $p = 2$ and $p' = 3$

$$\begin{cases} T.T' = (T_{ij}\,\underline{e}_i \otimes \underline{e}_j).(T'_{k\ell m}\,\underline{e}_k \otimes \underline{e}_\ell \otimes \underline{e}_m) \\[2mm] T.T' = T_{ij}T'_{k\ell m}\,\underline{e}_i\,(\underline{e}_j \cdot \underline{e}_k) \otimes \underline{e}_\ell \otimes \underline{e}_m \\[2mm] \Rightarrow (T.T')_{i\ell m} = T_{ij}T'_{j\ell m}\,\underline{e}_i \otimes \underline{e}_\ell \otimes \underline{e}_m\,. \end{cases} \qquad [\text{A}1.97]$$

A1.7.5. *Doubly contracted product of two second rank Euclidean tensors*

The doubly contracted product of two second rank Euclidean tensors $\underline{\underline{T}}$ and $\underline{\underline{T}}'$, denoted by $\underline{\underline{T}}:\underline{\underline{T}}'$, is the result of the contraction (section A1.5.2) of the contracted product $\underline{\underline{T}}.\underline{\underline{T}}'$

$$\underline{\underline{T}}:\underline{\underline{T}}' = \text{tr}\,(\underline{\underline{T}}.\underline{\underline{T}}') = (\underline{e}_i \cdot \underline{\underline{T}}).(\underline{\underline{T}}' \cdot \underline{e}_i) \qquad [\text{A}1.98]$$

in an orthonormal basis

$$\underline{\underline{T}}:\underline{\underline{T}}' = T_{ij}T'_{ji}\,. \qquad [\text{A}1.99]$$

We observe that this product is commutative and we note the straightforward identities

$$\text{tr}\,\underline{\underline{T}} = \underline{\underline{T}}:\underline{\underline{1}} \qquad [\text{A}1.100]$$

$$\text{tr}\,\underline{\underline{1}} = \underline{\underline{1}}:\underline{\underline{1}} = 3 \qquad [\text{A}1.101]$$

and, from [A1.55]

$$\underline{\underline{e}}_i:\underline{\underline{e}}_j = \varepsilon_{\ell m i}\,\varepsilon_{pqj}\,\delta_{mp}\delta_{\ell q} = -2\,\delta_{ij}\,. \qquad [\text{A}1.102]$$

A1.7.6. *Contraction of a Euclidean tensor of rank p*

In an orthonormal basis, consider as an example, the fifth rank Euclidean tensor written as $T = T_{ijk\ell m}\, \underline{e}_i \otimes \underline{e}_j \otimes \underline{e}_k \otimes \underline{e}_\ell \otimes \underline{e}_m$.

The contraction of this tensor may be defined on any pair of indices and results in a tensor of order $(5-2) = 3$. For instance, contraction on the indices j and ℓ yields the tensor

$$\underline{\underline{T_c}} = T_{ijk\ell m}\, \underline{e}_i \otimes \underline{e}_k \otimes \underline{e}_m\, (\underline{e}_j \cdot \underline{e}_\ell) = T_{ijkjm}\, \underline{e}_i \otimes \underline{e}_k \otimes \underline{e}_m \qquad\qquad [\text{A1.103}]$$

A1.8. Practicing

A1.8.1. *Problem 1*

$\forall \underline{\underline{T}}$ second rank Euclidean tensor

$$\operatorname{tr}\underline{\underline{T}} = \operatorname{tr}{}^t\underline{\underline{T}}$$

$$\det \underline{\underline{T}} = \det{}^t\underline{\underline{T}}$$

Solution

The results come directly from the explicit formulas

$$\operatorname{tr}\underline{\underline{T}} = T_{ii}$$

$$\det[T_{ij}] = \det \underline{\underline{T}} = \varepsilon_{ijk} T_{1i} T_{2j} T_{3k} = \varepsilon_{ijk} T_{i1} T_{j2} T_{k3}.$$

A1.8.2. *Problem 2*

$\forall \underline{\underline{T}}, \forall \underline{\underline{T'}}$ second rank Euclidean tensors

$$\underline{\underline{T}}:\underline{\underline{T'}} = \underline{\underline{T'}}:\underline{\underline{T}}$$

Solution

Explicitly:

$\underline{\underline{T}}:\underline{\underline{T}}' = T_{ij}T'_{ji}$ and $\underline{\underline{T}}':\underline{\underline{T}} = T'_{k\ell}:T_{\ell k}$ are proven identical by changing ℓ into i and k into j.

A1.8.3. *Problem 3*

$\forall \underline{\underline{T}}, \forall \underline{\underline{T}}', \underline{\underline{T}}''$ second rank Euclidean tensors

$$\text{tr}\,(\underline{\underline{T}}.\underline{\underline{T}}'.\underline{\underline{T}}'') = \text{tr}\,(\underline{\underline{T}}.\underline{\underline{T}}''.\underline{\underline{T}}') = \text{tr}\,(\underline{\underline{T}}'.\underline{\underline{T}}.\underline{\underline{T}}'') = ...$$

Solution

$\text{tr}\,(\underline{\underline{T}}.\underline{\underline{T}}'.\underline{\underline{T}}'') = T_{ij}T'_{jk}T''_{ki}$ and $\text{tr}\,(\underline{\underline{T}}.\underline{\underline{T}}''.\underline{\underline{T}}') = T_{i\ell}T''_{\ell k}T'_{ki}$ are proven identical by changing ℓ into i and k into j.

A1.8.4. *Problem 4*

$\forall \underline{\underline{S}}$ symmetric second rank Euclidean tensor, $\forall \underline{\underline{A}}$ antisymmetric second rank Euclidean tensor: $\underline{\underline{S}}:\underline{\underline{A}} = \underline{\underline{A}}:\underline{\underline{S}} = 0$.

Solution

Recall that $\underline{\underline{S}}:\underline{\underline{A}} = \text{tr}\,(\underline{\underline{S}}.\underline{\underline{A}})$. As a consequence of A1.8.2, we have

$$\text{tr}\,(\underline{\underline{S}}.\underline{\underline{A}}) = \text{tr}\,{}^t(\underline{\underline{S}}.\underline{\underline{A}}) = \text{tr}\,({}^t\underline{\underline{A}}.{}^t\underline{\underline{S}})$$

hence $\text{tr}\,(\underline{\underline{S}}.\underline{\underline{A}}) = -\text{tr}\,(\underline{\underline{A}}:\underline{\underline{S}})$.

But from A1.8.1, we also have $\underline{\underline{S}}:\underline{\underline{A}} = \underline{\underline{A}}:\underline{\underline{S}}$, thence

$$\text{tr}\,(\underline{\underline{S}}.\underline{\underline{A}}) = -\text{tr}\,(\underline{\underline{A}}:\underline{\underline{S}}) = \text{tr}\,(\underline{\underline{A}}:\underline{\underline{S}}) = 0.$$

A1.8.5. *Problem 5*

Let $\underline{\underline{T}}$ and $\underline{\underline{G}}$, second rank Euclidean tensors, be split into their symmetric and antisymmetric parts: $\underline{\underline{T}} = \underline{\underline{\sigma}} + \underline{\underline{\alpha}}$ and $\underline{\underline{G}} = \underline{\underline{d}} + \underline{\underline{\Omega}}$.

Then: $\underline{\underline{T}}:\underline{\underline{G}} = \underline{\underline{\sigma}}:\underline{\underline{d}} + \underline{\underline{\alpha}}:\underline{\underline{\Omega}}.$

Solution

$$\underline{\underline{T}}:\underline{\underline{G}} = (\underline{\underline{\sigma}}+\underline{\underline{\alpha}}):(\underline{\underline{d}}+\underline{\underline{\Omega}})$$

$$\underline{\underline{T}}:\underline{\underline{G}} = \underline{\underline{\sigma}}:\underline{\underline{d}} + \underline{\underline{\sigma}}:\underline{\underline{\Omega}} + \underline{\underline{\alpha}}:\underline{\underline{d}} + \underline{\underline{\alpha}}:\underline{\underline{\Omega}}$$

where the second and third terms are equal to zero from A1.8.4. Thence:

$$\underline{\underline{T}}:\underline{\underline{G}} = \underline{\underline{\sigma}}:\underline{\underline{d}} + \underline{\underline{\alpha}}:\underline{\underline{\Omega}}.$$

A1.8.6. *Problem 6*

Let $\underline{\underline{\Omega}}$ and $\underline{\underline{C}}$ be two antisymmetric second rank tensors defined on a three-dimensional Euclidean space, with $\underline{\Omega}$ and $\underline{C}$ their associated vectors such that $\forall \underline{u}, \underline{\Omega}.\underline{u} = \underline{\Omega} \wedge \underline{u}, \forall \underline{v}, \underline{C}.\underline{v} = \underline{C} \wedge \underline{v}.$

Then: $\underline{\underline{\Omega}}:\underline{\underline{C}} = -2\,\underline{\Omega}.\underline{C}.$

Solution

From [A1.55], we derive

$$\underline{\underline{\Omega}}:\underline{\underline{C}} = \Omega_i\,\underline{e}_i:C_j\,\underline{e}_j$$

and from [A1.102]

$$\underline{\underline{\Omega}}:\underline{\underline{C}} = -2\,\delta_{ij}\,\Omega_i\,C_j = -2\,\underline{\Omega}.\underline{C}.$$

A1.8.7. *Problem 7*

Let $\underline{\underline{\Omega}}$ be an antisymmetric second rank tensor and $\underline{\Omega}$ its associated vector through [A1.54]. With definition [A1.55], prove the following identities:

$$-(\underline{e}_i \otimes \underline{e}_j):\underline{e}_k = \varepsilon_{ijk} = [\underline{e}_i, \underline{e}_j, \underline{e}_k]$$

$$- (\underline{U} \otimes \underline{V}):\underline{\underline{\Omega}} = [\underline{U}, \underline{V}, \underline{\Omega}].$$

Solution

− Expanding $(\underline{e}_i \otimes \underline{e}_j):\underline{e}_{\underline{k}}$, we have

$$(\underline{e}_i \otimes \underline{e}_j):\underline{e}_{\underline{k}} = -(\underline{e}_i \otimes \underline{e}_j):\varepsilon_{\ell m k}\, \underline{e}_\ell \otimes \underline{e}_m = -\varepsilon_{\ell m k}\, \delta_{j\ell}\, \delta_{im}$$

thence

$$(\underline{e}_i \otimes \underline{e}_j):\underline{e}_{\underline{k}} = -\varepsilon_{jik} = \varepsilon_{ijk}.$$

With definition [A1.18] of the triple scalar product, this yields the general result:

$$(\underline{e}_i \otimes \underline{e}_j):\underline{e}_{\underline{k}} = \varepsilon_{ijk} = [\underline{e}_i, \underline{e}_j, \underline{e}_k].$$

$- (\underline{U} \otimes \underline{V}):\underline{\underline{\Omega}} = (U_i \underline{e}_i \otimes V_j \underline{e}_j):\Omega_k \underline{e}_{\underline{k}}$ thence

$$(\underline{U} \otimes \underline{V}):\underline{\underline{\Omega}} = (U_i \underline{e}_i \otimes V_j \underline{e}_j):\Omega_k \underline{e}_{\underline{k}} = \varepsilon_{ijk}\, U_i\, V_j\, \Omega_k$$

that is $(\underline{U} \otimes \underline{V}):\underline{\underline{\Omega}} = [\underline{U}, \underline{V}, \underline{\Omega}].$

A1.8.8. *Problem 8*

− Let $\underline{\underline{\Omega}}$ be an antisymmetric second rank tensor and $\underline{\Omega}$ its associated vector

through [A1.54]. With definition [A1.55], prove that $\underline{\Omega} = \dfrac{1}{2}({}^t\underline{\underline{\Omega}}:\underline{e}_{\underline{k}})\,\underline{e}_{\underline{k}}.$

− Let $\underline{\underline{T}} = \underline{\underline{S}} + \underline{\underline{\Omega}}$ be a second rank tensor with symmetric and antisymmetric

parts $\underline{\underline{S}}$ and $\underline{\underline{\Omega}}$ respectively. Prove that $\underline{\Omega} = \dfrac{1}{2}({}^t\underline{\underline{T}}:\underline{e}_{\underline{k}})\,\underline{e}_{\underline{k}}.$

Solution

− With $\underline{\underline{\Omega}} = \Omega_{ij}\, \underline{e}_i \otimes \underline{e}_j$, we have

$${}^t\underline{\underline{\Omega}}:\underline{e}_{\underline{k}} = \Omega_{ji}\, \underline{e}_i \otimes \underline{e}_j : \underline{e}_{\underline{k}}$$

and, from section A1.8.7,

$$\underline{\underline{\Omega}}_{ji}\underline{e}_i \otimes \underline{e}_j : \underline{e}_k = \varepsilon_{ijk}\underline{\underline{\Omega}}_{ji}.$$

Thence

$$({}^{t}\underline{\underline{\Omega}}:\underline{e}_k)\underline{e}_k = -\varepsilon_{ijk}\underline{\underline{\Omega}}_{ij}\underline{e}_k = 2\underline{\Omega} \text{ as defined in [A1.54].}$$

– Similarly, $\,{}^{t}\underline{\underline{T}}:\underline{e}_k = S_{ji}\underline{e}_i \otimes \underline{e}_j : \underline{e}_k + \underline{\underline{\Omega}}_{ji}\underline{e}_i \otimes \underline{e}_j : \underline{e}_k$

thence $\,{}^{t}\underline{\underline{T}}:\underline{e}_k = \varepsilon_{ijk}S_{ji} + \varepsilon_{ijk}\underline{\underline{\Omega}}_{ji}$

$\underline{\underline{S}}\,$ being symmetric implies $\varepsilon_{ijk}S_{ji} = 0$

and then $\dfrac{1}{2}({}^{t}\underline{\underline{T}}:\underline{e}_k)\underline{e}_k = \dfrac{1}{2}({}^{t}\underline{\underline{\Omega}}:\underline{e}_k)\underline{e}_k = \underline{\Omega}.$

Appendix 2

Differential Operators

A2.1. Derivative of a scalar field

A2.1.1. Affine space

Consider an affine space F with associated vector space E of dimension n. Chosen an origin O in F, any "point" M in F is characterized and represented by its "position" vector $\underline{OM} = \underline{x} \in E$.

A2.1.2. Gradient of a scalar field

Let f be a scalar field (or scalar function) over F

$$\begin{cases} F \xrightarrow{\;f\;} \mathbb{R} \\ M \in F \mapsto f(\underline{x}) \in \mathbb{R} \end{cases} \tag{A2.1}$$

The gradient of this scalar field at point M is the linear form, first rank Euclidean tensor, denoted by $\underline{\mathrm{grad}\, f(\underline{x})}$, defined over E as follows:

$$\begin{cases} E \xrightarrow{\;\mathrm{grad}\, f\;} \mathbb{R} \\ \underline{dx} \in E \mapsto \underline{\mathrm{grad}\, f(\underline{x})} . \underline{dx} = df(\underline{x}) \in \mathbb{R}, \end{cases} \tag{A2.2}$$

where $df(\underline{x})$ is the *differential* of the function f at point M.

In an orthonormal basis $\left\{ \underline{e}_i \right\}$ of E that defines *orthonormal Cartesian coordinates* x_i in F, we write the explicit formulas:

$$\begin{cases} \underline{dx} = dx_k \underline{e}_k \\[2mm] df(\underline{x}) = \dfrac{\partial f(\underline{x})}{\partial x_i} dx_i \\[2mm] \Rightarrow \underline{\mathrm{grad}}\, f(\underline{x}) = \dfrac{\partial f(\underline{x})}{\partial x_i} \underline{e}_i \, . \end{cases} \qquad \text{[A2.3]}$$

$\underline{\mathrm{grad}}\, f$ denotes the *gradient field* of the scalar field f over F.

A2.2. Derivative of a first rank Euclidean tensor field

A2.2.1. *Gradient of a vector field*

Let $\underline{U}$ be a vector field (first rank tensor field) over F. The gradient of this vector field at point M is the second rank Euclidean tensor, denoted by $\underline{\underline{\mathrm{grad}}}\, U(\underline{x})$, associated with the linear mapping of E onto E defined as follows:

$$\begin{cases} E \xrightarrow{\;\mathrm{grad}\,U\;} E \\[2mm] \underline{dx} \in E \mapsto \underline{\underline{\mathrm{grad}}}\, U(\underline{x}).\underline{dx} = \underline{dU}(\underline{x}) \in E, \end{cases} \qquad \text{[A2.4]}$$

with $\underline{dU}(\underline{x})$ the *differential* of the vector field $\underline{U}$ at point M. $\underline{\underline{\mathrm{grad}}}\, U$ is the *gradient field* of the vector field $\underline{U}$ over F.

Explicitly, in an orthonormal basis $\left\{ \underline{e}_i \right\}$ of E defining *orthonormal Cartesian coordinates* x_i

$$\begin{cases} \underline{dx} = dx_k \underline{e}_k \\[2mm] \underline{U}(\underline{x}) = U_j \underline{e}_j \\[2mm] \underline{dU}(\underline{x}) = dU_j \underline{e}_j = \dfrac{\partial U_j(\underline{x})}{\partial x_i} dx_i \underline{e}_j \Rightarrow \underline{\underline{\mathrm{grad}}}\, U(\underline{x}) = \dfrac{\partial U_j(\underline{x})}{\partial x_i} \underline{e}_j \otimes \underline{e}_i \, . \end{cases} \qquad \text{[A2.5]}$$

A2.2.2. *Divergence of a vector field*

The divergence of the vector field $\underline{U}$ at point M, denoted by $\mathrm{div}U(\underline{x})$, is the scalar obtained from the contraction of the gradient $\underline{\underline{\mathrm{grad}\,U(\underline{x})}}$

$$\mathrm{div}\,U(\underline{x}) = \mathrm{tr}\left(\underline{\underline{\mathrm{grad}\,U(\underline{x})}}\right) = \underline{\underline{\mathrm{grad}\,U(\underline{x})}}:\underline{\underline{1}}. \qquad [A2.6]$$

In *orthonormal Cartesian coordinates* x_i

$$\mathrm{div}\,\underline{U}(\underline{x}) = \frac{\partial U_i(\underline{x})}{\partial x_i}. \qquad [A2.7]$$

A2.2.3. *Curl of a vector field*

It is common practice to split the gradient $\underline{\underline{\mathrm{grad}\,U(\underline{x})}}$ into its symmetric and antisymmetric parts, $\underline{\underline{d}}(\underline{x})$ and $\underline{\underline{\Omega}}(\underline{x})$ respectively

$$\left\{ \begin{array}{l} \underline{\underline{\mathrm{grad}\,U(\underline{x})}} = \underline{\underline{d}}(\underline{x}) + \underline{\underline{\Omega}}(\underline{x}) \\[2mm] \underline{\underline{d}}(\underline{x}) = \frac{1}{2}\left(\underline{\underline{\mathrm{grad}\,U(\underline{x})}} +{}^{t}\underline{\underline{\mathrm{grad}\,U(\underline{x})}}\right) \\[2mm] \underline{\underline{\Omega}}(\underline{x}) = \frac{1}{2}\left(\underline{\underline{\mathrm{grad}\,U(\underline{x})}} -{}^{t}\underline{\underline{\mathrm{grad}\,U(\underline{x})}}\right) \end{array} \right. \qquad [A2.8]$$

hence

$$\mathrm{div}\,\underline{U}(\underline{x}) = \mathrm{tr}\,\underline{\underline{d}}(\underline{x}). \qquad [A2.9]$$

In *orthonormal Cartesian coordinates*

$$\left\{ \begin{array}{l} d_{ij}(\underline{x}) = \frac{1}{2}\left(\frac{\partial U_i(\underline{x})}{\partial x_j} + \frac{\partial U_j(\underline{x})}{\partial x_i}\right) \\[3mm] \Omega_{ij}(\underline{x}) = \frac{1}{2}\left(\frac{\partial U_i(\underline{x})}{\partial x_j} - \frac{\partial U_j(\underline{x})}{\partial x_i}\right). \end{array} \right. \qquad [A2.10]$$

In the case when E *is of dimension* $n = 3$, we know (Appendix 1, section A1.4.4) that the linear mapping from E onto E associated with $\underline{\underline{\Omega}}(\underline{x})$ can be expressed in the form of a vector product

$$\underline{\underline{\Omega}}(\underline{x}).\underline{dx} = \underline{\Omega}(\underline{x}) \wedge \underline{dx} \qquad [\text{A2.11}]$$

where the components of $\underline{\Omega}(\underline{x})$ in a right-handed basis are obtained from the six equations $\Omega_{ik} = \varepsilon_{ijk}\Omega_j$ that are proven compatible due to $\underline{\underline{\Omega}}(\underline{x})$ being antisymmetric.

We obtain

$$\underline{\Omega}(\underline{x}) = \frac{1}{2}[(\frac{\partial U_3}{\partial x_2} - \frac{\partial U_2}{\partial x_3})\underline{e}_1 + (\frac{\partial U_1}{\partial x_3} - \frac{\partial U_3}{\partial x_1})\underline{e}_2 + (\frac{\partial U_2}{\partial x_1} - \frac{\partial U_1}{\partial x_2})\underline{e}_3]. \qquad [\text{A2.12}]$$

The *curl* of the vector field $\underline{U}$ at point M is defined as *twice* this vector:

$$\underline{\text{curl}}\,\underline{U}(\underline{x}) = (\frac{\partial U_3}{\partial x_2} - \frac{\partial U_2}{\partial x_3})\underline{e}_1 + (\frac{\partial U_1}{\partial x_3} - \frac{\partial U_3}{\partial x_1})\underline{e}_2 + (\frac{\partial U_2}{\partial x_1} - \frac{\partial U_1}{\partial x_2})\underline{e}_3. \qquad [\text{A2.13}]$$

A2.3. Derivative of a second rank Euclidean tensor field

A2.3.1. *Gradient of a second rank Euclidean tensor field*

Consider $\underline{\underline{T}}$ a second rank Euclidean tensor field defined over F. The gradient of this tensor field at point M is defined in a similar way as in section A2.2.1. It is the third rank Euclidean tensor $\underline{\underline{\underline{\text{grad}\,T}}}(\underline{x})$

$$\begin{cases} E \xrightarrow{\ \underline{\underline{\text{grad}\,T}}\ } E \times E \\ \underline{dx} \in E \mapsto \underline{\underline{\underline{\text{grad}\,T}}}(\underline{x}).\underline{dx} = \underline{\underline{dT}}(\underline{x}) \in E \times E \end{cases} \qquad [\text{A2.14}]$$

with $\underline{\underline{dT}}(\underline{x})$ the *differential* of the tensor field $\underline{\underline{T}}$ at point M.

Using *orthogonal Cartesian coordinates*, we write

$$\left\{ \begin{array}{l} \underline{dx} = dx_k\, \underline{e}_k \\[2mm] \underline{\underline{T}}(\underline{x}) = T_{ij}\, \underline{e}_i \otimes \underline{e}_j \\[2mm] \underline{\underline{dT}}(\underline{x}) = dT_{ij}\, \underline{e}_i \otimes \underline{e}_j = \dfrac{\partial T_{ij}(\underline{x})}{\partial x_k}\, dx_k\, \underline{e}_i \otimes \underline{e}_j \\[4mm] \Rightarrow \underline{\underline{\mathrm{grad}\, T}}(\underline{x}) = \dfrac{\partial T_{ij}(\underline{x})}{\partial x_k}\, \underline{e}_i \otimes \underline{e}_j \otimes \underline{e}_k\, . \end{array} \right. \qquad [\text{A2.15}]$$

A2.3.2. *Divergence of a second rank Euclidean tensor field*

The *divergence* of the tensor field $\underline{\underline{T}}$ at point M is the vector (first rank tensor) obtained, in an *orthonormal basis* $\{\underline{e}_i\}$, from the contraction of the gradient $\underline{\underline{\mathrm{grad}\, T}}(\underline{x})$ on its *two last indices*. Using *orthogonal Cartesian coordinates*:

$$\mathrm{div}\,\underline{\underline{T}}(\underline{x}) = \dfrac{\partial T_{ij}(\underline{x})}{\partial x_k}\, \underline{e}_i\, (\underline{e}_j \cdot \underline{e}_k) \qquad [\text{A2.16}]$$

hence

$$\mathrm{div}\,\underline{\underline{T}}(\underline{x}) = \dfrac{\partial T_{ij}(\underline{x})}{\partial x_j}\, \underline{e}_i\, . \qquad [\text{A2.17}]$$

We may also define the "left-hand side" divergence (the former one being sometimes called "right-hand side divergence") obtained from the contraction on the first and last indices, which is just the divergence of $^t\underline{\underline{T}}(\underline{x})$

$$\left\{ \begin{array}{l} \mathrm{div}_L\,\underline{\underline{T}}(\underline{x}) = \dfrac{\partial T_{ij}(\underline{x})}{\partial x_k}\, \underline{e}_j\, (\underline{e}_i \cdot \underline{e}_k) \\[4mm] \mathrm{div}_L\,\underline{\underline{T}}(\underline{x}) = \dfrac{\partial T_{ij}(\underline{x})}{\partial x_i}\, \underline{e}_j = \mathrm{div}\,{}^t\underline{\underline{T}}(\underline{x}). \end{array} \right. \qquad [\text{A2.18}]$$

A2.4. Derivative of a Euclidean tensor field of rank p

A2.4.1. *Gradient of a Euclidean tensor field of rank p*

The definition given here above for the gradient of a second rank Euclidean tensor field can be generalized easily to Euclidean tensor fields of rank p. The gradient of the tensor field T at point M is the tensor $\mathrm{grad}\,T(\underline{x})$ of rank $p+1$

$$\begin{cases} E \xrightarrow{\mathrm{grad}\,T} E \times E...\times E \\ \underline{dx} \in E \mapsto \mathrm{grad}\,T(\underline{x}).\underline{dx} = dT(\underline{x}) \in E \times E...\times E \end{cases} \tag{A2.19}$$

where $dT(\underline{x})$ is the differential of the tensor field T at point M.

Using *orthogonal Cartesian coordinates*,

$$\begin{cases} T(\underline{x}) = T_{ij...p}(\underline{x})\,\underline{e}_i \otimes \underline{e}_j ... \otimes \underline{e}_p \\ \Rightarrow \mathrm{grad}\,T(\underline{x}) = \dfrac{\partial T_{ij...p}(\underline{x})}{\partial x_k}\,\underline{e}_i \otimes \underline{e}_j ... \otimes \underline{e}_p \otimes \underline{e}_k . \end{cases} \tag{A2.20}$$

A2.4.2. *Divergence of a Euclidean tensor field of rank p*

The *divergence* of the tensor field T at point M is the tensor of rank $(p-1)$ obtained, in an orthonormal basis $\{\underline{e}_i\}$, from the contraction of $\mathrm{grad}\,T(\underline{x})$ on its two last indices.

Thus, using *orthogonal Cartesian coordinates*,

$$\begin{cases} \mathrm{div}\,T(\underline{x}) = \dfrac{\partial T_{ij...p}(\underline{x})}{\partial x_k}\,\underline{e}_i \otimes \underline{e}_j ... \otimes \underline{e}_{p-1}\,(\underline{e}_p \cdot \underline{e}_k) \\ \mathrm{div}\,T(\underline{x}) = \dfrac{\partial T_{ij...p}(\underline{x})}{\partial x_p}\,\underline{e}_i \otimes \underline{e}_j ... \otimes \underline{e}_{p-1} . \end{cases} \tag{A2.21}$$

A2.5. The divergence theorem

For the sake of simplicity and *without any loss of generality*, the theorem will be stated in the three-dimensional case ($n = 3$) that makes geometrical representations possible.

A2.5.1. *The divergence theorem for a vector field*

– Continuous and continuously differentiable vector field

Let $\underline{U}$ be a continuous and continuously differentiable vector field over F. Consider a volume Ω in F with a regular boundary denoted by $\partial\Omega$. At any point M of the boundary, the outward unit normal is $\underline{n}(\underline{x})$.

The divergence theorem is classically expressed by the following equation:

$$\int_{\partial\Omega} \underline{U}(\underline{x}).\underline{n}(\underline{x})\,\mathrm{d}a = \int_{\Omega} \mathrm{div}\,\underline{U}(\underline{x})\,\mathrm{d}\Omega \qquad [A2.22]$$

– Piecewise continuous and continuously differentiable vector field

The theorem may be generalized to the case when $\underline{U}$ is a piecewise continuous and continuously differentiable vector field over F. The set of countable jump surfaces of $\underline{U}$ is denoted by Σ_U, with the jump (or discontinuity) $[\![\underline{U}(\underline{x})]\!]$ of the vector field at the point M being defined as

$$[\![\underline{U}(\underline{x})]\!] = \underline{U}(\underline{x})_2 - \underline{U}(\underline{x})_1 \qquad [A2.23]$$

where $\underline{U}(\underline{x})_1$ and $\underline{U}(\underline{x})_2$ are the values of the field $\underline{U}$ at point M, respectively "upstream" and "downstream" when crossing Σ_U *following* $\underline{n}(\underline{x})$, a unit normal to Σ_U at that point (Figure A2.1).

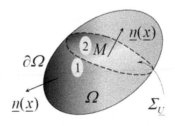

Figure A2.1. *Piecewise continuous and continuously differentiable vector field*

The divergence theorem now includes an integral over Σ_U taking the contribution of the jump $[\![\,\underline{U}(x)\,]\!]$ into account

$$\int_{\partial\Omega}\underline{U}(x).\underline{n}(x)\,da = \int_{\Omega} \operatorname{div}\underline{U}(x)\,d\Omega + \int_{\Sigma_U} [\![\,\underline{U}(x)\,]\!].\underline{n}(x)\,d\Sigma_U. \qquad [\text{A2.24}]$$

Note that this term does not depend on the orientation of $\underline{n}(x)$ at point M on Σ_U.

This result can be established simply by applying [A2.22] separately to regions 1 and 2. We obtain:

$$\begin{aligned} &\int_{\partial\Omega_1}\underline{U}(x).\underline{n}(x)\,da + \int_{\Sigma_U}\underline{U}_1(x).\underline{n}(x)\,d\Sigma_U \\ &+ \int_{\partial\Omega_2}\underline{U}(x).\underline{n}(x)\,da - \int_{\Sigma_U}\underline{U}_2(x).\underline{n}(x)\,d\Sigma_U = \int_{\Omega_1+\Omega_2}\operatorname{div}\underline{U}(x)\,d\Omega \end{aligned} \qquad [\text{A2.25}]$$

thence

$$\int_{\partial\Omega}\underline{U}(x).\underline{n}(x)\,da = \int_{\Omega}\operatorname{div}\underline{U}(x)\,d\Omega + \int_{\Sigma_U}(\underline{U}_2(x)-\underline{U}_1(x)).\underline{n}(x)\,d\Sigma_U. \qquad [\text{A2.26}]$$

A2.5.2. *The divergence theorem for any Euclidean tensor field*

The divergence theorem can be stated in a similar form for any piecewise continuous and continuously differentiable Euclidean tensor field T of rank $p \geq 1$.

With $[\![\,T(x)\,]\!] = T(x)_2 - T(x)_1$, the jump of the tensor $T(x)$ when crossing Σ_T at point M following $\underline{n}(x)$

$$\int_{\partial\Omega}T(x).\underline{n}(x)\,da = \int_{\Omega}\operatorname{div}T(x)\,d\Omega + \int_{\Sigma_T} [\![\,T(x)\,]\!].\underline{n}(x)\,d\Sigma_T \qquad [\text{A2.27}]$$

where the dots denote contracted products.

From the mathematical viewpoint, equations [A2.24] and [A2.27] are the expressions of the divergence theorem within the framework of *Distribution theory*: it means that the divergence field, in the case of a piecewise continuous and continuously differentiable tensor field T must be understood as composed of the *regular divergence field* div T over Ω to be integrated as a volume density *and the field* $[\![T]\!].\underline{n}$ over the jump surface Σ_T to be integrated as a surface density.

Within the same framework, the gradient field of a piecewise continuous and continuously differentiable tensor field T is composed of the *regular gradient field* grad T over Ω *and the field* $[\![T]\!]\otimes\underline{n}$ over the jump surface Σ_T.

A2.6. The curl theorem

E is supposed of dimension 3.

Let $\underline{U}$ be a continuous and continuously differentiable vector field over F. Consider a closed curve Γ that is spanned by a regular surface Σ (a "cap") transversally oriented ("two-sided"), with outward normal $\underline{n}(\underline{x})$, which yields the positive orientation of the curve Γ, called the boundary curve of Σ (Figure A2.2).

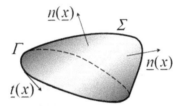

Figure A2.2. *Curl theorem*

With this orientation, the unit vector tangent to Γ at the current point is denoted as $\underline{t}(\underline{x})$ and the line integral $\oint_\Gamma \underline{U}(\underline{x}).\underline{t}(\underline{x})\,\mathrm{d}s$ is called the *circulation* of the vector field $\underline{U}$ along the closed curve Γ.

The *Curl theorem* is written as follows:

$$\oint_\Gamma \underline{U}(\underline{x}).\underline{t}(\underline{x})\,\mathrm{d}s = \int_\Sigma \mathrm{curl}\,\underline{U}(\underline{x}).\underline{n}(\underline{x})\,\mathrm{d}\Sigma. \qquad\qquad [\text{A2.28}]$$

It states that the circulation of the vector field along the closed curve Γ is equal to the flux of the curl through any regular surface Σ, transversally oriented, spanning this curve.

A2.7. Other noticeable results

A2.7.1. *Divergence of the curl*

A straightforward consequence of equation [A2.28] is that, under the same assumptions of continuity and continuous differentiability of the vector field $\underline{U}$, for any volume Ω in E with boundary $\partial\Omega$ transversally oriented, we have

$$\int_{\partial\Omega} \underline{\mathrm{curl}\,U}(\underline{x}).\underline{n}(\underline{x})\,\mathrm{d}a = 0. \qquad [\text{A2.29}]$$

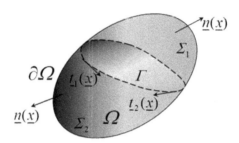

Figure A2.3. *Divergence of the curl*

This result comes from observing that any closed curve Γ drawn on $\partial\Omega$ delimitates two surfaces Σ_1 and Σ_2 that can be considered as caps on that curve but correspond to opposite orientations of Γ (Figure A2.3) and thus yield opposite values for the circulations in [A2.28]. Hence

$$\int_{\Sigma_1} \underline{\mathrm{curl}\,U}(\underline{x}).\underline{n}(\underline{x})\,\mathrm{d}\Sigma_1 = -\int_{\Sigma_2} \underline{\mathrm{curl}\,U}(\underline{x}).\underline{n}(\underline{x})\,\mathrm{d}\Sigma_2 \qquad [\text{A2.30}]$$

and

$$\int_{\Sigma} \underline{\mathrm{curl}\,U}(\underline{x}).\underline{n}(\underline{x})\,\mathrm{d}a = \int_{\Sigma_1} \underline{\mathrm{curl}\,U}(\underline{x}).\underline{n}(\underline{x})\,\mathrm{d}\Sigma_1 + \int_{\Sigma_2} \underline{\mathrm{curl}\,U}(\underline{x}).\underline{n}(\underline{x})\,\mathrm{d}\Sigma_2 = 0. \qquad [\text{A2.31}]$$

From the divergence theorem and Ω being arbitrary, it follows that:

$$\mathrm{div}\left(\underline{\mathrm{curl}\,U}(\underline{x})\right) = 0 \text{ over } F. \qquad [\text{A2.32}]$$

The result can also be obtained directly performing the differentiations explicitly.

A2.7.2. *Curl of a gradient field*

Consider now the case when the vector field $\underline{U}$ is the gradient of a continuous and continuously differentiable scalar field f defined over F. Then, from the gradient theorem (in fact, just the very definition of the gradient), the circulation of $\underline{U}$ along any oriented closed curve Γ is

$$\oint_\Gamma \underline{U}(\underline{x}).\underline{t}(\underline{x})\,\mathrm{d}s = \oint_\Gamma \mathrm{grad}\,f(\underline{x}).\underline{t}(\underline{x})\,\mathrm{d}s = 0. \qquad [\text{A2.33}]$$

Hence, for any oriented "cap" Σ spanning Γ

$$\int_\Sigma [\mathrm{curl}(\mathrm{grad}\,f(\underline{x}))].\underline{n}(\underline{x})\,\mathrm{d}\Sigma = 0, \qquad [\text{A2.34}]$$

and it follows:

$$\mathrm{curl}(\mathrm{grad}\,f(\underline{x})) = 0 \ \text{ over } \ F \qquad [\text{A2.35}]$$

which can also be obtained directly performing the differentiations explicitly.

Conversely, this *necessary* condition for a vector field to be a gradient field is also *sufficient*.

A2.7.3. *Laplacian of a tensor field*

In this section, E may be of any dimension n.

Let T be a continuous and continuously differentiable Euclidean tensor field T of rank $p \geq 1$. We consider the gradient field of the field T as defined in section A2.4.1:

$$\begin{cases} T(\underline{x}) = T_{ij\ldots p}(\underline{x})\,\underline{e}_i \otimes \underline{e}_j \ldots \otimes \underline{e}_p \\ \mathrm{grad}\,T(\underline{x}) = \dfrac{\partial T_{ij\ldots p}(\underline{x})}{\partial x_k}\,\underline{e}_i \otimes \underline{e}_j \ldots \otimes \underline{e}_p \otimes \underline{e}_k. \end{cases} \qquad [\text{A2.36}]$$

Then, the gradient of the gradient field $\operatorname{grad} T$ is written as

$$\operatorname{grad}(\operatorname{grad} T)(\underline{x}) = \frac{\partial}{\partial x_\ell}(\frac{\partial T_{ij\ldots p}(\underline{x})}{\partial x_k}\underline{e}_i \otimes \underline{e}_j \ldots \otimes \underline{e}_p \otimes \underline{e}_k) \otimes \underline{e}_\ell. \qquad \text{[A2.37]}$$

Taking the divergence (contraction on the two last indices), we obtain

$$\begin{cases} \operatorname{div}(\operatorname{grad} T)(\underline{x}) = \frac{\partial}{\partial x_\ell}(\frac{\partial T_{ij\ldots p}(\underline{x})}{\partial x_k}\underline{e}_i \otimes \underline{e}_j \ldots \otimes \underline{e}_p)(\underline{e}_k \cdot \underline{e}_\ell) \\[4mm] \operatorname{div}(\operatorname{grad} T)(\underline{x}) = \frac{\partial^2 T_{ij\ldots p}(\underline{x})}{\partial x_k \partial x_k}\underline{e}_i \otimes \underline{e}_j \ldots \otimes \underline{e}_p. \end{cases} \qquad \text{[A2.38]}$$

In other words, the *divergence of the gradient* is the Euclidean tensor of order p whose components in orthogonal Cartesian coordinates are the *Laplacian* of the components of the original tensor:

$$\operatorname{div}(\operatorname{grad} T)(\underline{x}) = \Delta T_{ij\ldots p}(\underline{x})\underline{e}_i \otimes \underline{e}_j \ldots \otimes \underline{e}_p. \qquad \text{[A2.39]}$$

This defines the *Laplacian of the Euclidean tensor field* T

$$\Delta T = \operatorname{div}(\operatorname{grad} T). \qquad \text{[A2.40]}$$

A2.8. Derivatives in an orthogonal curvilinear coordinate system

A2.8.1. *Curvilinear coordinates*

The definitions given in the preceding sections for the gradient of a scalar field or a Euclidean tensor field, whatever its rank refer to the contracted product that yields the differential of the concerned tensor field at a given point M in F: it is easily written in an orthonormal basis $\{\underline{e}_i\}$ of E at the considered point.

Additionally, we have seen that when *orthogonal Cartesian coordinates* are used, where the position vector $\underline{OM} = \underline{x}$ is expanded over the basis $\{\underline{e}_i\}$ as $\underline{x} = x_i \underline{e}_i$, simple explicit expressions of the gradient and derived quantities (divergence, etc.) at a point M are obtained in the orthogonal basis $\{\underline{e}_i\}$ in terms of the partial derivatives of the components of the tensor field with respect to the coordinates x_i. These expressions are the most commonly used in practice.

Nevertheless, it also proves useful to consider tensor fields over F where the point M, still to be referred to through its position vector $\underline{OM} = \underline{x}$, is labeled by n parameters η_i defining a system of *curvilinear coordinates*. The corresponding *coordinate lines* are defined giving fixed values to $(n-1)$ parameters and letting the last one free (for instance, given values to $\eta_2, \eta_3 ... \eta_n$ define a η_1 coordinate line).

A point M being defined by the values $(\eta_1, \eta_2 ..., \eta_n)$, we write

$$\underline{OM} = \underline{x}(\eta_1, \eta_2 ..., \eta_n). \tag{A2.41}$$

At that point M, a variation $d\eta_i$ of the parameter η_i, while the others are kept constant, defines a vector $\underline{dM} \in E$, which is tangent to the η_i coordinate line, in the form

$$\underline{dM} = \frac{\partial \underline{x}}{\partial \eta_i} d\eta_i = \underline{u}_i \, d\eta_i \text{ (no summation on repeated indices here)}. \tag{A2.42}$$

The vectors

$$\underline{u}_i = \frac{\partial \underline{x}}{\partial \eta_i} \ (i = 1, 2 ..., n) \tag{A2.43}$$

defined that way constitute a basis of E called the *local natural basis* of the curvilinear coordinate system at the point M. Unit vectors collinear with the $\underline{u}_i$ are $\underline{e}_i = \underline{u}_i / |\underline{u}_i|$.

Mathematical regularity conditions – e.g. diffeomorphism – are obviously imposed in equation [A2.41] for the validity of the description.

The most common curvilinear coordinate systems encountered in three-dimensional continuum mechanics are the cylindrical coordinate system about an axis Oz and the spherical coordinate system with center O. For these two systems, the local natural bases $\{\underline{u}_i\}$ are orthogonal: thus the $\{\underline{e}_i\}$ bases are orthonormal.

A2.8.2. *Derivatives in a cylindrical coordinate system*

A2.8.2.1. *Parametrization*

The position of a point M is described by parameters r, θ, z. The associated local orthonormal basis is $\underline{e}_r, \underline{e}_\theta, \underline{e}_z$ as shown in Figure A2.4.

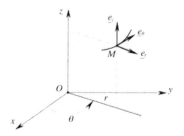

Figure A2.4.

$$\begin{cases} \underline{x} = \underline{OM} = r\,\underline{e}_r \\[2mm] \underline{dx} = \underline{dM} = dr\,\underline{e}_r + r\,d\theta\,\underline{e}_\theta + dz\,\underline{e}_z \\[2mm] \underline{e}_r = \dfrac{\partial \underline{x}}{\partial r}, \quad \underline{e}_\theta = \dfrac{1}{r}\dfrac{\partial \underline{x}}{\partial \theta}, \quad \underline{e}_z = \dfrac{\partial \underline{x}}{\partial z}. \end{cases} \qquad [\text{A2.44}]$$

The variation of the basis $\underline{e}_r, \underline{e}_\theta, \underline{e}_z$ as a function of the parameters r, θ, z is

$$\begin{cases} \dfrac{\partial \underline{e}_r}{\partial r} = 0, & \dfrac{\partial \underline{e}_\theta}{\partial r} = 0, & \dfrac{\partial \underline{e}_z}{\partial r} = 0, \\[3mm] \dfrac{\partial \underline{e}_r}{\partial \theta} = \underline{e}_\theta, & \dfrac{\partial \underline{e}_\theta}{\partial \theta} = -\underline{e}_r, & \dfrac{\partial \underline{e}_z}{\partial \theta} = 0, \\[3mm] \dfrac{\partial \underline{e}_r}{\partial z} = 0, & \dfrac{\partial \underline{e}_r}{\partial z} = 0, & \dfrac{\partial \underline{e}_z}{\partial z} = 0. \end{cases} \qquad [\text{A2.45}]$$

A2.8.2.2. *Derivative of a scalar field*

The scalar field (or scalar function) f is defined over F as a function of r, θ, z

$$\begin{cases} F \xrightarrow{\ f\ } \mathbb{R} \\ M \in F \mapsto f(\underline{x}) = f(r,\theta,z) \in \mathbb{R} \end{cases} \qquad \text{[A2.46]}$$

The definition of the gradient through the differential $df = \underline{\text{grad}}\, f(\underline{x}).d\underline{x}$ is now written as

$$\underline{\text{grad}}\, f(r,\theta,z).(dr\,\underline{e}_r + r\,d\theta\,\underline{e}_\theta + dz\,\underline{e}_z) = df = \frac{\partial f(\underline{x})}{\partial r}dr$$
$$+ \frac{\partial f(\underline{x})}{\partial \theta}d\theta + \frac{\partial f(\underline{x})}{\partial z}dz \qquad \text{[A2.47]}$$

which yields

$$\underline{\text{grad}}\, f(r,\theta,z) = \frac{\partial f(\underline{x})}{\partial r}\underline{e}_r + \frac{1}{r}\frac{\partial f(\underline{x})}{\partial \theta}\underline{e}_\theta + \frac{\partial f(\underline{x})}{\partial z}\underline{e}_z \qquad \text{[A2.48]}$$

According to [A2.40], the Laplacian is obtained as the divergence of the gradient, using the general expression that will be given in [A2.54] for a vector field. Hence:

$$\Delta f(r,\theta,z) = \frac{\partial^2 f(r,\theta,z)}{\partial r^2} + \frac{1}{r}\frac{\partial f(r,\theta,z)}{\partial r}$$
$$+ \frac{1}{r^2}\frac{\partial^2 f(r,\theta,z)}{\partial \theta^2} + \frac{\partial^2 f(r,\theta,z)}{\partial z^2}. \qquad \text{[A2.49]}$$

A2.8.2.3. Derivative of a vector field

$\underline{U}$ is a vector field (first rank Euclidean tensor field) over F. At the point M, it is expanded in terms of the local orthonormal basis $\underline{e}_r, \underline{e}_\theta, \underline{e}_z$ with U_r, U_θ, U_z as components:

$$\underline{U}(r,\theta,z) = U_r(r,\theta,z)\underline{e}_r + U_\theta(r,\theta,z)\underline{e}_\theta + U_z(r,\theta,z)\underline{e}_z \qquad \text{[A2.50]}$$

Expanding the definition of the gradient $\underline{\underline{\text{grad}}}\, U(\underline{x}).d\underline{x} = d\underline{U}(\underline{x})$, we have for the left-hand side

$$\underline{\underline{\text{grad}}}\, U(\underline{x}).d\underline{x} = \underline{\underline{\text{grad}}}\, U(r,\theta,z).(dr\,\underline{e}_r + r\,d\theta\,\underline{e}_\theta + dz\,\underline{e}_z) \qquad \text{[A2.51]}$$

and for the right-hand side

$$
\begin{aligned}
\underline{dU(x)} = &\frac{\partial}{\partial r}(U_r(r,\theta,z)\underline{e}_r + U_\theta(r,\theta,z)\underline{e}_\theta + U_z(r,\theta,z)\underline{e}_z)\,dr \\
&+\frac{\partial}{\partial r}(U_r(r,\theta,z)\underline{e}_r + U_\theta(r,\theta,z)\underline{e}_\theta + U_z(r,\theta,z)\underline{e}_z)\,dr \qquad [\text{A2.52}] \\
&+\frac{\partial}{\partial r}(U_r(r,\theta,z)\underline{e}_r + U_\theta(r,\theta,z)\underline{e}_\theta + U_z(r,\theta,z)\underline{e}_z)\,dr,
\end{aligned}
$$

where the variation of the basis according to [A2.45] must be taken into account.

Hence:

$$
\begin{aligned}
\underline{\underline{\text{grad}\,U}}(r,\theta,z) = &\frac{\partial U_r}{\partial r}\underline{e}_r \otimes \underline{e}_r + \frac{1}{r}\left(\frac{\partial U_r}{\partial \theta} - U_\theta\right)\underline{e}_r \otimes \underline{e}_\theta + \frac{\partial U_r}{\partial z}\underline{e}_r \otimes \underline{e}_z \\
&+\frac{\partial U_\theta}{\partial r}\underline{e}_\theta \otimes \underline{e}_r + \frac{1}{r}\left(\frac{\partial U_\theta}{\partial \theta} + U_r\right)\underline{e}_\theta \otimes \underline{e}_\theta + \frac{\partial U_\theta}{\partial z}\underline{e}_\theta \otimes \underline{e}_z \qquad [\text{A2.53}] \\
&+\frac{\partial U_z}{\partial r}\underline{e}_z \otimes \underline{e}_r + \frac{1}{r}\frac{\partial U_z}{\partial \theta}\underline{e}_z \otimes \underline{e}_\theta + \frac{\partial U_z}{\partial z}\underline{e}_z \otimes \underline{e}_z.
\end{aligned}
$$

The divergence is obtained from the contraction of the gradient:

$$
\text{div}\,\underline{U}(r,\theta,z) = \frac{\partial U_r}{\partial r} + \frac{U_r}{r} + \frac{1}{r}\frac{\partial U_\theta}{\partial \theta} + \frac{\partial U_z}{\partial z}. \qquad [\text{A2.54}]
$$

A2.8.2.4. *Derivative of a second rank Euclidean tensor field*

$\underline{\underline{T}}(\underline{x}) = \underline{\underline{T}}(r,\theta,z)$, a second rank Euclidean tensor, is expanded in terms of the local orthonormal basis $\underline{e}_r, \underline{e}_\theta, \underline{e}_z$:

$$
\begin{aligned}
\underline{\underline{T}}(r,\theta,z) = &T_{rr}\,\underline{e}_r \otimes \underline{e}_r + T_{r\theta}\,\underline{e}_r \otimes \underline{e}_\theta + T_{rz}\,\underline{e}_r \otimes \underline{e}_z \\
&+T_{\theta r}\,\underline{e}_\theta \otimes \underline{e}_r + T_{\theta\theta}\,\underline{e}_\theta \otimes \underline{e}_\theta + T_{\theta z}\,\underline{e}_\theta \otimes \underline{e}_z \qquad [\text{A2.55}] \\
&+T_{zr}\,\underline{e}_z \otimes \underline{e}_\theta + T_{z\theta}\,\underline{e}_z \otimes \underline{e}_\theta + T_{zz}\,\underline{e}_z \otimes \underline{e}_z.
\end{aligned}
$$

The gradient of the tensor field is obtained from the definition

$$\mathrm{grad}\,T(r,\theta,z).(\mathrm{d}r\,\underline{e}_r + r\,\mathrm{d}\theta\,\underline{e}_\theta + \mathrm{d}z\,\underline{e}_z) = \underline{\underline{\mathrm{d}T}}(\underline{x}) \qquad\qquad [\mathrm{A2.56}]$$

with

$$\underline{\underline{\mathrm{d}T}}(\underline{x}) = \frac{\partial T_{ij}(\underline{x})}{\partial x_k}\underline{e}_i \otimes \underline{e}_j\,\mathrm{d}x_k + T_{ij}(\underline{x})\frac{\partial(\underline{e}_i \otimes \underline{e}_j)(\underline{x})}{\partial x_k}\mathrm{d}x_k \qquad\qquad [\mathrm{A2.57}]$$

where the $\dfrac{\partial(\underline{e}_i \otimes \underline{e}_j)}{\partial x_k}$ are computed from [A2.45].

After rather tedious calculations, taking the trace of the gradient yields the divergence of the tensor field that will appear, in particular, in the equations of motion in continuum mechanics:

$$\begin{aligned}
\mathrm{div}\,\underline{\underline{T}}(r,\theta,z) = &\left(\frac{\partial T_{rr}}{\partial r} + \frac{1}{r}\frac{\partial T_{r\theta}}{\partial \theta} + \frac{\partial T_{rz}}{\partial z} + \frac{T_{rr} - T_{\theta\theta}}{r}\right)\underline{e}_r \\
&+\left(\frac{\partial T_{\theta r}}{\partial r} + \frac{1}{r}\frac{\partial T_{\theta\theta}}{\partial \theta} + \frac{\partial T_{\theta z}}{\partial z} + \frac{T_{\theta r} + T_{r\theta}}{r}\right)\underline{e}_\theta \\
&+\left(\frac{\partial T_{zr}}{\partial r} + \frac{1}{r}\frac{\partial T_{z\theta}}{\partial \theta} + \frac{\partial T_{zz}}{\partial z} + \frac{T_{zr}}{r}\right)\underline{e}_z.
\end{aligned} \qquad [\mathrm{A2.58}]$$

A2.8.3. Derivatives in a spherical coordinate system

A2.8.3.1. Parametrization

The position of a point M is described by parameters r,θ,φ as shown in Figure A2.5. The associated local orthonormal basis is $\underline{e}_r, \underline{e}_\theta, \underline{e}_\varphi$:

$$\left\{ \begin{aligned}
&\underline{x} = \underline{OM} = r\,\underline{e}_r \\
&\mathrm{d}\underline{x} = \underline{\mathrm{d}M} = \mathrm{d}r\,\underline{e}_r + r\,\mathrm{d}\theta\,\underline{e}_\theta + r\sin\theta\,\mathrm{d}\varphi\,\underline{e}_z \\
&\underline{e}_r = \frac{\partial \underline{x}}{\partial r},\ \underline{e}_\theta = \frac{1}{r}\frac{\partial \underline{x}}{\partial \theta},\ \underline{e}_\varphi = \frac{1}{r\sin\theta}\frac{\partial \underline{x}}{\partial \varphi}.
\end{aligned} \right. \qquad [\mathrm{A2.59}]$$

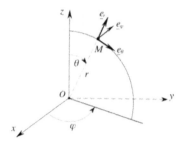

Figure A2.5.

The variation of the basis $\underline{e}_r, \underline{e}_\theta, \underline{e}_\varphi$ as a function of r, θ, φ is written as

$$
\begin{cases}
\dfrac{\partial \underline{e}_r}{\partial r} = 0, & \dfrac{\partial \underline{e}_\theta}{\partial r} = 0, & \dfrac{\partial \underline{e}_\varphi}{\partial r} = 0, \\[2mm]
\dfrac{\partial \underline{e}_r}{\partial \theta} = \underline{e}_\theta, & \dfrac{\partial \underline{e}_\theta}{\partial \theta} = -\underline{e}_r, & \dfrac{\partial \underline{e}_\varphi}{\partial \theta} = 0, \\[2mm]
\dfrac{\partial \underline{e}_r}{\partial \varphi} = \underline{e}_\varphi \sin\theta, & \dfrac{\partial \underline{e}_r}{\partial z} = -\underline{e}_\varphi \cos\theta, & \dfrac{\partial \underline{e}_\varphi}{\partial z} = -\underline{e}_r \sin\theta - \underline{e}_\theta \cos\theta.
\end{cases}
\qquad [\text{A2.60}]
$$

A2.8.3.2. Derivative of a scalar field

The scalar field f is defined over F as a function of r, θ, φ and the process for the calculation of the gradient is the same as described in [A2.47] and [A2.48], based upon

$$
\underline{\mathrm{grad}}\, f(r,\theta,\varphi).(dr\,\underline{e}_r + r\,d\theta\,\underline{e}_\theta + r\sin\theta\, d\varphi\,\underline{e}_z) = df = \frac{\partial f(x)}{\partial r}\,dr
$$
$$
+ \frac{\partial f(x)}{\partial \theta}\,d\theta + \frac{\partial f(x)}{\partial \varphi}\,d\varphi
\qquad [\text{A2.61}]
$$

hence

$$
\underline{\mathrm{grad}}\, f(r,\theta,\varphi) = \frac{\partial f(x)}{\partial r}\,\underline{e}_r + \frac{1}{r}\frac{\partial f(x)}{\partial \theta}\,\underline{e}_\theta + \frac{1}{r\sin\theta}\frac{\partial f(x)}{\partial \varphi}\,\underline{e}_\varphi.
\qquad [\text{A2.62}]
$$

Taking the divergence of this vector field according to [A2.65], we obtain the Laplacian:

$$\Delta f(r,\theta,\varphi) = \frac{\partial^2 f(\underline{x})}{\partial r^2} + \frac{2}{r}\frac{\partial f(\underline{x})}{\partial r} + \frac{1}{r^2}\frac{\partial^2 f(\underline{x})}{\partial\theta^2}$$
$$+ \frac{1}{r^2}\cot\theta\frac{\partial f(\underline{x})}{\partial\theta} + \frac{1}{r^2\sin^2\theta}\frac{\partial^2 f(\underline{x})}{\partial\varphi^2}.$$

[A2.63]

A2.8.3.3. Derivative of a vector field

The vector field $\underline{U}$, first rank Euclidean tensor field, is expanded at the point M in terms of the local orthonormal basis $\underline{e}_r, \underline{e}_\theta, \underline{e}_\varphi$ with U_r, U_θ, U_φ as components:

$$\underline{U}(r,\theta,\varphi) = U_r(r,\theta,\varphi)\underline{e}_r + U_\theta(r,\theta,\varphi)\underline{e}_\theta + U_\varphi(r,\theta,z)\underline{e}_\varphi.$$

[A2.64]

Taking [A2.60] into account, we obtain, in the same way as for [A2.53], the gradient

$$\underline{\underline{\text{grad}\,U}}(\underline{x}) = \frac{\partial U_r}{\partial r}\underline{e}_r\otimes\underline{e}_r + \frac{1}{r}(\frac{\partial U_r}{\partial\theta} - U_\theta)\underline{e}_r\otimes\underline{e}_\theta + \frac{1}{r}(\frac{1}{\sin\theta}\frac{\partial U_r}{\partial\varphi} - U_\varphi)\underline{e}_r\otimes\underline{e}_\varphi$$
$$+ \frac{\partial U_\theta}{\partial r}\underline{e}_\theta\otimes\underline{e}_r + \frac{1}{r}(\frac{\partial U_\theta}{\partial\theta} + U_r)\underline{e}_\theta\otimes\underline{e}_\theta + \frac{1}{r}(\frac{1}{\sin\theta}\frac{\partial U_\theta}{\partial\varphi} - U_\varphi\cot\theta)\underline{e}_\theta\otimes\underline{e}_\varphi$$
$$+ \frac{\partial U_\varphi}{\partial r}\underline{e}_\varphi\otimes\underline{e}_r + \frac{1}{r}\frac{\partial U_\varphi}{\partial\theta}\underline{e}_\varphi\otimes\underline{e}_\theta + \frac{1}{r}(\frac{1}{\sin\theta}\frac{\partial U_\varphi}{\partial\varphi} + U_\theta\cot\theta + U_r)\underline{e}_\varphi\otimes\underline{e}_\varphi.$$

[A2.65]

Hence, the divergence

$$\text{div}\,\underline{U}(r,\theta,\varphi) = \frac{\partial U_r}{\partial r} + \frac{2U_r}{r} + \frac{1}{r}\frac{\partial U_\theta}{\partial\theta} + \frac{U_\theta}{r}\cot\theta + \frac{1}{r\sin\theta}\frac{\partial U_\varphi}{\partial\varphi}.$$

[A2.66]

A2.8.3.4. Derivative of a second rank Euclidean tensor field

The second rank Euclidean tensor $\underline{\underline{T}}(\underline{x}) = \underline{\underline{T}}(r,\theta,\varphi)$ is expanded in terms of the local orthonormal basis $\underline{e}_r, \underline{e}_\theta, \underline{e}_\varphi$

$$\underline{\underline{T}}(r,\theta,\varphi) = T_{rr}\,\underline{e}_r \otimes \underline{e}_r + T_{r\theta}\,\underline{e}_r \otimes \underline{e}_\theta + T_{r\varphi}\,\underline{e}_r \otimes \underline{e}_\varphi$$
$$+ T_{\theta r}\,\underline{e}_\theta \otimes \underline{e}_r + T_{\theta\theta}\,\underline{e}_\theta \otimes \underline{e}_\theta + T_{\theta\varphi}\,\underline{e}_\theta \otimes \underline{e}_\varphi \qquad \text{[A2.67]}$$
$$+ T_{\varphi r}\,\underline{e}_\varphi \otimes \underline{e}_\theta + T_{\varphi\theta}\,\underline{e}_\varphi \otimes \underline{e}_\theta + T_{\varphi\varphi}\,\underline{e}_\varphi \otimes \underline{e}_\varphi$$

and the whole process for deriving the expressions of the gradient and divergence fields follows the same track as in section A2.8.2, using [A2.60] to compute the $\dfrac{\partial(\underline{e}_i \otimes \underline{e}_j)}{\partial x_k}$ in

$$\underline{\underline{dT}}(\underline{x}) = \frac{\partial T_{ij}(\underline{x})}{\partial x_k}\,\underline{e}_i \otimes \underline{e}_j\,dx_k + T_{ij}(\underline{x})\frac{\partial(\underline{e}_i \otimes \underline{e}_j)(\underline{x})}{\partial x_k}\,dx_k \qquad \text{[A2.68]}$$

The resulting equations are rather cumbersome and, for simplicity's sake, we only give here the expression of the divergence for a *symmetric* second rank Euclidean tensor field that is necessary to write the equations of motion for the classical three-dimensional continuum in a spherical coordinate system:

$$\operatorname{div}\underline{\underline{T}}(r,\theta,\varphi) = (\frac{\partial T_{rr}}{\partial r} + \frac{1}{r}\frac{\partial T_{r\theta}}{\partial \theta} + \frac{1}{r\sin\theta}\frac{\partial T_{r\varphi}}{\partial \varphi} + \frac{1}{r}(2T_{rr} - T_{\theta\theta} - T_{\varphi\varphi} + T_{r\theta}\cot\theta))\underline{e}_r$$
$$+ (\frac{\partial T_{\theta r}}{\partial r} + \frac{1}{r}\frac{\partial T_{\theta\theta}}{\partial \theta} + \frac{1}{r\sin\theta}\frac{\partial T_{\theta\varphi}}{\partial \varphi} + \frac{1}{r}[(T_{\theta\theta} - T_{\varphi\varphi})\cot\theta + 3T_{r\theta}])\underline{e}_\theta$$
$$+ (\frac{\partial T_{\varphi r}}{\partial r} + \frac{1}{r}\frac{\partial T_{\varphi\theta}}{\partial \theta} + \frac{1}{r\sin\theta}\frac{\partial T_{\varphi\varphi}}{\partial \varphi} + \frac{1}{r}(2T_{\theta\varphi}\cot\theta + 3T_{r\varphi}))\underline{e}_\varphi .$$

$$\text{[A2.69]}$$

A2.9. Practicing

Let $\underline{U}$ and $\underline{\underline{\sigma}}$ be continuous and differentiable vector and second rank tensor fields defined on a Euclidean three-dimensional affine space. Prove the following identities (referred to in Chapter 6, section 6.3.2):

$-\ \operatorname{grad}(\underline{\underline{\sigma}}.\underline{U}) = \underline{\underline{\sigma}}.\operatorname{grad}\underline{U} + \underline{U}.\operatorname{grad}{}^t\underline{\underline{\sigma}}\,;$

$-\ \operatorname{div}(\underline{\underline{\sigma}}.\underline{U}) = \underline{\underline{\sigma}}:\operatorname{grad}\underline{U} + \underline{U}.\operatorname{div}{}^t\underline{\underline{\sigma}}.$

Solution

$-$ $\forall \underline{dx}$, $\text{grad}\,(\underline{\underline{\sigma}}.\underline{U}).\underline{dx} = \underline{\underline{\sigma}}.\underline{dU} + \underline{\underline{d\sigma}}.\underline{U}$ may be written as

$$\forall \underline{dx},\ \text{grad}\,(\underline{\underline{\sigma}}.\underline{U}).\underline{dx} = \underline{\underline{\sigma}}.\underline{dU} + \underline{U}.\underline{\underline{d^t\sigma}} = \underline{\underline{\sigma}}.\text{grad}\,\underline{U}.\underline{dx} + \underline{U}.\text{grad}\,{}^t\underline{\underline{\sigma}}.\underline{dx}$$

thence, $\text{grad}\,(\underline{\underline{\sigma}}.\underline{U}) = \underline{\underline{\sigma}}.\text{grad}\,\underline{U} + \underline{U}.\text{grad}\,{}^t\underline{\underline{\sigma}}.$

$-$ $\text{div}\,(\underline{\underline{\sigma}}.\underline{U}) = \text{tr}\,(\text{grad}\,(\underline{\underline{\sigma}}.\underline{U}))$

$$\text{div}\,(\underline{\underline{\sigma}}.\underline{U}) = \underline{\underline{\sigma}}:\text{grad}\,\underline{U} + \text{tr}\,(\underline{U}.\text{grad}\,{}^t\underline{\underline{\sigma}})$$

$$\text{div}\,(\underline{\underline{\sigma}}.\underline{U}) = \underline{\underline{\sigma}}:\text{grad}\,\underline{U} + \text{tr}\,(\underline{U}.\frac{\partial \sigma_{ji}}{\partial x_k}\underline{e}_i \otimes \underline{e}_j \otimes \underline{e}_k)$$

$$\text{div}\,(\underline{\underline{\sigma}}.\underline{U}) = \underline{\underline{\sigma}}:\text{grad}\,\underline{U} + \text{tr}\,(\underline{U}.\frac{\partial \sigma_{ji}}{\partial x_k}\underline{e}_i \otimes \underline{e}_j \otimes \underline{e}_k)$$

$$\text{div}\,(\underline{\underline{\sigma}}.\underline{U}) = \underline{\underline{\sigma}}:\text{grad}\,\underline{U} + \underline{U}.\frac{\partial \sigma_{ji}}{\partial x_j}\underline{e}_i = \underline{\underline{\sigma}}:\text{grad}\,\underline{U} + \underline{U}.\text{div}\,{}^t\underline{\underline{\sigma}}.$$

Appendix 3

Distributors and Wrenches

A3.1. Distributors

A3.1.1. *Definition*

F is an affine space with associated vector space E of dimension $n = 3$. Chosen an origin O in F, any "point" M in F is characterized and represented by its "position" vector $\underline{OM} = \underline{x} \in E = \mathbb{R}^3$. Let us consider a vector field $\underline{U}$ defined by

$$\underline{U}(M) = \underline{U}(\underline{x}) = \underline{U}_0 + \underline{\omega}_0 \wedge \underline{OM} \qquad [A3.1]$$

with $\underline{U}_0$ and $\underline{\omega}_0$ two arbitrary vectors in $\mathbb{R}^3$. Introducing the antisymmetric tensor $\underline{\underline{\omega}}_0$ associated with $\underline{\omega}_0$ through the canonical identification formula

$$\forall \underline{v} \in \mathbb{R}^3, \ \underline{\omega}_0 \wedge \underline{v} = \underline{\underline{\omega}}_0 . \underline{v}, \qquad [A3.2]$$

namely[1]

$$\underline{\underline{\omega}}_0 = (\omega_0)_1 \underline{\underline{e}}_1 + (\omega_0)_2 \underline{\underline{e}}_2 + (\omega_0)_3 \underline{\underline{e}}_3, \qquad [A3.3]$$

equation [A3.1] can also be written as

$$\underline{U}(M) = \underline{U}(\underline{x}) = \underline{U}_0 + \underline{\underline{\omega}}_0 . \underline{OM}. \qquad [A3.4]$$

---

1 $\underline{\underline{e}}_i = -\varepsilon_{ijk} \underline{e}_j \otimes \underline{e}_k$.

We shall say that $O, \underline{U}_0, \underline{\omega}_0$ define the *Distributor* of the field $\underline{U}$ over F, denoted by

$$
\begin{cases}
\{\mathcal{D}\} = \{O, \underline{U}_0, \underline{\omega}_0\} \\
\text{or equivalently} \\
\{\mathcal{D}\} = \{O, \underline{U}_0, \underline{\underline{\omega}}_0\}.
\end{cases}
\qquad\text{[A3.5]}
$$

The vectors $\underline{U}_0$ and $\underline{\omega}_0$ are the reduced elements of the distributor $\{\mathcal{D}\}$ at point O.

Taking another reference point O', it comes out easily that the same vector field $\underline{U}$ is defined with respect to that point by the vectors $\underline{U}_{O'} = \underline{U}_0 + \underline{\omega}_0 \wedge \underline{OO'}$ and $\underline{\omega}_{O'} = \underline{\omega}_0$ substituted for $\underline{U}_0$ and $\underline{\omega}_0$ respectively. Thence, we write:

$$
\begin{cases}
\{\mathcal{D}\} = \{O, \underline{U}_0, \underline{\omega}_0\} = \{O', \underline{U}_0 + \underline{\omega}_0 \wedge \underline{OO'}, \underline{\omega}_0\} \\
\{\mathcal{D}\} = \{O, \underline{U}_0, \underline{\underline{\omega}}_0\} = \{O', \underline{U}_0 + \underline{\underline{\omega}}_0 \cdot \underline{OO'}, \underline{\underline{\omega}}_0\}
\end{cases}
\qquad\text{[A3.6]}
$$

or, with $\underline{\underline{O'O}}$ defined from $\underline{O'O}$ through [A3.2],

$$
\{\mathcal{D}\} = \{O, \underline{U}_0, \underline{\omega}_0\} = \{O', \underline{U}_0 + \underline{\underline{O'O}} \cdot \underline{\omega}_0, \underline{\omega}_0\}.
\qquad\text{[A3.7]}
$$

A3.1.2. *Rigid body motions*

Let us assume that the vector field $\underline{U}$ generated by the distributor $\{\mathcal{D}\}$ represents the velocity field of a set of geometrical points in F with respect to a given frame of reference so that: $\underline{U}(\underline{x}) = \dfrac{d\underline{x}}{dt}$. It comes out from [A3.1] that for two arbitrary points M and M', we have

$$
\underline{U}(\underline{x}') - \underline{U}(\underline{x}) = \underline{\omega}_0 \wedge \underline{MM'}.
\qquad\text{[A3.8]}
$$

which implies

$$\underline{MM'}.(\underline{U}(x')-\underline{U}(x)) = \frac{1}{2}\frac{d}{dt}(\underline{MM'})^2 = 0. \qquad [A3.9]$$

It means that in the instantaneous motion defined by the velocity field $\underline{U}$, the mutual separation of the points M and M' is conserved. This motion is called a *rigid body motion* defined over F.

A3.1.3. *Vector space of distributors*

Rigid body motions of the Euclidean space $\mathbb{R}^3$, and distributors as well, depend linearly on the vectors $\underline{U}_0$ and $\underline{\omega}_0$. They generate a six-dimensional vector space:

$$\begin{cases} \{\mathcal{D}^1\}=\{O, \underline{U}_0{}^1, \underline{\omega}_0{}^1\}, \{\mathcal{D}^2\}=\{O, \underline{U}_0{}^2, \underline{\omega}_0{}^2\} \\ \lambda\{\mathcal{D}^1\}=\{O, \lambda\underline{U}_0{}^1, \lambda\underline{\omega}_0{}^1\} \\ \{\mathcal{D}^1\}+\{\mathcal{D}^2\}=\{O, \underline{U}_0{}^1+\underline{U}_0{}^2, \underline{\omega}_0{}^1+\underline{\omega}_0{}^2\}. \end{cases} \qquad [A3.10]$$

A3.2. Wrenches

A3.2.1. *Definition*

Force modeling in mechanics makes it necessary to study the linear forms on the six-dimensional vector space of rigid body motions. The set of these linear forms is a six-dimensional vector space. Let $\mathcal{F}$ be any such linear form. Then, for any rigid body motion $\underline{U}$ defined by a distributor $\{\mathcal{D}\}=\{O, \underline{U}_0, \underline{\omega}_0\}$ through [A3.1], $\mathcal{F}(\underline{U})$ is a linear form in $\underline{U}_0$ and $\underline{\omega}_0$ which can be written as

$$\mathcal{F}(\underline{U}) = \underline{F}_0.\underline{U}_0 + \underline{C}_0.\underline{\omega}_0 \qquad [A3.11]$$

where $\underline{F}_0$ and $\underline{C}_0$ are two vectors.

These vectors define $\mathcal{F}$ with respect to O. We write

$$[\mathcal{F}]=[O, \underline{F}_0, \underline{C}_0] \qquad [A3.12]$$

and express the *duality product* in the form

$$[\mathcal{F}].\{\mathcal{D}\} = \underline{F}_0 . \underline{U}_0 + \underline{C}_0 . \underline{\omega}_0 = \{\mathcal{D}\}.[\mathcal{F}].$$ [A3.13]

The definition of $[\mathcal{F}]$ with respect to a different reference point O' proceeds from the invariance of the duality product. Thence:

$$[\mathcal{F}] = \left[O, \underline{F}_0, \underline{C}_0\right] = \left[O', \underline{F}_0, \underline{C}_0 + \underline{O'O} \wedge \underline{F}_0\right] = \left[O', \underline{F}_0, \underline{C}_0 + \underline{\underline{O'O}}.\underline{F}_0\right].$$ [A3.14]

$\underline{F}_0$ and $\underline{C}_0$ are the reduced elements of $[\mathcal{F}]$ at point O and the reduced elements at point O' are $\underline{F}_{0'} = \underline{F}_0$ and $\underline{C}_{0'} = \underline{C}_0 + \underline{O'O} \wedge \underline{F}_0 = \underline{C}_0 + \underline{\underline{O'O}}.\underline{F}_0$.

The linear form $[\mathcal{F}]$ is called a *Wrench*.

In the same way as for $\{\mathcal{D}\}$, we may introduce the antisymmetric tensor $\underline{\underline{C}}_0$ associated with $\underline{C}_0$ through [A3.2]; the duality product then takes the form

$$[\mathcal{F}].\{\mathcal{D}\} = \underline{F}_0 . \underline{U}_0 - \frac{1}{2}\underline{\underline{C}}_0 : \underline{\underline{\omega}}_0$$ [A3.15]

and

$$\underline{\underline{C}}_{0'} = \underline{\underline{C}}_0 + \underline{F}_0 \otimes \underline{O'O} - \underline{O'O} \otimes \underline{F}_0.$$ [A3.16]

A3.2.2. Wrench of a force system

As an important example of a linear form on the six-dimensional vector space of rigid body motions, we consider the case when $\mathcal{F}(\underline{U})$ is defined on a domain Ω of F as the integral of point, surface ($\Sigma \subset \Omega$) and volume (Ω) densities, which are linear functions of $\underline{U}(\underline{x})$ and $\underline{\omega}(\underline{x})$

$$
\begin{aligned}
\mathcal{F}(\underline{U}) = &\sum_i (\underline{F}_i . \underline{U}(\underline{x}_i) + \underline{C}_i . \underline{\omega}_0)) \\
&+ \int_{\Sigma} (\underline{f}_{\Sigma}(\underline{x}) . \underline{U}(\underline{x}) + \underline{c}_{\Sigma}(\underline{x}) . \underline{\omega}_0)) \, da \\
&+ \int_{\Omega} (\underline{f}_{\Omega}(\underline{x}) . \underline{U}(\underline{x}) + \underline{c}_{\Omega}(\underline{x}) . \underline{\omega}_0)) \, d\Omega
\end{aligned}
$$ [A3.17]

where $\underline{U}$ is a rigid body motion generated by an arbitrary distributor $\{\mathcal{D}\} = \{O, \underline{U}_0, \underline{\omega}_0\}$.

Identifying the corresponding wrench proceeds from [A3.13] and yields

$$\begin{cases} \underline{F}_0 = \sum_i \underline{F}_i + \int_\Sigma \underline{f}_\Sigma(\underline{x})\,da + \int_\Omega \underline{f}_\Omega(\underline{x})\,d\Omega \\[2mm] \underline{C}_0 = \sum_i \underline{c}_i + \int_\Sigma \underline{c}_\Sigma(\underline{x})\,da + \int_\Omega \underline{c}_\Omega(\underline{x})\,d\Omega \\[2mm] \quad + \sum_i \underline{OM}_i \wedge \underline{F}_i + \int_\Sigma \underline{OM} \wedge \underline{f}_\Sigma(\underline{x})\,da + \int_\Omega \underline{OM} \wedge \underline{f}_\Omega(\underline{x})\,d\Omega. \end{cases} \qquad [A3.18]$$

From a mechanical viewpoint, the point, surface and volume densities $\underline{F}_i$, $\underline{f}_\Sigma(\underline{x})\,da$ and $\underline{f}_\Omega(\underline{x})\,d\Omega$ on the one side, and $\underline{c}_i$, $\underline{c}_\Sigma(\underline{x})\,da$ and $\underline{c}_\Omega(\underline{x})\,d\Omega$ on the other side, can be interpreted as describing a system of forces and a system of couples acting on a mechanical system with volume Ω. $[\mathcal{F}]$ is then called the *Wrench of that force system*. $\underline{F}_0$ is the *resultant* of the force system and $\underline{C}_0$ the *resultant moment* with respect to O.

A3.3. Tensorial distributors and tensorial wrenches

The mechanical modeling of media with an underlying microstructure such as micropolar media, curvilinear media (beams, arcs) or plates calls for the consideration of fields of distributors and wrenches defined in the three-dimensional space or on a one- or two-dimensional subspace of this space. Provided the fields of their reduced elements are differentiable, the derivatives of such fields can be defined. It requires the introduction of the general concepts of tensorial distributors and tensorial wrenches.

A3.3.1. *Tensorial distributors*

Generally speaking, a *tensorial distributor* written $\{D\} = \{O, \underline{\underline{U}}, \underline{\underline{w}}\}$ is defined as a linear operator on $\mathbb{R}^3$ into the six-dimensional vector space of distributors:

$$\forall \underline{v} \in \mathbb{R}^3 \mapsto \{\mathbb{D}(\underline{v})\} = \left\{ O, \, \underline{U}_0 = \underline{\underline{U}}.\underline{v}, \, \underline{\omega}_0 = \underline{\underline{w}}.\underline{v} \right\} = \{\mathcal{D}\}, \qquad \text{[A3.19]}$$

also written as

$$\forall \underline{v} \in \mathbb{R}^3 \mapsto \{\mathbb{D}\}.\underline{v} = \left\{ O, \, \underline{U}_0 = \underline{\underline{U}}.\underline{v}, \, \underline{\omega}_0 = \underline{\underline{w}}.\underline{v} \right\}. \qquad \text{[A3.20]}$$

Applying [A3.7] to $\{\mathbb{D}(\underline{v})\}$, we derive

$$\{\mathbb{D}(\underline{v})\} = \left\{ O', \, (\underline{\underline{U}} + \underline{\underline{O'O}}.\underline{\underline{w}}).\underline{v}, \, \underline{\underline{w}}.\underline{v} \right\} \qquad \text{[A3.21]}$$

which proves that

$$\{\mathbb{D}\} = \left\{ O, \, \underline{\underline{U}}, \, \underline{\underline{w}} \right\} = \left\{ O', \, \underline{\underline{U}} + \underline{\underline{O'O}}.\underline{\underline{w}}, \, \underline{\underline{w}} \right\}. \qquad \text{[A3.22]}$$

A3.3.2. Tensorial wrenches

In the same way, a *tensorial wrench* $[\mathbb{T}] = \left[O, \, \underline{\underline{F}}, \, \underline{\underline{C}} \right]$ is defined as a linear operator on $\mathbb{R}^3$ into the six-dimensional vector space of wrenches:

$$\forall \underline{u} \in \mathbb{R}^3 \mapsto [\mathbb{T}(\underline{u})] = \left[O, \, \underline{\underline{F}}.\underline{u}, \, \underline{\underline{C}}.\underline{u} \right] = [\mathcal{F}] \qquad \text{[A3.23]}$$

or also

$$\forall \underline{u} \in \mathbb{R}^3 \mapsto [\mathbb{T}].\underline{u} = \left[O, \, \underline{\underline{F}}.\underline{u}, \, \underline{\underline{C}}.\underline{u} \right]. \qquad \text{[A3.24]}$$

Through [A3.14], we obtain

$$[\mathbb{T}(\underline{u})] = \left[O', \, \underline{\underline{F}}.\underline{u}, \, \underline{\underline{C}}.\underline{u} + \underline{\underline{O'O}}.\underline{\underline{F}}.\underline{u} \right] \qquad \text{[A3.25]}$$

hence:

$$[\mathbb{T}] = \left[O', \underline{\underline{F}}, \, \underline{\underline{C}} + \underline{\underline{O'O}}.\underline{\underline{F}} \right]. \qquad \text{[A3.26]}$$

A3.3.3. *Contracted products*

A3.3.3.1. *Contracted product of a tensorial distributor with a wrench*

The contracted product of the tensorial distributor $\{D\} = \{O, \underline{\underline{U}}, \underline{\underline{w}}\}$ with the wrench $[\mathcal{F}] = [O, \underline{F}_0, \underline{\underline{C}}_0]$ is the *vector* $[\mathcal{F}].\{D\} \in \mathbb{R}^3$ defined from the equation

$$\left\{\begin{array}{l} \forall \underline{v} \in \mathbb{R}^3, ([\mathcal{F}].\{D\}).\underline{v} = [\mathcal{F}].\{D(\underline{v})\} \\ \qquad = \underline{F}_0.\underline{\underline{U}}.\underline{v} + \underline{\underline{C}}_0.\underline{\underline{w}}.\underline{v} \\ \qquad = (\underline{F}_0.\underline{\underline{U}} + \underline{\underline{C}}_0.\underline{\underline{w}}).\underline{v} \end{array}\right. \qquad [\text{A3.27}]$$

hence

$$[\mathcal{F}].\{D\} = \underline{F}_0.\underline{\underline{U}} + \underline{\underline{C}}_0.\underline{\underline{w}} = {}^t\underline{\underline{U}}.\underline{F}_0 + {}^t\underline{\underline{w}}.\underline{\underline{C}}_0. \qquad [\text{A3.28}]$$

Note that defining $\{D\}.[\mathcal{F}]$ through $\forall \underline{v} \in \mathbb{R}^3, (\{D\}.[\mathcal{F}]).\underline{v} = \{D(\underline{v})\}.[\mathcal{F}]$ obviously yields the same result. Thence:

$$[\mathcal{F}].\{D\} = \{D\}.[\mathcal{F}] = \underline{F}_0.\underline{\underline{U}} + \underline{\underline{C}}_0.\underline{\underline{w}} = {}^t\underline{\underline{U}}.\underline{F}_0 + {}^t\underline{\underline{w}}.\underline{\underline{C}}_0. \qquad [\text{A3.29}]$$

A3.3.3.2. *Contracted product of a tensorial wrench with a distributor*

The contracted product of the tensorial wrench $[T] = [O, \underline{\underline{F}}, \underline{\underline{C}}]$ with the distributor $\{\mathcal{D}\} = \{O, \underline{U}_0, \underline{\omega}_0\}$ is the *vector* $[T].\{\mathcal{D}\}$ defined from the equation

$$\left\{\begin{array}{l} \forall \underline{u} \in \mathbb{R}^3, ([T].\{\mathcal{D}\}).\underline{u} = [T(\underline{u})].\{\mathcal{D}\} \\ \qquad = \underline{U}_0.\underline{\underline{F}}.\underline{u} + \underline{\omega}_0.\underline{\underline{C}}.\underline{u} \\ \qquad = (\underline{U}_0.\underline{\underline{F}} + \underline{\omega}_0.\underline{\underline{C}}).\underline{u} \end{array}\right. \qquad [\text{A3.30}]$$

hence

$$[T].\{\mathcal{D}\} = \underline{U}_0.\underline{\underline{F}} + \underline{\omega}_0.\underline{\underline{C}} = {}^t\underline{\underline{F}}.\underline{U}_0 + {}^t\underline{\underline{C}}.\underline{\omega}_0. \qquad [\text{A3.31}]$$

Here again, defining $\{\mathcal{D}\}.[T]$ through $\forall \underline{u} \in \mathbb{R}^3, (\{\mathcal{D}\}.[T]).\underline{u} = \{\mathcal{D}\}.[T(\underline{u})]$, we obtain

$$\{\mathcal{D}\}.[\mathrm{T}]=[\mathrm{T}].\{\mathcal{D}\}=\underline{U}_0.\underline{\underline{F}}+\underline{\omega}_0.\underline{\underline{C}}={}^t\underline{\underline{F}}.\underline{U}_0+{}^t\underline{\underline{C}}.\underline{\omega}_0. \qquad [\text{A3.32}]$$

A3.3.4. *Contracted product of a tensorial distributor with a tensorial wrench*

A contracted product between a tensorial distributor and tensorial wrench can be defined on the basis of the contracted product of a distributor and a wrench. From [A3.13], we derive:

$$\begin{cases} \forall \underline{u}\in\mathbb{R}^3, \forall \underline{v}\in\mathbb{R}^3, \\ [\mathrm{T}(\underline{u})].\{\mathrm{D}(\underline{v})\}=(\underline{\underline{F}}.\underline{u}).(\underline{\underline{U}}.\underline{v})+(\underline{\underline{C}}.\underline{u}).(\underline{w}.\underline{v}) \\ \qquad = \underline{u}.({}^t\underline{\underline{F}}.\underline{\underline{U}}+{}^t\underline{\underline{C}}.\underline{w}).\underline{v}, \end{cases} \qquad [\text{A3.33}]$$

where $({}^t\underline{\underline{F}}.\underline{\underline{U}}+{}^t\underline{\underline{C}}.\underline{w})$ is a bilinear form on $E\times E$.

The contracted product $\langle[\mathrm{T}]|\{\mathrm{D}\}\rangle$ is the trace of this bilinear form

$$\langle[\mathrm{T}]|\{\mathrm{D}\}\rangle=\langle\{\mathrm{D}\}|[\mathrm{T}]\rangle=\langle[0,\underline{\underline{F}},\underline{\underline{C}}]|\{0,\underline{\underline{U}},\underline{w}\}\rangle={}^t\underline{\underline{F}}:\underline{\underline{U}}+{}^t\underline{\underline{C}}:\underline{w}$$
$$={}^t\underline{\underline{U}}:\underline{\underline{F}}+{}^t\underline{w}:\underline{\underline{C}}, \qquad [\text{A3.34}]$$

which is obviously independent of the chosen point of reference, as it can be checked easily from [A3.22] and [A3.26].

A3.4. Distributor fields, wrench fields

A3.4.1. *Derivative of a distributor field*

Considering the field of distributors defined by

$$\{\mathcal{D}(\underline{x})\}=\{M,\underline{U}(\underline{x}),\underline{\omega}(\underline{x})\} \qquad [\text{A3.35}]$$

where $\underline{U}(\underline{x})$ and $\underline{\omega}(\underline{x})$, the reduced elements of $\{\mathcal{D}(\underline{x})\}$ at the field point M, are supposed to be *differentiable with respect to* $\underline{x}$, we write the fundamental equation

$$d\{\mathcal{D}(\underline{x})\} = \{\mathcal{D}(\underline{x}+\underline{dx})\} - \{\mathcal{D}(\underline{x})\}$$

$$= \left\{M+\underline{dx},\ \underline{U}(\underline{x}) + \operatorname{grad}\underline{U}(\underline{x}).\underline{dx},\ \underline{\omega}(\underline{x}) + \operatorname{grad}\omega(\underline{x}).\underline{dx}\right\}$$

$$-\{M,\ \underline{U}(\underline{x}),\ \underline{\omega}(\underline{x})\}$$

[A3.36]

which yields, through [A3.6],

$$d\{\mathcal{D}(\underline{x})\} = \left\{M,\ \operatorname{grad}\underline{U}(\underline{x}).\underline{dx} - \underline{\omega}(\underline{x}) \wedge \underline{dx},\ \operatorname{grad}\omega(\underline{x}).\underline{dx}\right\}$$

$$= \left\{M,\ \operatorname{grad}\underline{U}(\underline{x}).\underline{dx} - \underline{\underline{\omega}}(\underline{x}).\underline{dx},\ \operatorname{grad}\omega(\underline{x}).\underline{dx}\right\}.$$

[A3.37]

Referring to [A3.2], this equation identifies the *gradient of the field* $\{\mathcal{D}\}$ at the field point M as the *tensorial distributor*

$$\{\operatorname{grad}\{\mathcal{D}\}(\underline{x})\} = \left\{M,\ \operatorname{grad}\underline{U}(\underline{x}) - \underline{\underline{\omega}}(\underline{x}),\ \operatorname{grad}\omega(\underline{x})\right\}.$$

[A3.38]

Note that in the case of a distributor field defined on a curve in $\mathbb{R}^3$ as a function of arc length s (as considered in the mechanical modeling of curvilinear media), we obtain the derivative

$$\frac{d}{ds}\{\mathcal{D}(s)\} = \left\{M,\ \frac{d\underline{U}(s)}{ds} - \underline{\omega}(s) \wedge \underline{t}(s),\ \frac{d\underline{\omega}(s)}{ds}\right\}$$

$$= \left\{M,\ \frac{d\underline{U}(s)}{ds} - \underline{\underline{\omega}}(s).\underline{t}(s),\ \frac{d\underline{\omega}(s)}{ds}\right\}$$

[A3.39]

where $\underline{t}(s)$ denotes the unit tangent vector to the curve at the field point M.

It is clear from the above differentiation formulas that a distributor field being constant *does not imply*, but for some particular cases, that the fields $\underline{U}$ and $\underline{\omega}$, which define it through [A3.35], are constant.

A3.4.2. Derivative of a wrench field

Considering now the case of a wrench field $[\mathcal{F}]$ supposed to be differentiable with respect to $\underline{x}$

$$[\mathcal{F}(\underline{x})] = [M, \underline{F}(\underline{x}), \underline{C}(\underline{x})]$$

[A3.40]

we write, in the same way as in the preceding section,

$$d[\mathcal{F}(\underline{x})] = \left[M + d\underline{x}, \underline{F}(\underline{x}) + \operatorname{grad} F(\underline{x}).d\underline{x}, \underline{C}(\underline{x}) + \operatorname{grad} C(\underline{x}).d\underline{x} \right]$$
$$- [M, \underline{F}(\underline{x}), \underline{C}(\underline{x})]$$

[A3.41]

and, through [A3.14],

$$d[\mathcal{F}(\underline{x})] = \left[M, \operatorname{grad} F(\underline{x}).d\underline{x}, \operatorname{grad} C(\underline{x}).d\underline{x} + d\underline{x} \wedge \underline{F}(\underline{x}) \right]$$
$$= \left[M, \operatorname{grad} F(\underline{x}).d\underline{x}, \operatorname{grad} C(\underline{x}).d\underline{x} - \underline{\underline{F}}(\underline{x}).d\underline{x} \right]$$

[A3.42]

with

$$\underline{\underline{F}}(\underline{x}) = F_1(\underline{x})\underline{\underline{e}}_1 + F_2(\underline{x})\,\underline{\underline{e}}_2 + F_3(\underline{x})\,\underline{\underline{e}}_3$$

[A3.43]

denoting the antisymmetric tensor associated with $\underline{F}(\underline{x})$.

Thus, the *gradient of the field* $[\mathcal{F}]$ at the field point M is the *tensorial wrench*

$$\left[\operatorname{grad}[\mathcal{F}](\underline{x}) \right] = \left[M, \operatorname{grad} F(\underline{x}), \operatorname{grad} C(\underline{x}) - \underline{\underline{F}}(\underline{x}) \right]$$

[A3.44]

according to the definition given in section A3.3.2.

In the case of a wrench field defined on a curve in $\mathbb{R}^3$ as a function of arc length s as in section A3.4.1, we obtain the derivative

$$\frac{d}{ds}[\mathcal{F}(s)] = \left[M, \frac{dF(s)}{ds}, \frac{dC(s)}{ds} - \underline{F}(s) \wedge \underline{t}(s) \right]$$
$$= \left[M, \frac{d\underline{F}(s)}{ds}, \frac{dC(s)}{ds} - \underline{\underline{F}}(s).\underline{t}(s) \right].$$

[A3.45]

In the same way as for a distributor field, a wrench field being constant *does not necessarily imply* that the corresponding $\underline{F}$ and $\underline{C}$ fields in [A3.40] are constant.

A3.4.3. *Derivative of the duality product between a distributor and a wrench*

$[\mathcal{F}]$ and $\{\mathcal{D}\}$ being a wrench field and a distributor field, the duality product [A3.13] of $[\mathcal{F}(\underline{x})]$ and $\{\mathcal{D}(\underline{x})\}$ at the field point M is written as

$$[\mathcal{F}(\underline{x})].\{\mathcal{D}(\underline{x})\} = \underline{F}(\underline{x}).\underline{U}(\underline{x}) + \underline{C}(\underline{x}).\underline{\omega}(\underline{x}). \qquad [\text{A3.46}]$$

Its differential is

$$d\big([\mathcal{F}(\underline{x})].\{\mathcal{D}(\underline{x})\}\big) = \underline{U}(\underline{x}).(\mathrm{grad}\,F(\underline{x}).d\underline{x}) + \underline{F}(\underline{x}).(\mathrm{grad}\,U(\underline{x}).d\underline{x})$$
$$+ \underline{\omega}(\underline{x}).(\mathrm{grad}\,C(\underline{x}).d\underline{x}) + \underline{C}(\underline{x}).(\mathrm{grad}\,\omega(\underline{x}).d\underline{x}), \qquad [\text{A3.47}]$$

which defines the gradient

$$\mathrm{grad}\big([\mathcal{F}].\{\mathcal{D}\}\big)(\underline{x}) = \underline{U}(\underline{x}).\mathrm{grad}\,F(\underline{x}) + \underline{F}(\underline{x}).\mathrm{grad}\,U(\underline{x})$$
$$+ \underline{\omega}(\underline{x}).\mathrm{grad}\,C(\underline{x}) + \underline{C}(\underline{x}).\mathrm{grad}\,\omega(\underline{x}) \qquad [\text{A3.48}]$$

or, equivalently,

$$\mathrm{grad}\big([\mathcal{F}].\{\mathcal{D}\}\big)(\underline{x}) = {}^t\mathrm{grad}\,F(\underline{x}).\underline{U}(\underline{x}) + \underline{F}(\underline{x}).\mathrm{grad}\,U(\underline{x})$$
$$+ {}^t\mathrm{grad}\,C(\underline{x}).\underline{\omega}(\underline{x}) + \underline{C}(\underline{x}).\mathrm{grad}\,\omega(\underline{x}). \qquad [\text{A3.49}]$$

With [A3.38] and [A3.44], this expression can be considered as the result of the *formal differentiation* of the scalar field $[\mathcal{F}].\{\mathcal{D}\}$ in the form:

$$\mathrm{grad}\big([\mathcal{F}].\{\mathcal{D}\}\big)(\underline{x}) = \{\mathcal{D}(\underline{x})\}.\big[\mathrm{grad}[\mathcal{F}](\underline{x})\big] + [\mathcal{F}(\underline{x})].\{\mathrm{grad}\{\mathcal{D}\}(\underline{x})\}. \qquad [\text{A3.50}]$$

When the wrench and distributor fields are defined on a curve in $\mathbb{R}^3$ as a function of arc length s along the curve, we obtain the derivative:

$$\frac{d}{ds}\big([\mathcal{F}].\{\mathcal{D}\}\big) = \{\mathcal{D}(s)\}.\frac{d}{ds}[\mathcal{F}(s)] + [\mathcal{F}(s)].\frac{d}{ds}\{\mathcal{D}(s)\}. \qquad [\text{A3.51}]$$

A3.4.4. *Divergence of the contracted product of a tensorial wrench with a distributor*

Let $[T]$ and $\{\mathcal{D}\}$ be a tensorial *wrench field* and a *distributor field* defined by $[T(\underline{x})] = \left[M, \underline{\underline{F}}(\underline{x}), \underline{\underline{C}}(\underline{x}) \right]$ and $\{\mathcal{D}(\underline{x})\} = \{M, \underline{U}(\underline{x}), \underline{\omega}(\underline{x})\} = \{M, \underline{U}(\underline{x}), \underline{\omega}(\underline{x})\}$. Through [A3.31], the contracted product $[T(\underline{x})].\{\mathcal{D}(\underline{x})\}$ defines the vector field

$$[T(\underline{x})].\{\mathcal{D}(\underline{x})\} = \underline{U}(\underline{x}).\underline{\underline{F}}(\underline{x}) + \underline{\omega}(\underline{x}).\underline{\underline{C}}(\underline{x}) = {}^{t}\underline{\underline{F}}(\underline{x}).\underline{U}(\underline{x}) + {}^{t}\underline{\underline{C}}(\underline{x}).\underline{\omega}(\underline{x}) \qquad \text{[A3.52]}$$

whose gradient field is the *second rank tensor field*

$$\underline{\underline{\text{grad}}}\left([T(\underline{x})].\{\mathcal{D}(\underline{x})\} \right) = {}^{t}\underline{\underline{F}}(\underline{x}).\underline{\underline{\text{grad}}}\, \underline{U}(\underline{x}) + {}^{t}\underline{\underline{C}}(\underline{x}).\underline{\underline{\text{grad}}}\, \underline{\omega}(\underline{x})$$
$$+ \underline{U}(\underline{x}).\underline{\underline{\text{grad}}}\, \underline{\underline{F}}(\underline{x}) + \underline{\omega}(\underline{x}).\underline{\underline{\text{grad}}}\, \underline{\underline{C}}(\underline{x}). \qquad \text{[A3.53]}$$

Taking the trace of [A3.53], we derive the corresponding divergence field in the form of the scalar field

$$\text{div}\left([T(\underline{x})].\{\mathcal{D}(\underline{x})\} \right) = {}^{t}\underline{\underline{F}}(\underline{x}):\underline{\underline{\text{grad}}}\, \underline{U}(\underline{x}) + {}^{t}\underline{\underline{C}}(\underline{x}):\underline{\underline{\text{grad}}}\, \underline{\omega}(\underline{x})$$
$$+ \underline{U}(\underline{x}).\text{div}\, \underline{\underline{F}}(\underline{x}) + \underline{\omega}(\underline{x}).\text{div}\, \underline{\underline{C}}(\underline{x}). \qquad \text{[A3.54]}$$

The right-hand term of the first line in this equation can be transformed using definition [A3.34] of the contracted product of $[T(\underline{x})]$ with $\{\underline{\underline{\text{grad}}}\{\mathcal{D}\}(\underline{x})\}$ and becomes

$$\text{div}\left([T(\underline{x})].\{\mathcal{D}(\underline{x})\} \right) = \left\langle [T(\underline{x})] \middle| \{\underline{\underline{\text{grad}}}\{\mathcal{D}\}(\underline{x})\} \right\rangle$$
$$+ \underline{\omega}(\underline{x}):{}^{t}\underline{\underline{F}}(\underline{x}) + \underline{U}(\underline{x}).\text{div}\, \underline{\underline{F}}(\underline{x}) + \underline{\omega}(\underline{x}).\text{div}\, \underline{\underline{C}}(\underline{x}). \qquad \text{[A3.55]}$$

Introducing $\underline{\underline{F}}_{a}(\underline{x})$, the antisymmetric part of $\underline{\underline{F}}(\underline{x})$, and $\underline{F}_{a}(\underline{x})$ its associated vector, the term $\underline{\omega}(\underline{x}):{}^{t}\underline{\underline{F}}(\underline{x})$ in this equation can be transformed into

$$\underline{\omega}(\underline{x}):{}^{t}\underline{\underline{F}}(\underline{x}) = -\underline{\omega}(\underline{x}):\underline{\underline{F}}_{a}(\underline{x}) = 2\,\underline{\omega}(\underline{x}).\underline{F}_{a}(\underline{x}) \qquad \text{[A3.56]}$$

and we know, from Appendix 1 (sections A1.8.6 and A1.8.8), that

$$\begin{cases} 2\,\underline{F}_a(\underline{x}) = ({}^t\underline{\underline{F}}(\underline{x})\!:\!\underline{e}_i)\,\underline{e}_i \\ \text{with } \underline{\underline{e}}_k = -\varepsilon_{ijk}\,\underline{e}_i \otimes \underline{e}_j . \end{cases} \qquad \text{[A3.57]}$$

Thence, [A3.55] may be written as

$$\mathrm{div}\big([\mathrm{T}(\underline{x})].\{\mathcal{D}(\underline{x})\}\big) = \big\langle [\mathrm{T}(\underline{x})]\,\big|\,\{\mathrm{grad}\{\mathcal{D}\}(\underline{x})\}\big\rangle$$
$$+\underline{\omega}(\underline{x}).({}^t\underline{\underline{F}}(\underline{x})\!:\!\underline{e}_i)\,\underline{e}_i + \underline{U}(\underline{x}).\mathrm{div}\,\underline{\underline{F}}(\underline{x}) \qquad \text{[A3.58]}$$
$$+\underline{\omega}(\underline{x}).\mathrm{div}\,\underline{\underline{C}}(\underline{x}),$$

or, equivalently,

$$\mathrm{div}\big([\mathrm{T}(\underline{x})].\{\mathcal{D}(\underline{x})\}\big) = \big\langle [\mathrm{T}(\underline{x})]\,\big|\,\{\mathrm{grad}\{\mathcal{D}\}(\underline{x})\}\big\rangle$$
$$+\Big[\,M,\ \mathrm{div}\,\underline{\underline{F}}(\underline{x}),\ \mathrm{div}\,\underline{\underline{C}}(\underline{x}) + ({}^t\underline{\underline{F}}(\underline{x})\!:\!\underline{e}_i)\,\underline{e}_i\,\Big].\{\mathcal{D}(\underline{x})\}. \qquad \text{[A3.59]}$$

From this equation, we can interpret $\Big[\,M,\ \mathrm{div}\,\underline{\underline{F}}(\underline{x}),\ \mathrm{div}\,\underline{\underline{C}}(\underline{x}) + ({}^t\underline{\underline{F}}(\underline{x})\!:\!\underline{e}_i)\,\underline{e}_i\,\Big]$ as the divergence of the tensorial wrench field $[\mathrm{T}]$

$$\mathrm{div}\big[\mathrm{T}(\underline{x})\big] = \Big[\,M,\ \mathrm{div}\,\underline{\underline{F}}(\underline{x}),\ \mathrm{div}\,\underline{\underline{C}}(\underline{x}) + ({}^t\underline{\underline{F}}(\underline{x})\!:\!\underline{e}_i)\,\underline{e}_i\,\Big] \qquad \text{[A3.60]}$$

and write

$$\mathrm{div}\big([\mathrm{T}(\underline{x})].\{\mathcal{D}(\underline{x})\}\big) = \big\langle [\mathrm{T}(\underline{x})]\,\big|\,\{\mathrm{grad}\{\mathcal{D}\}(\underline{x})\}\big\rangle + \{\mathcal{D}(\underline{x})\}.\mathrm{div}\big[\mathrm{T}(\underline{x})\big]. \qquad \text{[A3.61]}$$

This equation is the counterpart of the equation $\mathrm{div}(\underline{\underline{\sigma}}.\underline{U}) = \underline{\underline{\sigma}}\!:\!\mathrm{grad}\,\underline{U} + \underline{U}.\mathrm{div}\,{}^t\underline{\underline{\sigma}}$ obtained in Appendix 2 (section A2.9), taking into account the presence of the transpose signs in [A3.34].

Bibliography

[ALE 43] ALEMBERT J. (LE ROND D'), *Traité de dynamique*, David l'aîné, Paris, 1743.

[ARC] ARCHIMEDES, *De Planorum Æquilibriis* (A Treatise on the Equilibrium of Planes or Their Centres of Gravity).

[ARI 24] ARISTOTLE, *Metaphysics*, Book I, translated by W.D. Ross, available at: http://izt.ciens.ucv.ve/ecologia/Archivos/Filosofia-I/Aristotle%20-%20Metaphysics.pdf, 1924.

[ARI 36] ARISTOTLE, *Quaestionae Mechanicae* (*Mechanical Problems*), Loeb Classical Library Edition, available at: http://penelope.uchicago.edu/Thayer/E/Roman/Texts /Aristotle/Mechanica*.html, 1936.

[ARI 62] ARIS R., *Vectors, Tensors and the Basic Equations of Fluid Mechanics*, Dover, New York, 1962.

[ARI 09] ARISTOTLE, *Physicae Auscultationes* (*Lectures on Nature*), translated by R.P. Hardie and R.K. Gaye, available at: http://classics.mit.edu/Aristotle/physics.7.vii.html, 2009.

[BAR 55] BARRE DE SAINT-VENANT A., *De la torsion des prismes avec des considérations sur leur flexion ainsi que sur l'équilibre des solides élastiques en général...*, Imprimerie impériale, Paris, 1855.

[BAR 64] BARRE DE SAINT-VENANT A., "Établissement élémentaire des formules et équations générales de la théorie des corps solides", in NAVIER C.-L.-M.-H. (ed.), *De la résistance des corps solides*, 3rd ed., Paris, 1864.

[BEL 86] BELTRAMI E., "Sull'interpretazione meccanica delle formole di Maxwell", *Memorie della Reale Accademia delle Scienze dell'Istituto di Bologna*, Series IV, vol. VII, pp. 1–38, 1886.

[BEL 89] BELTRAMI E., "Sur la théorie de la déformation infiniment petite d'un milieu", *Comptes rendus de l'Académie des Sciences Paris*, vol. 108, pp. 344–347, 1889.

[BEN 81] BENVENUTO E., *La Scienza delle Costruzioni e il suo Sviluppo Storico*, Sansoni, Florence, 1981.

[BEN 91] BENVENUTO E., *An Introduction to the History of Structural Mechanics. Part I, Statics and Resistance of Solids*, Springer-Verlag, New York, 1991.

[BIS 53] BISHOP J.F.W., "On the complete solution to problems of deformations of a plastic-rigid material", *Journal of the Mechanics and Physics of Solids*, vol. 2, no. 1, pp. 43–53, 1953.

[BON 72] BONNET, MORIO G., Travail de fin d'études, LCPC, Paris, 1972.

[BRA 07] BRAESTRUP M.W., "Yield line theory and concrete plasticity", *Proceedings of Morley Symposium on Concrete Plasticity and its Application*, University of Cambridge, pp. 43–48, 23 July 2007.

[CAP 12] CAPECCHI D., *History of Virtual Work Laws: A History of Mechanics Perspective*, Springer-Verlag, Italy, 2012.

[CAU 23] CAUCHY A., "Recherches sur l'équilibre et le mouvement des corps solides ou fluides, élastiques ou non élastiques", *Bulletin de la Société philomathique*, Paris, pp. 9–13, 1823.

[CAU 27a] CAUCHY A., "Addition à l'article précédent", *Exercices de Mathématiques*, de Bure, Paris, 1827.

[CAU 27b] CAUCHY A., "De la pression ou tension dans un corps solide", *Exercices de Mathématiques*, de Bure, Paris, 1827.

[CAU 27c] CAUCHY A., "Sur les relations qui existent dans l'état d'équilibre d'un corps solide ou fluide entre les pressions ou tensions et les forces accélératrices", *Exercices de Mathématiques*, de Bure, Paris, 1827.

[CAU 29] CAUCHY A., "Sur les pressions ou tensions supportées en un point donné d'un corps solide par trois plans perpendiculaires entre eux", *Exercices de Mathématiques*, de Bure, Paris, 1829.

[CAU 30] CAUCHY A., "Sur les diverses méthodes à l'aide desquelles on peut établir les équations qui représentent les lois d'équilibre ou le mouvement intérieur des corps solides ou fluides", *Bulletin de Férussac*, vol. XIII, pp. 169–176, 1830.

[CHA 59] CHARNES A., LEMKE C.E., ZIENKIEWICZ O.C., "Virtual work, linear programming and plastic limit analysis", *Proceedings of the Royal Society A*, vol. 251, pp. 110–116, 1959.

[CHA 99] CHADWICK P., *Continuum Mechanics: Concise Theory and Problems*, Dover, New York, 1999.

[CHU 96] CHUNG T.J., *Applied Continuum Mechanics*, Cambridge University Press, Cambridge, 1996.

[COH 99] COHEN I.B., WHITMAN A. (trans.), *The Principia*, University of California Press, Berkeley, 1999.

[COU 65] COURBON J., *Résistance des matériaux*, vol. 2, Dunod, Paris, 1965.

[DAU 90/00] DAUTRAY R., LIONS J.-L., *Mathematical Analysis and Numerical Methods for Science and Technology*, vols 1–6, Springer-Verlag, 1990-2000.

[DEC 15] DE CAUS S., *Les Raisons des forces mouvantes, avec diverses machines tant utiles que plaisantes*, Jan Norton, Frankfurt, 1615.

[DEG 08] DE GROOT J., "Dunamis and the science of mechanics: Aristotle on animal motion", *Journal of the History of Philosophy*, vol. 46, no. 1, pp. 43–68, 2008.

[DEL 09] DEL PIERO G., "On the method of virtual power in continuum mechanics", *Journal of Mechanics of Materials and Structures*, vol. 4, pp. 281–292, 2009.

[DES 68] DESCARTES R., *Œuvres de Descartes. Correspondance.1* (April 1622–February 1638); *Correspondance.2* (March 1638–December 1639); *Correspondance.4* (July 1643–April 1647), (Adam–Tannery, Amsterdam, 1969; J. Vrin, Paris, 1988), 1668.

[DES 86] DESRUES J., BRAULT G., "Strain localisation in a biaxial test", available at: https://www.researchgate.net/publication/321623908, 1986.

[DRU 54] DRUCKER D.C., HOPKINS H.G., "Combined concentrated and distributed load on ideally plastic circular plates", *Proceedings of the 2nd U.S. National Congress of Applied Mechanics*, pp. 517–520, A.S.M.E., Ann Arbor, 1954.

[DUG 50] DUGAS R., *Histoire de la Mécanique*, Éditions du Griffon, Neuchatel, 1950.

[DUH 05] DUHEM P.M.M., *Les origines de la Statique*, vol. 1, Hermann, Paris, 1905.

[DUH 06] DUHEM P.M.M., *Les origines de la Statique*, vol. 2, Hermann, Paris, 1906.

[ERI 67] ERINGEN A.C., *Mechanics of Continua*, John Wiley & Sons, New York, 1967.

[EUG 14] EUGSTER S.R., On the Foundations of Contiunuum Mechanics and its Applications to Beam Theories, Doctoral thesis, ETH, Zurich, 2014.

[EUG 15] EUGSTER S.R., *Geometric Continuum Mechanics and Induced Beam Theories*, Springer, New York, 2015.

[FOS 96] FOSSOMBRONI V., *Memoria sul principio delle velocità virtuali*, Florence, 1796.

[FOU 98] FOURIER J., "Mémoire sur la statique contenant la démonstration du principe des Vitesses virtuelles, et la théorie des Moments", *Journal de l'école polytechnique, V^e cahier, Tome II, prairial an VI*, pp. 20–60, Imprimerie de la République, Paris, 1798.

[FUN 94] FUNG Y.C., *A First Course in Continuum Mechanics*, 3rd ed., Prentice-Hall, Englewood Cliffs, 1994.

[GAL 99] GALILEO GALILEI, *Le Mecaniche. Della vite*, available at: http://it.wikisource.org/wiki/Le_mecaniche/Della_vite, 1599.

[GAL 34] GALILEO GALILEI, *Les Méchaniques* (translation by Mersenne of Della Scienza Meccanica, Ravenna (1649)), 1634.

[GAL 38] GALILEO GALILEI, *Discorsi e dimostrazioni matematiche intorno a duo nuove scienze attenenti alla Meccanica, ed ai movimenti locali (Dialogues Concerning Two New Sciences*, translation by Crew H. and de Salvio A., Dover, New York, 1954), 1638.

[GER 72] GERMAIN P., "Sur l'application du principe des puissances virtuelles en mécanique des milieux continus", *Comptes rendus de l'Académie des Sciences*, Series A, vol. 274, pp. 1051–1055, 1972.

[GER 73a] GERMAIN P., "La méthode des puissances virtuelles en mécanique des milieux continus. 1ère partie: Théorie du second gradient", *Journal de Mécanique*, vol. 12, pp. 235–274, 1973.

[GER 73b] GERMAIN P., "The method of virtual power in continuum mechanics. Part 2: Microstructure", *SIAM Journal on Applied Mathematics*, vol. 25, pp. 556–575, 1973.

[GER 86] GERMAIN P., *Mécanique*, Ellipses, Paris, 1986.

[GUR 81] GURTIN M.E., *An Introduction to Continuum Mechanics*, Academic Press, New York, 1981.

[HAB 84] HABIB P., "Les surfaces de glissement en mécanique des sols", *Revue française de géotechnique*, vol. 27, pp. 7–21, 1984.

[HEY 66] HEYMAN J., "The stone skeleton", *International Journal of Solids and Structures*, vol. 2, no. 2, pp. 249–279, 1966.

[HEY 98] HEYMAN J., *Structural Analysis*, 3rd ed., Prentice-Hall, Upper Saddle River, 1998.

[HJE 97] HJELMSTAD K.D., *Fundamentals of Structural Mechanics*, Prentice-Hall, Upper Saddle River, 1997.

[HOP 53] HOPKINS H.G., PRAGER W., "The load-carrying capacity of circular plates", *Journal of the Mechanics and Physics of Solids*, vol. 2, no. 1, pp. 1–13, 1953.

[HOP 54] HOPKINS H.G., WANG A.J., "Load-carrying capacities for circular plates of perfectly-plastic material with arbitrary yield conditions", *Journal of the Mechanics and Physics of Solids*, vol. 3, pp. 117–129, 1954.

[JOH 31] JOHANSEN K.W., "Beregning af krydsarmerede jernbetonpladers brudmoment", *Bygningsstatiske Meddelelser*, vol. 3, no. 1, pp. 1–18, 1931.

[JOH 52] JOHANSEN K.W., *Brudlinieteorier*, Gjellerup, Copenhagen (English translation: *Yield-Line Theory*, Cement and Concrete Association, London, 1962), 1952.

[KEM 65] KEMP K.O., "The yield criterion for orthotropically reinforced concrete slabs", *International Journal of Mechanical Sciences*, vol. 7, no. 11, pp. 737–746, 1965.

[KIR 50] KIRCHHOFF G., "Über das Gleichgewicht and die Bewegung einer elastischen Scheibe", *Journal für reine und angewandte Mathematik*, vol. 40, pp. 51–88, 1850.

[KLI 63] KLINE S.J., *Flow visualization*, available at: https://searchworks.stanford.edu/view /7861399, 1963.

[LAG 88a] LAGRANGE J.-L., *Méchanique Analitique*, Courcier, Paris, 1788.

[LAG 97] LAGRANGE J.-L., "Sur le principe des vitesses virtuelles", *Journal de l'école polytechnique, V^e cahier, vol. II, prairial an VI*, pp. 115–118, Imprimerie de la République, Paris, 1797.

[LAG 88b] LAGRANGE J.-L., *Œuvres complètes*, vol. 11, Gauthier-Villars, Paris, 1888.

[LEO 87/08] LEONARDO DA VINCI, *Les Manuscrits de Léonard de Vinci*, Manuscripts A–M, Bibliothèque de l'Institut de France, Paris, 1487–1508.

[LOV 88] LOVE A.E.H., "On the small free vibrations and deformations of elastic shells", *Philosophical Transactions of the Royal Society A*, vol. 17, pp. 491–549, 1888.

[LOV 44] LOVE A.E.H., *A Treatise on the Mathematical Theory of Elasticity*, Dover, New York, 1944.

[LUB 90] LUBLINER J., *Plasticity Theory*, MacMillan Publ. Company, New York, 1990.

[MAL 65] MALVERN L.E., *Introduction to the Mechanics of a Continuous Medium*, Prentice-Hall, Englewood Cliffs, 1965.

[MAN 66] MANDEL J., *Cours de mécanique des milieux continus*, Gauthier-Villars, Paris, 1966.

[MAN 74] MANDEL J., *Introduction à la mécanique des milieux continus déformables*, Éditions scientifiques de Pologne, Warsaw, 1974.

[MAS 63] MASSONNET C.E., SAVE M., *Calcul plastique des constructions, II, Structures spatiales*, CBLIA, Brussels, 1963.

[MAS 67] MASSONNET C.E., "Complete solutions describing the limit state of reinforced concrete slabs", *Magazine of Concrete Research*, vol. 19, no. 58, pp. 13–32, 1967.

[MAS 70] MASE G.E., *Theory and Problems of Continuum Mechanics*, Schaum's Outline Series, McGraw-Hill, New York, 1970.

[MIN 51] MINDLIN R.D., "Influence of rotatory inertia and shear on flexural motions of isotropic, elastic plates", *Journal of Applied Mechanics*, vol. 18, pp. 31–38, 1951.

[MOR 66] MORLEY C.T., "On the yield criterion of an orthogonally reinforced concrete slab element", *Journal of the Mechanics and Physics of Solids*, vol. 14, no. 1, pp. 33–47, 1966.

[MOT 29] MOTTE A. (trans.), *The Principia*, available at: http://www.archive.org/stream /newtonspmathema00newtrich#page/n7/mode/2up, 1729.

[NEW 87] NEWTON I., *Philosophiae naturalis principia mathematica*, available at: http://www.gutenberg.org/files/28233/28233-pdf.pdf, 1687.

[NIE 64] NIELSEN M.P., "Limit analysis of reinforced concrete slabs", *Acta Polytechnica Scandinavia Ci 26*, Copenhagen, 1964.

[NOL 59] NOLL W., "The foundations of classical mechanics in the light of recent advances in continuum mechanics", *The Axiomatic Method, with Special Reference to Geometry and Physics*, North-Holland Publishing, Amsterdam, 1959.

[PEL 87] PELTZER G., Contribution à l'étude de la collision Inde-Asie, PhD thesis, University Paris 7, 1987.

[PRA 61] PRAGER W., *Introduction to Mechanics of Continua*, Gin & Co, New York, 1961.

[REI 45] REISSNER E., "The effect of transverse shear deformations on the bending of elastic plates", *Journal of Applied Mechanics*, vol. 12, pp. A69–A76, 1945.

[SAL 01] SALENÇON J., *Handbook of Continuum Mechanics*, Springer-Verlag, 2001.

[SAL 02] SALENÇON J., *De l'Élasto-plasticité au Calcul à la rupture*, Éditions de l'École polytechnique, Palaiseau, France, 2002.

[SAL 13] SALENÇON J., *Yield Design*, ISTE Ltd, London and John Wiley & Sons, New York, 2013.

[SAL 16] SALENÇON J., *Mécanique des milieux continus*, vol. III, Éditions de l'École polytechnique, Palaiseau, 2016.

[SAV 73] SAVE M., MASSONNET C.E., *Calcul plastique des constructions*, 2nd ed., CBLIA, Brussels, 1973.

[SAV 85] SAVE M., PRAGER W., *Structural Optimization, vol. 1: Optimality Criteria*, Plenum Press, 1985.

[SAV 95] SAVE M., *Atlas of Limit Loads of Metal Plates, Shells and Disks*, Elsevier, Amsterdam, 1995.

[SAV 98] SAVE M.A., MASSONNET C.E., DE SAXCE G., *Plastic Limit Analysis of Plates, Shells and Disks*, Elsevier, Amsterdam, 1998.

[SAW 63] SAWCZUK A., JAEGER T., *Grenztragfähigkeits-Theorie der Platten*, Springer, 1963.

[SEG 87] SEGEL L.E., *Mathematics Applied to Continuum Mechanics*, Dover, New York, 1987.

[SPE 04] SPENCER A.J.M., *Continuum Mechanics*, Dover, New York, 2004.

[STE 86] STEVIN S., *Beghinselen der Weeghconst*, Leyden (E.J. DISKSTERHUIS (ed.), *The Principal Works of Simon Stevin, vol. I, General Introduction, Mechanics*, C.V. Swets & Zeitlinger, Amsterdam, 1955), 1586.

[STE 05/08] STEVIN S., *Hypomnemata mathematica (Additamenti Staticæ pars secunda: de Trochleostatica)*, Lugodini Batavorum, ex officina Ioannis Patii, academiæ Typographi, 1605-1608.

[TIM 59] TIMOSHENKO S., WOINOWSKY-KRIEGER S., *Theory of Plates and Shells*, McGraw-Hill, New York, 1959.

[TRU 66] TRUESDELL C., *The Elements of Continuum Mechanics*, Springer, 1966.

[TRU 68] TRUESDELL C., *Essays in the History of Mechanics*, Springer, 1968.

[VAR 25] VARIGNON P., *Nouvelle Mécanique ou Statique*, vol. II, Claude Jombert, Paris, 1725.

[WAG 97] WAGONER R.H., CHENOT J.L., *Fundamentals of Metal Forming*, John Wiley & Sons, New York, 1997.

[WAL 70] WALLIS J., *Mechanica: sive, De Motu, Tractatus Geometricus, Pars Prima*, 1670.

Index

Other titles from

in

Mechanical Engineering and Solid Mechanics

2017

BOREL Michel, VÉNIZÉLOS Georges
Movement Equations 2: Mathematical and Methodological Supplements
(Non-deformable Solid Mechanics Set – Volume 2)

Movement Equations 3: Dynamics and Fundamental Principle
(Non-deformable Solid Mechanics Set – Volume 3)

BOUVET Christophe
Mechanics of Aeronautical Solids, Materials and Structures

BRANCHERIE Delphine, FEISSEL Pierre, BOUVIER Salima,
IBRAHIMBEGOVIC Adnan
From Microstructure Investigations to Multiscale Modeling:
Bridging the Gap

CHEBEL-MORELLO Brigitte, NICOD Jean-Marc, VARNIER Christophe
From Prognostics and Health Systems Management to Predictive
Maintenance 2: Knowledge, Traceability and Decision
(Reliability of Multiphysical Systems Set – Volume 7)

EL HAMI Abdelkhalak, DELAUX David, GRZESKOWIAK Henry
Reliability of High-Power Mechatronic Systems 1
Reliability of High-Power Mechatronic Systems 2

EL HAMI Abdelkhalak, RADI Bouchaib
Dynamics of Large Structures and Inverse Problems
(Mathematical and Mechanical Engineering Set – Volume 5)

Fluid-Structure Interactions and Uncertainties: Ansys and Fluent Tools
(Reliability of Multiphysical Systems Set – Volume 6)

KHARMANDA Ghias, EL HAMI Abdelkhalak
Biomechanics: Optimization, Uncertainties and Reliability
(Reliability of Multiphysical Systems Set – Volume 5)

LEDOUX Michel, EL HAMI Abdelkhalak
Compressible Flow Propulsion and Digital Approaches in Fluid Mechanics
(Mathematical and Mechanical Engineering Set – Volume 4)

Fluid Mechanics: Analytical Methods
(Mathematical and Mechanical Engineering Set – Volume 3)

MORI Yvon
Mechanical Vibrations: Applications to Equipment

2016

BOREL Michel, VÉNIZÉLOS Georges
Movement Equations 1: Location, Kinematics and Kinetics
(Non-deformable Solid Mechanics Set – Volume 1)

BOYARD Nicolas
Heat Transfer in Polymer Composite Materials

CARDON Alain, ITMI Mhamed
New Autonomous Systems
(Reliability of Multiphysical Systems Set – Volume 1)

DAHOO Pierre Richard, POUGNET Philippe, EL HAMI Abdelkhalak
Nanometer-scale Defect Detection Using Polarized Light
(Reliability of Multiphysical Systems Set – Volume 2)

DE SAXCÉ Géry, VALLÉE Claude
Galilean Mechanics and Thermodynamics of Continua

DORMIEUX Luc, KONDO Djimédo
Micromechanics of Fracture and Damage
(Micromechanics Set – Volume 1)

EL HAMI Abdelkhalak, RADI Bouchaib
Stochastic Dynamics of Structures
(Mathematical and Mechanical Engineering Set – Volume 2)

GOURIVEAU Rafael, MEDJAHER Kamal, ZERHOUNI Noureddine
From Prognostics and Health Systems Management to Predictive
Maintenance 1: Monitoring and Prognostics
(Reliability of Multiphysical Systems Set – Volume 4)

KHARMANDA Ghias, EL HAMI Abdelkhalak
Reliability in Biomechanics
(Reliability of Multiphysical Systems Set –Volume 3)

MOLIMARD Jérôme
Experimental Mechanics of Solids and Structures

RADI Bouchaib, EL HAMI Abdelkhalak
Material Forming Processes: Simulation, Drawing, Hydroforming and
Additive Manufacturing
(Mathematical and Mechanical Engineering Set – Volume 1)

2015

KARLIČIĆ Danilo, MURMU Tony, ADHIKARI Sondipon, MCCARTHY Michael
Non-local Structural Mechanics

SAB Karam, LEBÉE Arthur
Homogenization of Heterogeneous Thin and Thick Plates

2014

ATANACKOVIC M. Teodor, PILIPOVIC Stevan, STANKOVIC Bogoljub, ZORICA Dusan
Fractional Calculus with Applications in Mechanics: Vibrations and Diffusion Processes

ATANACKOVIC M. Teodor, PILIPOVIC Stevan, STANKOVIC Bogoljub, ZORICA Dusan
Fractional Calculus with Applications in Mechanics: Wave Propagation, Impact and Variational Principles

CIBLAC Thierry, MOREL Jean-Claude
Sustainable Masonry: Stability and Behavior of Structures

ILANKO Sinniah, MONTERRUBIO Luis E., MOCHIDA Yusuke
The Rayleigh–Ritz Method for Structural Analysis

LALANNE Christian
Mechanical Vibration and Shock Analysis – 5-volume series – 3rd edition
Sinusoidal Vibration – Volume 1
Mechanical Shock – Volume 2
Random Vibration – Volume 3
Fatigue Damage – Volume 4
Specification Development – Volume 5

LEMAIRE Maurice
Uncertainty and Mechanics

2013

ADHIKARI Sondipon
Structural Dynamic Analysis with Generalized Damping Models: Analysis

ADHIKARI Sondipon
Structural Dynamic Analysis with Generalized Damping Models: Identification

BAILLY Patrice
Materials and Structures under Shock and Impact

BASTIEN Jérôme, BERNARDIN Frédéric, LAMARQUE Claude-Henri
*Non-smooth Deterministic or Stochastic Discrete Dynamical Systems:
Applications to Models with Friction or Impact*

EL HAMI Abdelkhalak, BOUCHAIB Radi
Uncertainty and Optimization in Structural Mechanics

KIRILLOV Oleg N., PELINOVSKY Dmitry E.
Nonlinear Physical Systems: Spectral Analysis, Stability and Bifurcations

LUONGO Angelo, ZULLI Daniele
Mathematical Models of Beams and Cables

SALENÇON Jean
Yield Design

2012

DAVIM J. Paulo
Mechanical Engineering Education

DUPEUX Michel, BRACCINI Muriel
Mechanics of Solid Interfaces

ELISHAKOFF Isaac *et al.*
*Carbon Nanotubes and Nanosensors: Vibration, Buckling and Ballistic
Impact*

GRÉDIAC Michel, HILD François
Full-Field Measurements and Identification in Solid Mechanics

GROUS Ammar
*Fracture Mechanics – 3-volume series
Analysis of Reliability and Quality Control – Volume 1
Applied Reliability – Volume 2
Applied Quality Control – Volume 3*

RECHO Naman
Fracture Mechanics and Crack Growth

2011

KRYSINSKI Tomasz, MALBURET François
Mechanical Instability

SOUSTELLE Michel
An Introduction to Chemical Kinetics

2010

BREITKOPF Piotr, FILOMENO COELHO Rajan
Multidisciplinary Design Optimization in Computational Mechanics

DAVIM J. Paulo
Biotribolgy

PAULTRE Patrick
Dynamics of Structures

SOUSTELLE Michel
Handbook of Heterogenous Kinetics

2009

BERLIOZ Alain, TROMPETTE Philippe
Solid Mechanics using the Finite Element Method

LEMAIRE Maurice
Structural Reliability

2007

GIRARD Alain, ROY Nicolas
Structural Dynamics in Industry

GUINEBRETIÈRE René
X-ray Diffraction by Polycrystalline Materials